Hybrid Enhanced Oil Recovery Processes for Heavy Oil Reservoirs

DEVELOPMENTS IN PETROLEUM SCIENCE 73

Hybrid Enhanced Oil Recovery Processes for Heavy Oil Reservoirs

Xiaohu Dong
Associate professor
College of Petroleum Engineering
China University of Petroleum-Beijing

Huiqing Liu
Professor
College of Petroleum Engineering
China University of Petroleum-Beijing

Zhangxin Chen
Professor
Department of Chemical and Petroleum Engineering;
Director of the Foundation CMG/Frank and Sarah Meyer Collaboration Centre
University of Calgary, Alberta, Canada

ELSEVIER

Elsevier
Radarweg 29, PO Box 211, 1000 AE Amsterdam, Netherlands
The Boulevard, Langford Lane, Kidlington, Oxford OX5 1GB, United Kingdom
50 Hampshire Street, 5th Floor, Cambridge, MA 02139, United States

Notices
Knowledge and best practice in this field are constantly changing. As new research and experience broaden our understanding, changes in research methods, professional practices, or medical treatment may become necessary.

Practitioners and researchers must always rely on their own experience and knowledge in evaluating and using any information, methods, compounds, or experiments described herein. In using such information or methods they should be mindful of their own safety and the safety of others, including parties for whom they have a professional responsibility.

To the fullest extent of the law, neither the Publisher, nor the authors, contributors, or editors, assume any liability for any injury and/or damage to persons or property as a matter of products liability, negligence or otherwise, or from any use or operation of any methods, products, instructions, or ideas contained in the material herein.

Library of Congress Cataloging-in-Publication Data
A catalog record for this book is available from the Library of Congress

British Library Cataloguing-in-Publication Data
A catalogue record for this book is available from the British Library

ISBN: 978-0-12-823954-4
ISSN: 0376-7361

Publisher: Candice Janco
Acquisitions Editor: Amy Shapiro
Editorial Project Manager: Naomi Robertson
Production Project Manager: Swapna Srinivasan
Cover Designer: Miles Hitchen

Typeset by TNQ Technologies

Contents

7. Microscale experiments for hybrid enhanced oil recovery processes

8. Macroscale experiments for hybrid enhanced oil recovery processes

9. Challenges in application of hybrid enhanced oil recovery processes

10 Other enhanced oil recovery processes and future trends

Chapter 1

Introduction to hybrid enhanced oil recovery processes

1.1 Introduction to heavy oil and oil sands reservoirs

1.1.1 Distribution of heavy oil resources

Heavy crude oil refers to liquid petroleum whose American Petroleum Institute (API) gravity is less than 20° or viscosity is higher than 100 cp under reservoir conditions. Oil sands resources (tar sands) are another kind of heavy oil resources whose API gravity is usually less than 10 degrees and whose viscosity is higher than 10,000 cp under reservoir conditions. However, there is no explicit difference between heavy oil and oil sands [1,2]. Heavy oil and oil sands resources are distributed in 32 basins in 16 countries [3]. Total crude oil resources are approximately 9−11 trillion barrels (bbls) in the world, among which more than two-thirds are heavy oil and bitumen (the crudes contained in oil sands). Out of the total 8 trillion bbls of heavy oil and bitumen resources, Canada and Venezuela possess about 2−3 trillion bbls each [4−6].

Canada has the third largest proven oil reserves in the world, most of which are heavy oil and oil sands. Total oil sands reserves in Canada are 166.3 billion bbls, accounting for about 97% of Canada's 171 billion bbls of proven oil reserves [7]. In Canada, almost all heavy oil and oil sands deposits lie in Alberta. In Alberta, oil sands resources are mainly distributed in Athabasca, Cold Lake, and Peace River [8]. In Athabasca, the total oil-bearing area is about 40,000 km^2. About 20% of oil sands reserves in Fort McMurray in Athabasca are recovered by the method of surface mining. In Cold Lake, the oil-bearing area is about 22,000 km^2. It has the second largest heavy oil reserves in Alberta. The burial depth is about 300−600 m. In Peace River, the oil-bearing area is about 8000 km^2. The burial depth is about 300−770 m. For oil reserves in Cold Lake and Peace River, considering the burial depth, in situ recovery technology is the primary exploitation method.

Hybrid Enhanced Oil Recovery Processes for Heavy Oil Reservoirs.
https://doi.org/10.1016/B978-0-12-823954-4.00007-2

In Venezuela, heavy oil resources mainly lie parallel to the northern bank of the Orinoco River and extend from east to west along the Orinoco petroleum belt. The main producing areas are Cerro Negro (Carabobo), Hamaca (Ayacucho), Zuata (Junín), and Machete (Boyacá) [4]. The total length is about 500–600 km. Proven total heavy oil reserves in these four areas are about 1200 billion bbls. The reservoir burial depth in the Orinoco Oil Belt is 240–1200 m, the reservoir temperature is 37–60°C, the initial reservoir pressure is 3.1–8.2 MPa, the formation oil viscosity is 1000–5000 cp, the reservoir thickness is 6–90 m, the porosity is 28%–34%, and permeability is 1–20 μm^2 [9]. Different from recovery processes in Canada, considering the properties of foamy oil, cold heavy oil production with sands (CHOPS) is the primary recovery method in the Orinoco Oil Belt.

In China, more than 70 heavy oil fields have been discovered and distributed within 12 basins. Proven onshore heavy oil reserves in China are about 28 billion bbls. Proven offshore heavy oil reserves are about 18 billion bbls and are mainly distributed in Bohai Bay (operated by China National Offshore Oil Corporation [CNOOC]) [5,10]. Liaohe oil field, Shengli oil field, Karamay oil field, and Henan oil field are typical onshore heavy oil fields, as shown in Fig. 1.1. Compared with heavy oil resources in Canada and

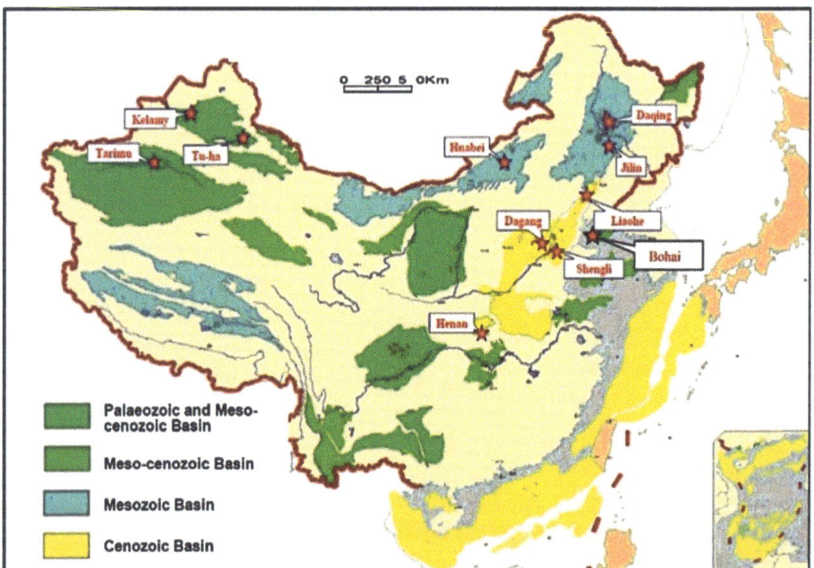

FIGURE 1.1 Heavy oil field locations in China.

Venezuela, the burial depth of heavy oil reservoirs in China is deeper. Proven heavy oil reserves whose burial depth is greater than 900 m account for more than 60%; some even reach 1300–1700 m. The maximum burial depth of proven heavy oil reservoirs in China is located in the Tuha oil field, at about 3300 m. Heavy oil resources in the Henan and Xinjiang oil fields are relatively shallow (200–600 m).

1.1.2 Characteristics of heavy crude oil

Heavy crude oil is different from conventional light oil, It is usually characterized by high viscosity and high density in the original formation temperature condition. Therefore, to recover it effectively, reducing its viscosity (μ_o) and improving its mobility (k/μ_o) are top priorities. Considering the temperature sensitivity of heavy oil or bitumen viscosity, a thermal recovery process is introduced. For thermal recovery, a hot fluid such as steam is cyclically or continuously injected into a formation. Then, both the formation rock and fluids around wells are heated and the temperature increases. Thus, the oil viscosity is reduced and the mobility of heavy oil and bitumen is improved. As shown in Fig. 1.2, as the temperature increases, the oil viscosity is reduced by orders of magnitude. A thermal recovery technique was first started in Trintopec's operations in 1966, with a small cyclic pilot project in the Palo Seco field [11]. It remains the main method of exploitation for recovering heavy oil and bitumen resources all over the world. Considering the high heat-carrying capacity of steam, it is the most commonly used and ideal hot fluid for a thermal recovery project [12,13]. In situ steam-based technology has long been

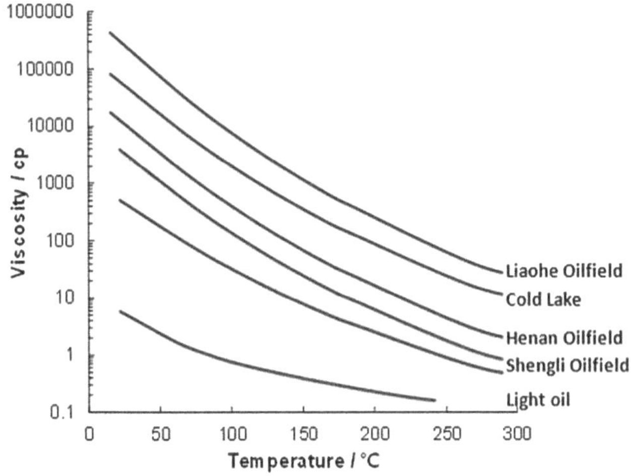

FIGURE 1.2 Schematic of oil viscosity versus temperature.

widely applied as an enhanced oil recovery (EOR) process for heavy oil and oil sands reservoirs. It is also the most advanced of all EOR processes in terms of field experience and thus has the least uncertainty in estimating performance [6].

Another unique characteristic of heavy oil is its rheologic behavior in porous media. Compared with conventional light crude oil, heavy oil is usually considered to be a temperature-sensitive non-Newtonian fluid, especially for superheavy oil (10,000–50,000 cp) and extraheavy oil (>50,000 cp) [14]. It is caused by the unique compositions of heavy oil, especially of asphaltene and resin. One important rheologic characteristic of heavy oil is the behavior of shear thinning. However, as the temperature increases, the behavior of shear thinning gradually vanishes, and the behavior of a Newtonian fluid appears. The transition temperature from a non-Newtonian fluid to a Newtonian fluid is called the critical temperature. It is the important turning point of the rheologic behavior of heavy oil. Theoretically, only when the formation temperature is higher than the critical temperature can heavy oil flow continuously in the formation. Therefore, for thermal recovery, a minimal requirement for a steam injection volume is that the total steam volume should guarantee the formation temperature to be higher than the critical temperature.

1.1.3 New classification of heavy oil reservoirs

The purpose of classifying heavy oil reservoirs is to find an effective exploitation method for these reservoirs. From the performance characteristics of different thermal recovery processes, the classification criterion of heavy oil reservoirs is based on the sensitive factors of their recovery performance. It usually includes the reservoir structure, lithology, formation thickness, oil quality, and associated aquifers. Among them, the formation thickness and oil quality are two important factors.

Based on the theory of fluid flow in porous media, a steady displacement front in a tilted formation meets the requirement:

$$(M_0 - 1)v_c < \frac{K_1}{\mu_1}(\rho_1 - \rho_2)g \sin\alpha \tag{1.1}$$

$$M_0 = \frac{\mu_2}{\mu_1}\frac{K_1}{K_2} \tag{1.2}$$

where K_1 and K_2 are effective permeabilities of a displacing fluid and a displaced fluid in the formation, $10^{-3}\mu m^2$; μ_1 and μ_2 are the viscosity of the displacing and displaced fluids, cp; ρ_1 and ρ_2 are their densities, kg/m^3, respectively; M_0 is their mobility ratio; v_c is a critical flow rate, m/s; and α is the formation angle with the horizontal direction, deg.

For a horizontal formation ($\sin \alpha = 0$), because $v_c \neq 0$, Eq. (1.1) can be simplified as:

$$M_0 < 1 \tag{1.3}$$

In water flooding, water is the displacing fluid and oil is the displaced fluid, and thus:

$$M_0 = \frac{\mu_o}{\mu_w}\frac{K_w}{K_o} \tag{1.4}$$

Eq. (1.4) shows that reducing the mobility ratio can be achieved by increasing the water viscosity (e.g., by polymer flooding), reducing the oil viscosity (by a thermal recovery process), or reducing the water-phase effective permeability (by water shut-off or profile control). Increasing the viscosity of the displacing fluid and reducing the viscosity of the displaced fluid are two effective EOR approaches. Thus, with a viscosity difference between the displacing fluid and displaced fluid, recovery performance can be improved. The methods of polymer flooding and gel injection are proposed based on increasing the displacing fluid viscosity, and thermal recovery methods are proposed based on reducing the displaced fluid viscosity. For heavy oil reservoirs with a relatively low oil viscosity, the capacity reducing oil viscosity by increasing the temperature is limited. Thus, increasing the viscosity of a displacing fluid will be the EOR direction, by water and chemical flooding, for example. From water flooding to polymer flooding and gel injection is an EOR process of gradually increasing the displacing fluid viscosity. They are suited to heavy oil reservoirs with different oil viscosities. Especially for offshore heavy oil reservoirs, exploiting an easily operated cold oil production method (water and chemical flooding) is highly recommended. For heavy oil reservoirs with an oil viscosity of 50* cp to 350* cp, water or hot water flooding is commonly used, such as in the B21 heavy oil block ($\mu_o = 95$ cp) in the Shengli oil field, Sinopec (China Petroleum and Chemical Corporation) and the Du66 heavy oil block ($\mu_o = 108$ cp) in the Liaohe oil field, China National Petroleum Corporation (CNPC) [16,18,19].

Fig. 1.3 shows a criterion for the development methods for heavy oil reservoirs. For heavy oil reservoirs with a relatively low oil viscosity range of 50–350* cp, chemical flooding (polymer flooding and/or gel injection) can yield good recovery performance. According to the field tests in Bohai Bay, CNOOC, the original heavy oil classification can be further subdivided, as shown in Fig. 1.3. Previous heavy oil reservoirs with an oil viscosity of 150* cp to 10,000 cp are subdivided into two subtypes, including type I with an oil viscosity of 150–350* cp (gel injection) and type II with an oil viscosity of 350* cp to 10,000 cp (steam injection).

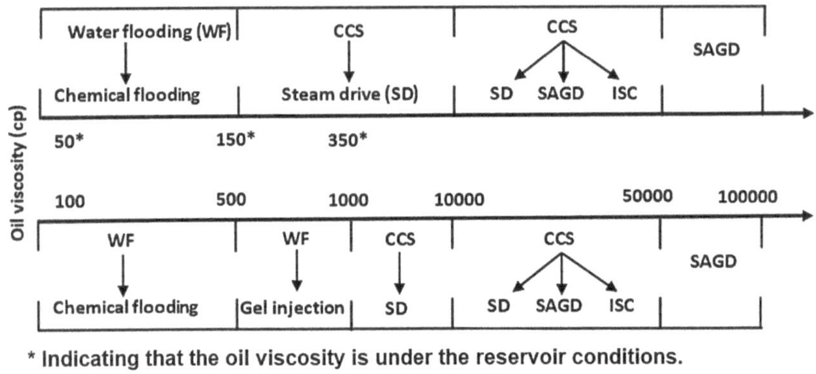

FIGURE 1.3 Development methods of different heavy oil reservoirs. *ISC*, in situ combustion; *SAGD*, steam-assisted gravity drainage; *CCS*, cyclic steam stimulation.

1.2 Steam-based recovery processes

Exploitation processes for heavy oil reservoirs include surface mining and in situ recovery technology. Surface mining is usually applied for shallow oil sands resources (<75 m). In a region of about 500 km^2 in northern Alberta, oil sands resources are recovered by surface mining [7]. The operation procedure of a surface mining process includes oil sands extraction and collection, conditioning, separation, secondary separation, and froth treatment. In comparison, in situ recovery technology is usually applied for deep heavy oil reservoirs. It includes cold oil production and thermal oil recovery. Cold oil production is used for heavy oil reservoirs with a low oil viscosity; it includes natural depletion, water flooding, chemical flooding, CHOPS, and vapor extraction (VAPEX). Commonly used thermal oil recovery processes include cyclic steam stimulation (CSS), steam flooding, and steam-assisted gravity drainage (SAGD) [12]; they account for about 90% of total heavy oil production in the world [15]. In this section, the status of these three typical steam-based recovery processes is presented.

1.2.1 Cyclic steam stimulation (huff n' puff)

CSS is the simplest form of steam injection operation; it has been widely used to recover heavy oil and oil sands reservoirs in the world. This operation includes three major phases: steam injection (several weeks), soaking (3–5 days), and oil production (tens or hundreds of days) [12,16]. All of these phases are performed within the same well, as shown in Fig. 1.4. After the phase of oil production, steam is reinjected to start a new cycle. The recovery mechanisms of CSS essentially consist of oil viscosity reduction, heat swelling, and solution gas driving. They are predominantly performed in vertical wells [6]. Typical oil recovery factors are 20–35% with steam–oil

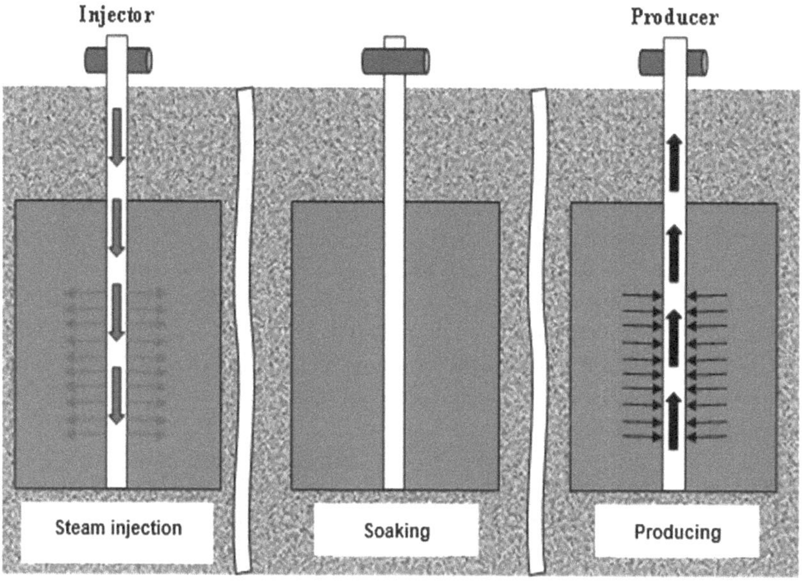

FIGURE 1.4 Three phases of cyclic steam stimulation process.

ratios (SORs) of 3.0–5.22. Alvarez and Han conducted a critical review of the current status of CSS technology and discussed commercial projects in the world [17]. It is usually applied to heavy oil reservoirs whose pay thickness is greater than 30 ft, reservoir depth is less than 3000 ft, porosity is higher than 0.3, and oil saturation is greater than 40% [12,16]. It is still the main recovery method for most heavy oil fields in the world. In China, almost 75% of heavy oil production comes from CSS projects [16,18,19].

On the other hand, thin heavy oil reservoirs have become important in boosting oil production. For this type of heavy oil reservoir, the conventional vertical well-based CSS technique is no longer economical because of huge overburden and underburden heat losses. The horizontal well-based CSS technique (HW-CSS) has been performed in this type of heavy oil reservoir [20,21]. Also, because of a reduction in directional drilling costs and the improvement in sweep efficiency, HW-CSS has been considered a successful EOR process in heavy oil reservoirs.

In a conventional CSS process, specifically at a later stage of CSS operations, the thermal efficiency of steam is dramatically reduced. The problems of steam fingering and steam breakthrough become top priorities. Considering this situation, many attempts have been made to improve CSS, followed up by, for example, combining the CSS process with multiple wells, steam flooding, or a steam-additive process [15,16,22]. First, for the combined CSS process with multiple wells, a single well group or several well groups are considered as an operation unit, as shown in Fig. 1.5. These well groups generally possess a higher steam breakthrough degree or higher recovery factor within certain layers. They use a steam channeling path to perform steam injection with a well-group operation mode instead of the conventional single well operation. This mode can effectively recover remaining oil between wells. In addition, retarding steam overlay is another important mechanism of this process. Other follow-up techniques that use steam flooding or steam-additive processes will be discussed in detail in later sections.

On the other hand, CSS has been applied to improve a steam chamber expansion in SAGD [23]. It is called a fast-SAGD process or hybrid CSS/SAGD process, and it is proposed based on CSS and SAGD processes. In the fast-SAGD, an offset well is drilled between adjacent SAGD well pairs and is performed under CSS to accelerate steam chamber growth sideways [60]. In hybrid CSS/SAGD, a CSS well is placed between the SAGD well pairs. This well is operated in a CSS mode until steam chambers are in contact with each other, and then it is switched to an SAGD operation. In comparison, a hybrid CSS/SAGD process can recover a greater amount of bitumen with lower steam injection than CSS, SAGD, and fast-SAGD [23]. Xu et al. numerically investigated and optimized the performance of hybrid CSS/SAGD in Long Lake heavy oil reservoirs with lean zones. The authors observed that hybrid CSS/SAGD performed better than conventional SAGD in oil sands reservoirs with lean zones [24].

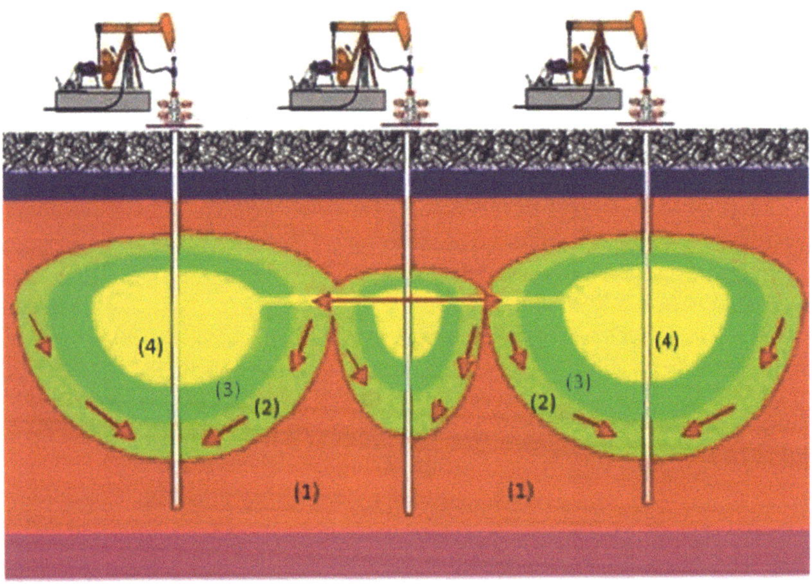

FIGURE 1.5 Combined cyclic steam stimulation process with multiple wells: ① original crude oil; ② heated area; ③ condensate area; ④ steam zone.

For a field operation, CSS is one of the most widely used in-situ recovery techniques for heavy oil and bitumen resources. It was first applied in Venezuela in 1959. Since then, this method has been applied in many oil fields across the world, such as the San Joaquin Valley and the Los Angeles Basin in the United States, Cold Lake in Canada, Lake Maracaibo in Venezuela, and the Liaohe oil field in China [12,25,26]. Nehring et al. (1983) provided a list of significant heavy oil fields and pools in the United States [27]. There are 89 fields and 219 pools. Among them, based on a reservoir screening process, 41 pools are qualified for CSS, 101 pools are qualified for steam flooding, 32 pools are qualified for in situ combustion, and 45 pools are qualified for any thermal recovery methods. Except for a few thin pools, nearly all that are qualified for steam flooding are also qualified for a CSS operation.

For the oil sands resources in Alberta, Canada, CSS has evolved more than 45 years in Athabasca, Cold Lake, Peace River, and Grosmont [28–30]. The Athabasca oil sands possess a large deposit of oil-rich bitumen located in northern Alberta, and almost all oil resources in Athabasca are deposited in the Fort McMurray formation [31]. The commercial production of oil from Athabasca oil sands began in 1967 by Great Canadian Oil Sands Limited (currently Suncor Energy) using surface mining. In Athabasca, except for surface mining, CSS provides the highest oil production output among in situ recovery techniques [32]. Furthermore, CSS has been applied in Cold Lake by Imperial Oil since the 1980s and is used by Canadian Natural Resources at Primrose and Wolf Lake and by Shell Canada at Peace River. Another application is in fractured carbonate heavy oil reservoirs, such as the Grosmont Formation in northern Alberta. It is a dolomitized, karsted, and fractured carbonate reservoir that contains a massive bitumen accumulation [30,33]. The CSS operation in Grosmont is the most widely and successful in situ recovery process. One of the best performing CSS wells in Grosmont, well 10A-5-88-19W4, has recovered about 100,000 bbls of oil through over 10 cycles, with a cumulative steam-oil ratio (CSOR) of 6 [33,34]. In addition, the Lake Maracaibo area in Venezuela is a successful application of the CSS process for heavy crude oil [25,35]. CSS operations in the Lake Maracaibo area began in 1971, and up to 1995 a total of 325 wells were stimulated with 860 cycles [25]. In China, CSS operations for heavy oil recovery began in the Karamay oil field, CNPC, in the 1960s. Then, based on successful operations in pilot tests of Karamay, the CSS technique was expanded to recover heavy oil reservoirs in the Liaohe, Shengli, and Henan oil fields. In 1995, the total heavy oil production in China increased to over 10 million tons. China has become a major country for heavy oil production [10,36]. Furthermore, CSS and steam flooding have been widely applied to recover heavy oil resources in Russia [37], Indonesia, Colombia [38,39], Oman, and Mexico [40–42].

1.2.2 Steam flooding (steam drive)

Steam flooding is a logical follow-up stage of CSS. In this process, steam is continuously injected into a reservoir using a vertical or horizontal well. A proper well pattern is usually required for successful steam flooding, including injectors and producers. Steam flooding is similar in concept to water flooding [16,27]. Steam is continuously injected into a reservoir through an injector and travels through the reservoir. It effectively heats the heavy crude oil within a formation and its surrounding rocks, as shown in Fig. 1.6. Then, the condensed water flows to a producer, which efficiently displaces the crude oil. The recovery mechanisms of steam flooding include the improvement of an oil—-water mobility ratio, and changes in relative permeability curves, steam distillation, and emulsification behavior [12,16]. Based on a screening criterion, steam flooding is usually suited for a reservoir whose thickness is greater than 10 m, porosity is greater than 20%, permeability is greater than 200×10^{-3} μm^2, oil viscosity is less than 20,000 cp under reservoir conditions, and oil saturation is higher than 0.5 [15,16]. On the other hand, with progress in science and technology and changes in oil prices, this screening criterion is also changing.

For a field operation, the projects of Kern River in the United States and Cold Lake in Alberta are two successful steam flooding cases in North America [43—45]. First, for the steam flooding project in Kern River, it was initially performed in a 10-pattern area. In this area, the formation average depth is 213—243 m, the reservoir original pressure is about 1.55 MPa, oil viscosity is 2710 cp at 85°F, and the formation thickness is 29.5 m. For the performance of steam flooding in this area, the cumulative production—injection ratio of this test was 0.81 and the oil recovery factor of steam flooding reached about 37% (based on oil reserves before steam

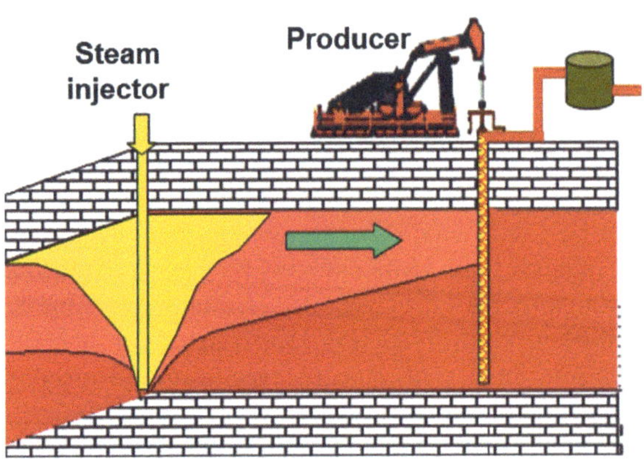

FIGURE 1.6 Steam flooding process.

flooding) [46,47]. However, as steam injection continued, steam breakthrough in some production wells hindered continuing this test. Therefore, some adjustments were conducted, including a steam injection rate reduction, a water flooding process, well repairing, infill drilling, and profile control [48,49]. The Cold Lake project in Alberta has been operated by Imperial Oil since 1975. About 55 infills are located in 34 producing pads (almost 700 wells in total) [45]. In addition, the Qi40 reservoir in the Liaohe oil field is successful steam flooding project in China. Involving the previous CSS stage, the total oil recovery factor after steam flooding reached about 55.7% [36,50]. Except for the operation in Qi40, steam flooding has been performed in many heavy oil reservoirs in China, such as Jin45 (a reservoir with a boundary aquifer) and the Wa38 reservoir in the Liaohe oil field, Shan83 (a reservoir with a boundary aquifer) and the Shan56 reservoir in the Shengli oil field, and the BQ10 reservoir in the Henan Oil field [19,51,52].

1.2.3 Steam-assisted gravity drainage

The SAGD recovery process was initially proposed by Dr. Butler and colleagues in the 1980s [53,54]. In this process, a horizontal well pair is placed in parallel in a bottom section of a reservoir. The upper horizontal well performs as a steam injector and the bottom one is a producer. The vertical distance between them is about 5–7 m, as shown in Fig. 1.7. A normal SAGD project involves two phases: a preheating (startup) phase and an SAGD phase. The preheating phase aims to create a thermal connection between a producer and an injector. CSS, steam circulation, fracturing, electrical heating, and solvent coinjection are commonly used startup approaches for SAGD [55–57]. After that, steam is continuously injected from the upper injector, and a steam

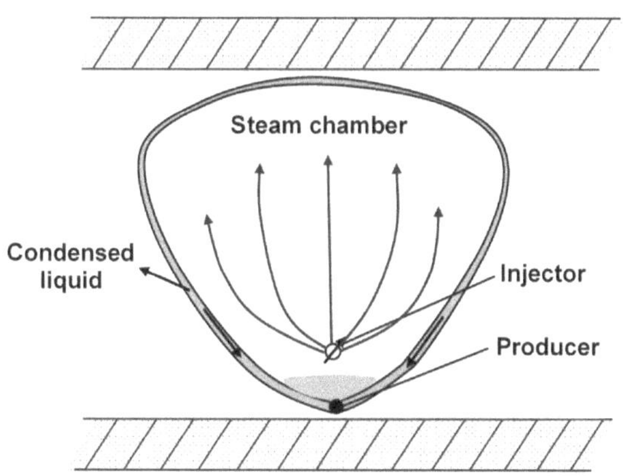

FIGURE 1.7 Schematic of steam-assisted gravity drainage process.

chamber is created. According to the different stages of steam chamber expansion, the SAGD phase includes the stages of vertical expansion, horizontal expansion, and exhaustion [12,31,58]. Once the steam chamber front reaches the top of a reservoir, it expands horizontally. Analogously, once the steam chamber arrives at the reservoir boundary, the steam chamber is depleted and the oil production rate starts to decrease [31]. During SAGD, the heated oil and condensate water flow downward along the boundary of a steam chamber and finally are produced from the bottom producer [6,58].

Different from the techniques of CSS and steam flooding, in SAGD, the dominating driving force for oil drainage is gravity. Al Bahlani and Babadagli critical reviewed the current status and future trends of SAGD [59]. For EOR mechanisms, different from conventional mechanisms of steam injection, multiphase fluid flow and emulsification phenomena at the edge of a steam chamber are important [61,62]. First, for fluid flow behavior in a steam chamber boundary, based on an assumption of single oil-phase flow in the steam chamber boundary, and without considering the effect of the heat convection mechanism, Butler proposed an ideal productivity model for SAGD [31]. However, after a comparison of Butler's model and field data, it was found that his model could not match the data well. Therefore, many researchers proposed some modified models for SAGD recovery performance based on Butler's model. Based on the experimental observation that a steam-zone shape is an inverted triangle, Akin proposed a mathematical model for SAGD [63]. In this model, both the effects of steam distillation and asphaltene-deposition are considered. Sharma and Gates considered the impact of oil saturation and relative permeability on an oil mobility profile at the edge of a steam chamber to propose a novel model for SAGD performance [64]. On the other hand, the heat transmission mechanism in a steam chamber boundary is another popular topic for SAGD. Butler and Reis believed that heat conduction is the primary heat transfer mechanism of steam chamber expansion [31,65]. However, the effect of heat convection is a nonnegligible issue. Irani and Ghannadi, Zhang et al., and Keshavarz et al. considered the effects of heat conduction and heat convection and proposed some modified prediction models for SAGD [66–68]. The effect of heat convection on oil drainage was a function of temperature. At the side of a steam chamber, water saturation was higher, temperature was higher, and the effect of heat convection was enhanced. However, for the side of a cold oil reservoir, because of low temperature, the effect of heat convection was significantly reduced. Second, for the emulsification phenomenon, Noik et al. used a differential scanning calorimeter technique, microscopy, and image analysis to characterize the water-in-oil emulsion and the reverse emulsion characteristics in produced fluids in SAGD [69]. Using a Computer Modeling Group Ltd. simulator, Ezeuko et al. numerically discussed emulsification at the boundary of a steam chamber in SAGD [70]. The presence of emulsion can further improve oil mobility, promote SAGD performance, and increase oil-steam ratio (OSR).

Hascakir et al. comparatively analyzed the emulsion features of steam flooding and SAGD from experimental observations [71]. Because of the increased interaction time of asphaltene with water in a steam chamber, SAGD can produce more water-in-oil emulsions than steam flooding. From the experimental point of view, there have been many related investigations into the recovery performance of SAGD since it was proposed by Butler [72,73]. Experimental setups include two-dimensional (2D), 3D, and visualized experiments in SAGD. For a suitable reservoir type, SAGD can be applied to conventional thick, heavy oil reservoirs and also reservoirs with a bottom aquifer [56,74]. It can be considered a recovery method with a constant pressure drop. Therefore, application of the SAGD process in reservoirs with a bottom aquifer does not cause the issue of water coning. Compared with the performance in conventional thick, heavy oil reservoirs, cumulative oil recovery in reservoirs with a bottom aquifer is not significantly reduced.

For field applications, the SAGD technique has been widely applied as an EOR process for heavy oil reservoirs, especially for oil sands reservoirs in Alberta. More than 15 commercial SAGD projects have been operated or remain in operation in Canada, mainly in Athabasca, Cold Lake, and Lloydminster [55,75,76]. In China, operation sites of SAGD projects are located in the Xinjiang oil field in Karamay and Liaohe oil field in Panjin. Especially in the Liaohe oil field, a modified SAGD process using a combination of vertical and horizontal wells was proposed as a follow-up process to CSS [77]. To predict the recovery performance (recovery factor and CSOR) of SAGD quickly in oil sands reservoirs, an empirical correlation was proposed using numerical simulation and gray relational analysis [75]. Through this correlation, SAGD recovery performance can be quickly obtained. Furthermore, for the operation parameters of SAGD, operation pressure is important for SAGD. It is generally around the reservoir pressure. The operation pressure in most SAGD projects is about 2–4 MPa, as shown in Table 1.1 [78–80]. However, for an operation in reservoirs with a bottom aquifer, a pressure-decline analysis is generally required before the SAGD stage [56].

1.3 Concepts of hybrid enhanced oil recovery processes

For heavy oil reservoirs that have gone through long-term stimulated steam-based recovery processes, their underground fluid distributions have changed significantly [15,17]. Therefore, continuing their original steam-based recovery processes is no longer effective. In this situation, hybrid EOR processes are proposed in the poststeam injection era. In hybrid EOR processes, multifluid injection processes are performed. For heavy oil reservoirs, steam and some additives are coinjected into reservoirs. For different purposes, hybrid EOR processes include three kinds of processes: hybrid thermo-solvent, hybrid thermal-noncondensable gas (NCG), and hybrid thermochemical. All three techniques have been tested to enhance heavy oil recovery. In field, they can be operated by a cyclic injection mode, a continuous injection mode, or even a

TABLE 1.1 Reservoir properties of some steam-assisted gravity drainage projects.

Steam-assisted gravity drainage projects	Depth (m)	φ (%)	K_h (μm^2)	NTG	H (m)	μ_o (10^4cp)	ΔS_o (%)	Operation pressure (kPa)
UTF project	180	33	3–5	0.8	21	200–300	65	2250
Hangingstone	255–280	30	2.4	0.8	30–50	>100	65	4500
Foster Creek	180–225	33	3–5	0.8	25	6–8	65	2700
MacKay River	98–145	31	3–5	0.8	15–25	>100	64	1700
Long Lake	210	30	2	0.8	30		60	2500
Peace River	550–600	28	1.5	0.8	29	1–100	57	3000
Firebag	270–325	32	7	0.8	37		64	2000
Celtic field	470	35	2–13	0.8	20	1.5	60	
Tia Juana field	305	38	1.5	0.8	26	2	65	2000
Fengcheng oil field	100–230	33	1.8	0.8	33	1.4	55	2200–3000
Du 84 block	530–640	36.3	5.5	0.8	91	23.19	45	3000

NTG: Net-to-gross

gravity drainage mode. Accordingly, these three operation modes have great potential for improving the recovery performance of the three basic steam-based recovery processes for heavy oil reservoirs (CSS, steam flooding, and SAGD).

First, for hybrid thermo-solvent processes, a small amount of vaporized but condensable hydrocarbon solvent is added to steam. Solvent and steam are coinjected simultaneously or periodically into a reservoir to improve the recovery performance of steam-based processes for heavy oil resources [81,82]. After injection, the solvent condenses along with steam at a bitumen–vapor interface and mixes with bitumen to reduce the oil viscosity and enhance the oil production rate. There are generally five types of hybrid thermo-solvent processes: liquid addition to steam for enhancing recovery (LASER), steam-alternating solvent (SAS), expanding solvent-SAGD (ES-SAGD), solvent-aided process (SAP), and solvent-enhanced steam flooding (SESF). The main mechanisms of hybrid thermo-solvent processes improve the properties of heavy oil and increase the sweep efficiency.

Second, for hybrid thermal-NCG processes, steam and NCG are coinjected simultaneously or periodically into a reservoir to improve the recovery performance of steam-based processes for heavy oil resources. In this process, commonly used NCGs include nitrogen, carbon dioxide, air, flue gas, and methane. Different from the mechanisms of the hybrid thermo-solvent process, the addition of NCG further reduces oil viscosity, improves steam injectivity, increases the size of heated areas, recovers reservoir energy, and provides additional drive energy [15,83,84].

For hybrid thermochemical processes, the main purpose is to control steam injection profiles effectively [85–87]. Commonly used chemical additives include alkali, surfactant, polymer, NCG (N_2 and CO_2) foam, gels, and solid particles. Especially for NCG foam, gels, and solid particles, they can effectively plug a steam breakthrough path and improve the recovery performance of steam-based processes for heavy oil reservoirs. On the other hand, combining the advantages of NCG and chemical additives, a hybrid thermal-NCG-chemical process has been proposed and field-tested in [88,89]. Specific operation methods include horizontal wells, dissolver, nitrogen, and steam (HDNS), horizontal wells, dissolver, CO_2, and steam (HDCS), and horizontal wells, dissolver, air, and steam (HDAS). Dissolver is a kind of surfactant used for heavy crude oil. It can effectively reduce heavy oil viscosity by changing the colloid structure of heavy oil in porous media. This hybrid process has been widely applied in the Shengli oil field and Sinopec and Xinjiang oil fields, CNPC.

1.4 Multicomponent and multiphase fluids

Multicomponent and multiphase fluids (MMFs) refer to a multicomponent multifluid mixture with the characteristics of multiple phases. Multiple phases indicate that different types of fluids in MMFs have different fluid properties. In petroleum reservoirs, this kind of fluid mixture has been widely applied in EOR processes in both light oil and heavy oil reservoirs. For heavy oil reservoirs, a steam injection operation is a typical EOR process with two-component three-phase fluids, as shown in Table 1.2. Hybrid thermal-NCG, thermo-solvent, and thermochemical EOR processes are typical multicomponent three-phase EOR processes. A hybrid thermochemical-NCG EOR process is a four-component three-phase EOR process. In addition, to address the problem of steam breakthrough, solid particles have been used for profile control in heavy oil reservoirs. This process is a three-component four-phase EOR process. For hybrid EOR processes in heavy oil reservoirs, fluid flow behavior in porous media involves a multicomponent multiphase flow. In addition, during operations, the purposes of different additives are distinct.

For hybrid EOR processes, different types of fluids can be injected simultaneously or separately. The operation procedure depends on the purpose of a specific operation [15]. For a hybrid thermal-NCG EOR process, steam and NCG are usually injected separately to recover reservoir energy (e.g., a hybrid steam-CH_4/N_2 process), especially when it is applied in a multi-CSS-cycled heavy oil reservoir. In some cases, however, they can also be injected simultaneously to reduce oil viscosity further (e.g., a hybrid steam-CO_2 process). For hybrid thermo-solvent EOR processes, steam and solvent additives are injected simultaneously (e.g., LASER, SESF, and ES-SAGD). For hybrid thermochemical EOR processes, chemical additives are usually injected first and then steam is injected. That is because of the low temperature endurance of chemical additives.

TABLE 1.2 Multicomponent and multiphase fluids in hybrid enhanced oil recovery processes for heavy oil reservoirs.

Multiphase fluids	Two phases (water and oil)	Three phases (oil, water, and vapor/gas)	Four phases (oil, water, vapor/gas, and solid)
Multicomponent fluids	Liquid additive injection (water flooding, chemical flooding, etc.)	Steam injection	Steam and solid-particle coinjection
		NCG (steam) injection	
		Steam and chemical coinjection	Steam, chemical and solid-particle coinjection
		Steam and solvent coinjection	

1.5 Hybrid thermo-solvent processes

Solvents are an environmentally friendly additive for EOR processes. Compared with other additives, solvent-based and solvent-assisted recovery processes have a lower carbon footprint. Hybrid thermo-solvent processes in heavy oil reservoirs can reduce steam requirements. The EOR mechanisms of a hybrid thermo-solvent process include the mechanisms of conventional steam injection and also the extra effects of solvent additives. A condensed solvent fraction can dissolve into bitumen to improve fluid flowability in a reservoir. It further reduces oil viscosity. Specifically, the phase equilibrium or pressure-volume-temperature (PVT) behavior of a heavy oil/solvent/steam system in hybrid thermo-solvent processes has attracted much attention [90,91]; equation of state modeling is a promising method that has been applied to characterize this behavior [92,93]. Second, the mechanism of emulsion breaking is an important aspect during this process. Kar et al. experimentally characterized the emulsions of SAGD and ES-SAGD [94]. The produced oil in the ES-SAGD process had lower emulsion stability. In this process, the coinjection of solvent can control the wettability alteration owing to its interaction with asphaltenes in heavy crude oil. In the field, to decrease the effect of emulsification on steam-based recovery performance, an asphaltene-soluble solvent is recommended and can be applied in operation [71].

1.5.1 Liquid addition to steam for enhancing recovery

LASER is a cyclic steam injection with the addition of a C5+ condensate to the steam during injection. In the late stage of CSS heavy oil reservoirs, LASER is a preferable EOR process. In this process, a small fraction of a light hydrocarbon is applied. As shown in Fig. 1.8, the solvent (~6% by volume) will be added to steam in a well. During the operation, the vaporized solvent transports with steam in reservoirs. Around the displacement front, because of heat conduction, steam and solvent condensation occurs. On the other hand, the solvent can dissolve in bitumen to improve the flowability of heavy oil

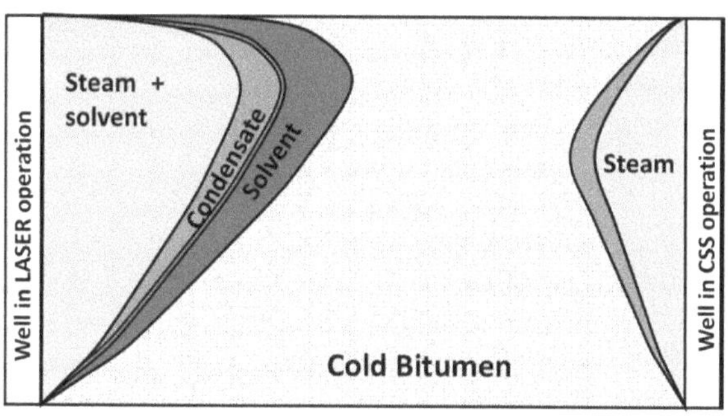

FIGURE 1.8 Concept of liquid addition to steam for enhancing recovery (LASER) process. *CSS*, cyclic steam stimulation.

[95,96]. Based on experimental studies, the addition of solvent can further reduce in situ oil viscosity and improve recovery by more than 5%. Moreover, during the operation, solvent condensate can be produced. Thus, the produced solvent can be also reinjected into reservoirs to continue the LASER operation.

1.5.2 Solvent enhanced steam flooding

SESF was proposed based on the extension of a solvent-aided CSS process and a solvent-aided SAGD process. In some other cases, this process is also called SAP, as shown in Fig. 1.9. The main mechanisms of SESF are an enhanced gas drive and solvent bank miscible displacement [97]. In particular, for thin, heavy oil reservoirs whose thickness is less than 5 m, SESF performs better than a pure steam injection operation [98]. For thin, heavy oil reservoirs, the problems of heat loss and steam breakthrough are top concerns for thermal recovery processes. In SESF, based on steam injection, the addition of solvent can further improve the thermal efficiency of steam and reduce SOR. For an operation in a thin, heavy oil reservoir, a solvent-rich channel can be observed at the top of the reservoir after solvent breakthrough at a producer. Under the effect of solvent-rich channeling, the oil-solvent mixing behavior at the periphery of the channel and the heat transfer behavior of oil beyond it will benefit the recovery performance [98]. On the other hand, the mechanisms of wettability alteration and solvent diffusion during SESF are also important [99].

In a field operation of SESF, a suitable solvent type and a technically feasible operation strategy are top concerns. Medium-alkane solvents usually perform better than heavy-alkane solvents [99]. A pilot test was performed at the Senlac thermal project by EnCana in 2002. In this project, butane was used as the solvent for coinjection with steam [100]. To analyze the performance of the SESF process, the cumulative-energy-to-produced-oil ratio (cEOR) can be used [98]. The optimal solvent-aided steam operation process should apply an operating strategy with a lower cEOR and a cumulative water-to-oil ratio.

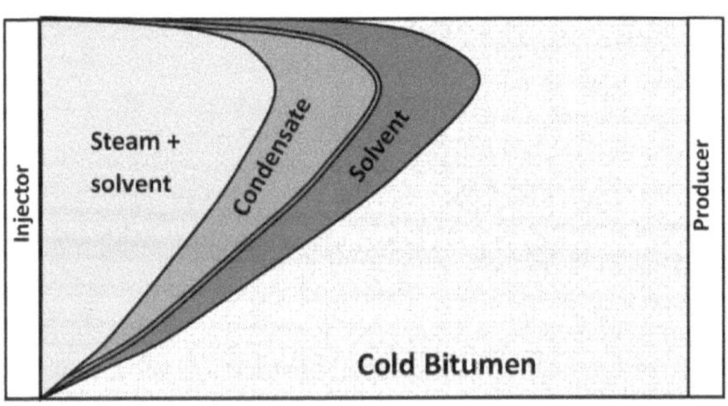

FIGURE 1.9 Schematic of solvent-enhanced steam flooding process.

1.5.3 Expanding solvent–steam-assisted gravity drainage

The third steam-solvent process is ES-SAGD or the solvent-aided SAGD process. Different from the LASER process with a cyclic operation and the SESF process with a continuous injection operation, ES-SAGD aims to enhance the performance of SAGD. In this process, solvent (hexane, heptane or octane) and steam are coinjected into a reservoir to assist oil drainage, as shown in Fig. 1.10. The ideal solvent concentration in ES-SAGD is about 4%–8%. The solvent recovery in ES-SAGD can be over 70% [101]. In ES-SAGD, the condensation and diffusion of liquid solvent into bitumen have an important role in the successful operation of this process. Because of the performance of liquid solvent, the operating temperature in ES-SAGD is often much lower than in SAGD, so that heat loss is reduced. In addition, ES-SAGD can significantly improve the oil production rate and decrease the SOR. In ES-SAGD, because of the performance of heat and mass transfer behavior and solvent diffusion, the addition of solvent can further reduce oil viscosity, increase oil phase permeability, improve oil mobility, and reduce the SOR. On the other hand, the injection of solvent can also improve the configuration of steam chamber expansion [102]. Compared with SAGD, ES-SAGD can increase oil recovery by about 30%, and the temperature of a steam chamber in ES-SAGD is also much lower than that in SAGD [103]. The oil production rate in ES-SAGD is highly related to the concentration of solvent in a steam chamber front. The higher the solvent concentration in the steam chamber front, the higher the oil production rate.

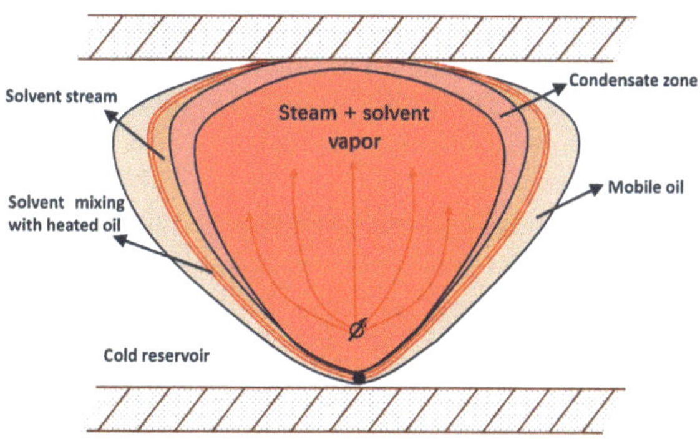

FIGURE 1.10 Schematic of expanding solvent–steam-assisted gravity drainage process.

1.5.4 Steam-alternating solvent

SAS is another modification of SAGD for heavy oil reservoirs. SAS involves injecting steam and solvent alternately [91,104]. It can be seen as an alternative operation of a pure steam or pure solvent process. The same as

ES-SAGD, the operation temperature of SAS is lower than that of SAGD. During SAS, a conventional pure steam-based SAGD operation is performed. When a stem chamber grows to the cap rock and contacts the top surface of a reservoir, steam injection is stopped and solvent injection is started. When a reduction in the temperature of the steam chamber is observed, the solvent injection is stopped and steam injection is restarted. Then, alternating steam and solvent injection are carried out until it is no longer economical. Compared with SAGD, the cumulative steam injection volume in SAS is lower. Moreover, during SAS, the dissolution of solvent in heavy crude oil can further reduce oil viscosity. On the other hand, alternating steam and solvent injection in SAS benefits the mixing behavior between solvent and heavy oil. The interfacial area will increase [104]. Similar to the three hybrid thermo-solvent processes, a solvent type is an important factor that dominates the success of SAS. For the solvent type, the dew point temperature of the used solvent should be between the reservoir temperature and steam temperature. This temperature range can guarantee a liquid state of solvent is maintained during the SAS operation. Zhao et al. experimentally compared the difference between SAGD and SAS using a 2D high-temperature and high-pressure model [105]. In this SAS experiment, a mixture of propane and methane was used as the solvent. The SAS process took advantage of SAGD and VAPEX processes to minimize the energy input in heavy oil recovery.

1.6 Hybrid thermal–noncondensable gas processes

NCG is another important additive for hybrid thermal recovery processes. For a hybrid thermal-NCG process, NCG is usually coinjected with steam into a formation to assist oil drainage. In other processes, NCG additives and steam are injected separately. An NCG slug is injected before or after steam injection. Compared with solvent and chemical additives, NCG additives are more economical and easily operated. A hybrid thermal-NCG process can be also operated by a cyclic injection mode, a continuous injection mode, or even a gravity drainage mode. The EOR mechanisms of hybrid thermal-NCG processes were discussed in Section 1.3. For the EOR of heavy oil reservoirs after steam injection, hybrid thermal-NCG processes are attractive.

1.6.1 Noncondensable gas–cyclic steam stimulation processes

In NCG-CSS, the NCG components are used to improve the recovery performance of post-CSS heavy oil reservoirs. It usually includes N_2-CSS, CO_2-CSS, flue gas-CSS, CH_4-CSS, and air-CSS. The most effective operation process among them is CO_2-CSS, because of the high solubility of CO_2 in heavy oil and the performance of miscible gas injection [106]. Compared with conventional CSS, the heat energy required for a hybrid CO_2-CSS process is lower. On the other hand, because of the lower saturation temperature of CO_2, the addition of CO_2 reduces the injection temperature. Srivastava et al. experimentally assessed the suitability and effectiveness of three gases for heavy oil recovery: pure CO_2, flue-gas (15 mol% CO_2 in N_2), and produced-gas (15 mol% CO_2 in CH_4) [107]. CO_2 was the best-suited gas to recover heavy oils. In addition, in the pure CO_2 case, the solubilization mechanism of CO_2 can dominate the process, whereas in the produced-gas and flue-gas cases, except for the solubilization mechanism, the free-gas drive was also important. Specifically, among several hybrid processes, for cyclic steam-air injection, after air injection, it can react with heavy oil in a formation through a low-temperature oxidization (LTO) reaction. Then, the produced mixture gases (including CO_2, CO, and CH_4) after LTO and the unreacted N_2 can enhance recovery. Compared with conventional CSS, hybrid thermal-air injection can significantly increase oil production.

Another type of heat carrier, multiple thermal fluids (MTFs), has been introduced into recovery for heavy oil reservoirs [108–110]. MTFs are based on the combustion and jetting mechanisms of a rocket engine. As a new heat carrier, MTFs are different from a conventional gas mixture of steam and NCG. First, MTFs are directly produced from a combustion process in a high-temperature, high-pressure generator. The NCG fraction in MTFs is a wide mixture of N_2, CO_2, CH_4, and CO. Therefore, an MTF-based process can also be considered a steam-solvent-gas coinjection process. Second, in field operations, MTFs are always injected into a reservoir directly after generation. It is different from a separate injection method in the conventional case. Since 2009, many MTF-based recovery projects have been carried out in heavy oil fields in China, and an obvious increase in oil production is achieved [110,111].

On the other hand, to reduce oil viscosity and improve the mobility ratio based on the hybrid thermal-NCG process, a surfactant, an oil viscosity reducer (VR) was used. Typical processes include HDCS, HDNS, and HDAS, as mentioned earlier. These EOR processes can comprehensively apply multiple EOR mechanisms of different additives [89]. Based on NCG, the addition of a VR additive can further destroy the micellar structure of resin and asphaltene in heavy oil and improve the flowability of heavy crude oil in porous media. Horizontal wells can reduce steam injection pressure and increase sweep efficiency, especially for extraheavy oil reservoirs. This technique has been applied to recover extraheavy crude oil reserves effectively in the Shengli oil field, Sinopec.

1.6.2 Hybrid steam–noncondensable gas process as poststeam flooding process

A hybrid steam-NCG process can be applied to improve the recovery performance of poststeam flooding reservoirs. In the later stage of steam flooding, steam breakthrough and a high steam-oil ratio are the main problems that hinder the continuous development of heavy oil reservoirs. Compared with continuous steam injection, because of the performance of a continuous NCG stream, a hybrid steam-NCG process can further increase displacement efficiency and sweep efficiency. Commonly used continuous injection methods of hybrid steam-NCG include CO_2-assisted steam flooding, N_2-assisted steam flooding, and flue gas–assisted steam flooding. Similarly, because of high solubility, a steam-CO_2 mixture is superior to either steam-N_2 or steam–flue gas combinations. Alnoaimi experimentally and numerically investigated the effect of gas additives on the recovery performance of a steam flooding process in naturally fractured carbonate heavy oil reservoirs [112]. The addition of NCG to steam flooding accelerated oil production at an early stage. Furthermore, for the hybrid steam-CO_2 process, the behaviors of extraction and diffusion are also important. During the steam-CO_2 process, lighter oil compositions in heavy crude oil were extracted, which indicates that the oil compositions of produced oil are significantly different from the original oil compositions of heavy oil in porous media. On the other hand, for a hybrid steam-N_2 process, because of the low solubility and high swelling factor of N_2, it can also be applied to increase the heating range of steam stimulation. N_2 injection has been also used to control water-coning behavior during thermal recovery for heavy oil reservoirs with a bottom aquifer, and good improvement in recovery was observed [113]. For flue gas–assisted steam flooding, recovery performance is between the steam-CO_2 process and steam-N_2 process. It combines their EOR advantages. For field operations, a gas source is usually the top priority. Another consideration for a continuous steam-NCG injection process is to adopt a suitable steam–NCG ratio. A low steam–NCG ratio can result in a low heat energy requirement and reduce the percentage of oil viscosity reduction. A high steam–NCG ratio can waste heat energy and increase operation costs.

1.6.3 Noncondensable gas–steam-assisted gravity drainage process

NCG-SAGD is another important type of hybrid thermal-NCG process. It is also called the steam and gas push process [114]. It is a modification of the conventional SAGD process in which a small amount of NCG, such as N_2, CO_2, or CH_4, is added to steam. During the operation, the injected NCG accumulates in an upper part of a steam chamber to reduce the temperature in the chamber and the heat loss rate to overburden. For some thin, heavy oil reservoirs, compared with SAGD, NCG-SAGD tremendously increases the thermal efficiency of steam. For some heavy oil reservoirs with top water, NCG accumulation behavior in the top of a reservoir benefits the recovery performance. Furthermore, in NCG-SAGD, because of the dissolution of NCG in heavy oil, the configuration of a steam chamber is different from that in SAGD, and the operation pressure is usually lower than that in SAGD. The addition of NCG can further reduce the steam requirement and improve the steam–oil ratio [115,116].

During this type of oil drainage process, NCG can gather in a steam chamber front. It is beneficial for steam chamber expansion. On the other hand, because of the performance of NCG, a high-temperature distribution area in a steam chamber cannot reflect the actual oil drainage area [117,118]. Although this hybrid process can reduce heat loss rate to overburden, a reduction in the oil production rate and the oil recovery factor negates the benefits of such a heat loss reduction, especially for N_2-SAGD and CH_4-SAGD processes [112,119]. This is because of the low solubility of these gases in heavy crude oil. Therefore, most of the injected gases accumulate in the vicinity of a steam chamber and reduce the heat transmission into cold bitumen at the steam chamber boundary. However, for CO_2-SAGD, because of its high solubility, CO_2 acts as a solvent. Therefore, it corresponds to an ES-SAGD process. For an air-SAGD process (combustion-assisted gravity drainage) or oxygen-SAGD process, in situ upgrading and in situ combustion are the most important mechanisms during the operation [120,121].

In addition, to improve the performance of post-SAGD reservoirs, the recovery performance of multithermal fluid-assisted gravity drainage (MFAGD) has was tested [122,123]. During SAGD, once a steam chamber front contacts with cap rock, the steam chamber expands laterally. At this stage, heat transmission between the steam chamber and the cap rock also significantly reduces the thermal efficiency of the injected steam. In this situation, an MFAGD process is proposed to enhance the recovery performance of SAGD. It fully combines the advantages of SAGD and gas-assisted-gravity-drainage and tremendously improves the development effect of thick, heavy oil reservoirs. Especially for offshore thick, heavy oil reservoirs, compared with conventional SAGD, this technique has a higher recovery rate and is highly efficient. Within the limited service life of an offshore platform, through the operation of the MFAGD process, we can get higher oil recovery.

Fig. 1.11 compares pure SAGD and SAGD-MFAGD. After MFAGD, the steam chamber is shaped like a liquid drop instead of a conventional inverted

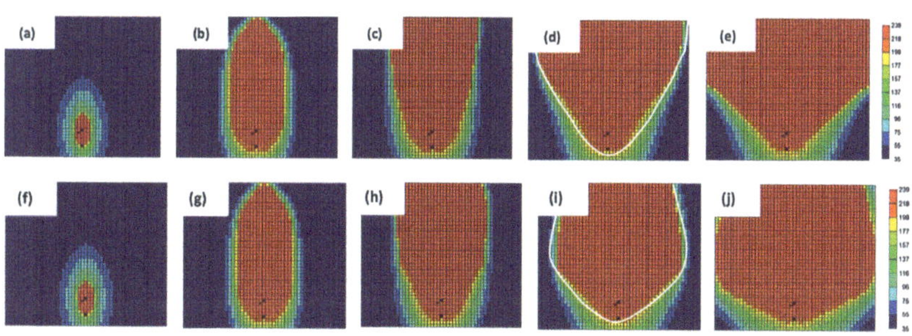

FIGURE 1.11 Temperature distribution of pure steam-assisted gravity drainage (SAGD) and SAGD-multithermal fluid-assisted gravity drainage (MFAGD) processes: (A–E) Different stages of pure SAGD processes; (F–J) Different stages of SAGD-MFAGD processes.

triangle in SAGD. For SAGD-MFAGD, MFAGD was operated once the steam chamber of SAGD reached the reservoir top. Comparatively, the top of the steam chamber in the SAGD process reached the reservoir boundary first. For MFAGD, the middle part of the steam chamber reached the reservoir boundary first, mainly because of the effect of an NCG fraction, especially the N_2 fraction in MTFs. Because N_2 is an insoluble gas in heavy oil, after injection it rises to the reservoir top rapidly and hinders the lateral expansion of a steam chamber.

1.7 Hybrid thermochemical processes

Chemical additives can be also applied to improve the performance of thermal recovery. Different from the chemical additives we usually employ in light oil reservoirs, for heavy oil reservoirs, chemical additives must afford the high temperature condition of steam injection operation. Compared with the two hybrid processes, a hybrid thermochemical process is usually applied to control a steam injection profile, especially with NCG foam and gel systems.

Alkaline, surfactant, and polymer are the three main chemical additives used in a thermal recovery process. First, for a hybrid steam-alkaline process, in addition to conventional steam-based EOR mechanisms, extra mechanisms are emulsification, wettability alteration, interface tension (IFT) reduction, and rigid film breaking. Commonly used alkali include Na_2CO_3 and NaOH. When remaining oil saturation after primary recovery is low, steam-alkaline flooding is more effective and can recover more original oil-in-place than conventional steam flooding under a similar condition [124,125]. However, considering the problem of scaling inside a reservoir, the application of HASP has always been severely restricted.,

Second, a hybrid steam-surfactant process (HSSP) uses a small amount of surfactant coinjected with steam to enhance the oil recovery of a steam-based recovery process. The mechanisms involve IFT reduction, wettability alteration, oil relative permeability enhancement, and in situ emulsification [126]. Compared with the recovery performance of SAGD, HSSP can further increase an oil production rate, reduce the CSOR and enhance the ultimate oil recovery [127]. Biodiesel (BD) has been also applied as a surfactant additive in HSSP. Based on an experimental investigation, the application of BDs (fatty acid methyl esters) can significantly increase the efficiency of bitumen recovery in SAGD and CSS processes [128]. Srivastava and Castro provided a successful field application of surfactant additives (thin film spreading agents) to enhance the recovery performance of thermal processes in heavy oil reservoirs [129]. HSSP has been also applied to enhance heavy oil recovery in some heavy oil fields.

Third, for the polymer additive, different from alkaline and surfactant, it is generally used for a nonthermal recovery process, especially for a waterflooded heavy oil reservoir because of the low-temperature resistance of polymer. Therefore, polymer is well-suited for an EOR process for offshore heavy oil fields, such as the Bressay Field and Bently Field in North Sea, UK and Bohai oil field in China [130,131]. The SZ36-1 reservoir in the Bohai oil field is one of the most successful polymer-based heavy oil EOR projects [132]. Some high-temperature polymers have been proposed and tested. Considering the high-viscosity characteristic of polymer, polymer injection is technically feasible to improve SAGD performance in oil sands reservoirs with top water. Based on a numerical investigation, it was observed that by injecting a polymer, a stable high viscosity layer can be developed at the bottom of top water [133]. On the other hand, another hybrid thermochemical process, alkali-co-solvent-polymer, was developed [134,135]. In this process, different additives perform differently during operations. Alkali is used to reduce interfacial tension, polymer is used to increase water viscosity for mobility control, and cosolvent is used to optimize phase behavior and prevent the formation of highly viscous emulsions [134]. Then, through a combination of electrical resistance preheating and hot water flooding, this process can well handle the challenges of injectivity, heating, and oil displacement and production [135].

1.7.1 Noncondensable gas–foam

NCG-foam in enhanced heavy-oil recovery processes can be applied to change the flow direction of steam in porous media and improve a steam injection profile. Especially for heavy oil reservoirs with a serious steam breakthrough path, NCG-foam is an important plugging agent to improve a steam injection profile. Therefore, it is usually applied to enhance the recovery performance of a steam flooding process. On the other hand, considering its sound plugging capability, NCG-foam can be also applied to control the water-coning behavior of a bottom/edge aquifer. Commonly used NCGs to generate an NCG-foam system include N_2, CO_2, and CH_4. The applied hydrosoluble surfactant (foaming agent) is usually high-temperature resistant. There are many types of foaming agents, including anionic, cationic, nonionic, and amphoteric. To form an effective foam system in a formation, the selection of a foaming agent is the dominating factor. The main EOR mechanisms of NCG-foam in a steam flooding process are that:

(1) The addition of surfactant additive can reduce the oil–water IFT and improve the displacement efficiency.
(2) NCG-foam can increase residual gas saturation and gas apparent viscosity and reduce gas mobility.
(3) NCG-foam can improve a steam injection profile in a formation.
(4) NCG-foam can increase the thermal efficiency of injected steam.

For thermal recovery processes in heavy oil reservoirs, through the injection of a foam system, steam viscosity increases and steam mobility is reduced. NCG-foam can effectively control viscosity fingering, gravity override, and steam breakthrough. NCG-foam can effectively plug a steam channeling path and improve sweep efficiency in heterogeneous heavy oil reservoirs. On the other hand, some other chemical additives have also been applied to enhance the performance of foam in thermal recovery processes, including alkaline and polymer. For an alkaline-enhanced foam process, the addition of alkaline can further increase the foam-propagation rate and improve foam mobility. Also, the presence of alkaline can reduce the oil—-water interfacial tension to enable the formation of an oil-in-water emulsion. Thus, the residual oil saturation is reduced [136]. For a polymer-enhanced foam (PEF) process, the addition of polymer can further increase the viscosity of the liquid membrane and reduce the liquid drainage rate in the liquid membrane. Thus, gas diffusivity is significantly reduced. Compared with a conventional foam system, the stability and plugging performance of a PEF system are more remarkable [137].

1.7.2 High-temperature gel

A high temperature gel (HTG) agent is another effective chemical additive to control a steam injection profile in heavy oil reservoirs. Compared with an NCG-foam system, HTG is a blocking agent with higher plugging capacity. For an HTG operation process in a poststeam-stimulated heavy oil reservoir with a widely distributed steam breakthrough path, the gel solution first enters a high permeability path during the injection process. Then, after a certain time, the gel status with a large strength is achieved. The viscosity of gel will increase tremendously. For some gel systems, it can reach 25,000 mPa·s. Therefore, a gel wall can be formed and a chief zone will be effectively plugged. Thus, the next steam injected can reenter the low-permeability path and improve the steam injection profile. However, different from the conventional gel system for water-flooded reservoirs, HTG in heavy oil reservoirs has higher thermal stability. The tolerable temperature of HTG is generally about 200°C. To increase the properties of HTG in thermal recovery processes, a number of HTG systems have been proposed [138,139]. For some HTG systems, after adding some chemical agents, the tolerable temperature can even be 250—300°C.

For field operations, the selection of a suitable HTG system for a specific reservoir is related to the formation temperature, salinity, and hardness level of the water used as well as the lithology of the reservoir [140]. The cost factor is also nonnegligible, especially when oil prices are low. A thermo-reversible gel is a good candidate for in-depth conformance control in steam-stimulated wells [141]. A novel HTG has been proposed to control a steam breakthrough path in heavy oil reservoirs, and a parallel sand-pack experiment was also conducted to evaluate its performance [140]. Based on experimental observation, this gelling system has strong salt and dilution resistance [139]. It can effectively plug a steam breakthrough path and force the subsequent steam to enter into a low-permeability path.

1.7.3 Surfactant assisted—steam-assisted gravity drainage

A surfactant additive in a surfactant assisted-SAGD (SA-SAGD) process usually refers to a VR. It is a special hybrid steam-surfactant process. Similar to the MFAGD process discussed previously, SA-SAGD is also applied at the later stage of SAGD, when a steam chamber front contacts the reservoir top. The addition of VR improves the expansion of a steam chamber. Once it is coinjected with steam, a higher oil drainage rate can be observed. In a steam chamber front, a mixture of condensate and surfactant can further reduce oil viscosity, and the downward expansion rate of the steam chamber is accelerated. Thus, the total recovery period is expanded. Based on a 3D experiment investigation, SA-SAGD can increase the recoverable oil reserve by about 17% [142].

1.7.4 Chemical additive and foam-assisted steam-assisted gravity drainage

The recovery performance of SAGD highly depends on steam chamber conformance and the heating reservoir volume. However, considering long horizontal wellbores, reservoir heterogeneity tends to affect steam chamber expansion greatly. For example, top water can work as an enthalpy sink; it absorbs heat and reduces thermal efficiency. Low permeable zones are less accessible to steam; they cause the reserved oil to be bypassed and decrease oil recovery. Based on the discussion in Section 1.7.1, an NCG-foam system can improve gas apparent viscosity and reduce gas relative permeability, which controls steam mobility and preserves steam chamber conformance. The successful application of NCG-foam in CSS and steam flooding processes can provide some new insights into a foam-assisted SAGD process in terms of chemical additive and foam-assisted SAGD (CAFA-SAGD).

Considering foam generation, foam collapse, and foam movement mechanisms, Chen adopted a local-equilibrium foam simulator to analyze the performance of foam-assisted SAGD [143]. The simulation results showed that foam was conducive to improving steam efficiency in SAGD. Li analyzed EOR mechanisms with a homogeneous simulation model and further applied a foam model into a heterogenous Athabasca simulation model to check the impact of foam on SAGD by comparing oil production and an SOR [144]. She considered the mechanisms of steam mobility control, IFT reduction, and emulsification. The injection of surfactants and generation of in situ steam foam at a steam chamber boundary controlled steam vertical development and promoted steam lateral movement, which benefited uniform steam chamber growth and heating area expansion (Fig. 1.12). Moreover, steam foam was

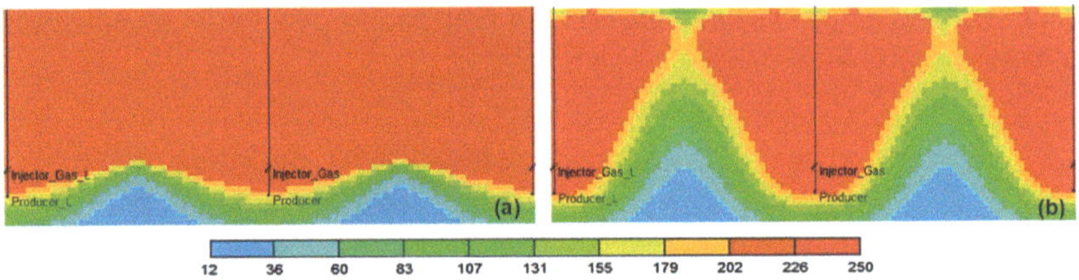

FIGURE 1.12 Temperature distribution comparison between steam-assisted gravity drainage (SAGD) and chemical additive and foam-assisted (CAFA)-SAGD model: (A) SAGD; (B) CAFA-SAGD.

likely to exist in high-permeable layers to preserve steam chamber conformance. Abdul Ghani et al. adopted a homogeneous reservoir simulation model with typical Athabasca oil sands properties to find the best injection strategy for SAGD production [145]. They maintained that an SAGD/surfactant slug alternative injection method is the best way to preserve surfactants in a steam chamber and further steam foam. Adetunji et al. tested 12 surfactants as foam agents at 250°C for SAGD production [146]. They used methane as a gas phase and found that methane—steam coinjection brought about the highest foam stability. The effects of surfactant concentration, oil saturation, and an amount of methane were also studied.

Most CAFA-SAGD investigations are conducted with numerical simulations. For the numerical simulation of foam, mechanism models and empirical models are two commonly used methods [144,147,148]. CAFA-SAGD experiments mainly focus on selecting foaming agents. Both the foaming ability and high-temperature stability of the surfactant candidates with different concentrations are measured. Other chemical additives such as cosurfactants or nanoparticles are also tested as a potential foam stabilizer [149].

CAFA-SAGD is a potential technology to improve SAGD performance; it relies on high-temperature bubbles to control steam mobility to preserve steam chamber conformance and increase heat efficiency. Foam stability in a high-temperature and high-pressure environment is the prerequisite for CAFA-SAGD. More investigations need to be carried out to find the best foaming agent to extend the lifetime of steam bubbles to control steam chamber growth better.

1.8 Field implementation of hybrid enhanced oil recovery processes

All three hybrid EOR processes have been applied to enhance the recovery of some poststeamed heavy oil reservoirs in the world. In this section, several field pilot tests of these three hybrid processes are presented.

1.8.1 Field tests of hybrid thermo-solvent processes

Most of the hybrid thermo-solvent processes discussed earlier were been tested in the field [150]. These processes are not sensitive to oil viscosity, in that they can be applied to heavy oil reservoirs with a very high viscosity. There are many field implementations of these hybrid processes in oil sands reservoirs in Canada, including LASER, SAP, and ES-SAGD, as shown in Table 1.3. The oil viscosity in most tested oil sands reservoirs in Table 1.3 is over 100×10^4 mPa·s. Differently, in China, a cyclic steam-CO_2 coinjection process (SAP) and a continuous steam-CO_2 coinjection process (SESF) were applied to improve the recovery performance of poststeamed heavy oil

TABLE 1.3 Hybrid thermo-solvent process field implementations in Canada.

Process	Field	Operation corporation	Solvent type	Steam–oil ratio reduction
Liquid addition to steam for enhancing recovery	Cold Lake	Imperial Oil, 2002	C_{5+} (6%)	32%
Solvent-enhanced steam flooding/ solvent-aided process	Senlac	EnCana, 2002	C_4	40%
	Christina Lake	—	C_4 (10%)	31%
Expanding solvent–steam-assisted gravity drainage	Firebag	Suncor, 2003	Naptha (Ave. C_8)	—
	Christina Lake	EnCana, 2004	C_4	66% (from 5 to 1.6)
	Long Lake	Nexen, 2006	Jet B, C_7–C_{12} (5%–10%)	7%
	Grosmont	Laricina Energy, 2013	C_3, >C_{5+}	25%–30%
	Algar	Connacher, 2012	Diluent (10%–15%)	16%
	Cold Lake	ExxonMobil and Imperial Oil Resources, 2013	Light hydrocarbon (~20%)	—
	Surmont	ConocoPhillips, 2014	NGL mixture	14%

reservoirs in Liaohe, Shengli, and Xinjiang oil fields [151,152]. The oil viscosity in the tested heavy oil reservoirs in China was relatively lower than in the oil sands reservoirs in Canada. Furthermore, for the SAGD operation test in the Fengcheng oil field in Xinjiang, a startup approach of xylene-steam coinjection was to accelerate a preheating process between an injector and a producer [153]. Results indicated that the startup time of SAGD was reduced by about 60 days compared with pure steam circulation. Some typical hybrid thermo-solvent processes will be discussed next.

1.8.1.1 Liquid addition to steam for enhancing recovery process

In field applications, LASER has been pilot-tested in the H22 pad in the Cold Lake area by Imperial Oil [150]. This pad was a standard 20-well pad. In Mar. 2002, about six CSS cycles were completed in pad H22 and recovered 17% original bitumen in place (OBIP). Then, eight wells in pad H22 were selected for coinjection of solvent into steam to start a LASER operation. In addition, the other 12 CSS wells remained a CSS recovery process. In this field test, a 6% volume fraction of C5+ was coinjected. After the first LASER cycle, about 80% of the injected solvent was recovered.

1.8.1.2 Expanding solvent–steam-assisted gravity drainage process

Compared with the LASER and SESF processes, ES-SAGD had more field pilot tests in Canada. There were many field implementations of ES-SAGD or SA-SAGD in oil sands resources in Canada, including the EnCana-Senlac pilot (butane coinjection, ES-SAGD), Nexen Long Lake pilot (ES-SAGD), Laricina Energy-Grosmont carbonate reservoir pilot (solvent cyclic SAGD, SC-SAGD), and Imperial Oil Resources-Cold Lake pilot (SA-SAGD). As shown in Table 1.3, both light and heavy hydrocarbons have been exploited in the ES-SAGD process.

Nexen's Long Lake ES-SAGD test was performed from Feb. 13 to Apr. 16, 2006. The solvent used in this test was a mixture of C_7–C_{12} (Jet B). It was coinjected with steam at a concentration of 10%. During the process, an operating pressure of 1400 kPa was applied. Under this pressure condition, the solvent mixture and steam were maintained in a vapor phase. In this process, once a steam chamber was filled with about 5% solvent, the solvent concentration in the coinjection was reduced to 5%. The 5% solvent concentration was maintained in the steam chamber [101]. After the operation, a slight increase in the oil rate and a small decrease in SOR were observed.

1.8.2 Field tests of hybrid thermal–noncondensable gas processes

Compared with other hybrid processes, hybrid thermal-NCG is more attractive in heavy oil reservoirs with a low oil viscosity ($\mu_o < 10{,}000$ mPa·s). That is because that as the oil viscosity increases, the solubility of NCG in heavy crude oil is gradually reduced. On the other hand, based on experimental investigations, the swelling factor of a heavy oil-NCG system is also reduced as the oil viscosity increases. Some typical hybrid thermal-NCG processes will be discussed next.

1.8.2.1 N_2-cyclic steam stimulation process

Compared with other NCG components, N_2 is the most commonly used NCG in hybrid thermal-NCG processes, especially when operation costs are considered. N_2 is usually characterized by low heat conduction efficiency and large swelling. Through the coinjection of N_2 and steam, the flowability of heavy crude oil can be significantly improved. For heavy oil reservoirs after multiple CSS cycles, N_2-CSS can increase the heating range, recover formation energy, and improve recovery performance. Since the 1970s, some pilot tests of N_2-CSS have been performed in heavy oil reservoirs in the United States and Canada. In China, this process has been applied to enhance heavy oil recovery in Liaohe, Shengli, Xinjiang, and Henan oil fields.

The JL, YL, YQ, BQ and XQ blocks in Henan oil field, Sinopec, China are heavy oil reservoirs. Their geological features are usually summarized by four words, "shallow, thin, viscous, and scattered." Their reservoir depth is about 500 m and the formation thickness is only 3–5 m. For these complicated heavy oil reservoirs, CSS was the first exploitation method. However, after long-term steam stimulation, continuing CSS cycles was no longer economical and the N_2-CSS process was applied to some blocks. Field test results indicated that after the operation, the average water-cut was reduced from 72% to 68% and the oil rate increased from 1.8 to 2.9 t/d. Compared with pure steam stimulation, the SOR of N_2-CSS was reduced by 20% and good recovery was obtained.

The Leng 42 heavy oil block in the Liaohe oil field, CNPC, China is another typical heavy oil reservoir. The reservoir permeability is 765×10^{-3} μm^2 and its oil viscosity is 1000–5000 mPa·s. Since its large-scale development in 1995, CSS has been the method of exploitation. After long-term CSS stimulation, the recovery performance was gradually unsatisfactory and a follow-up recovery technique was required. In 2001, N_2-CSS was applied in 11 CSS wells. After the operation, the recovery performance of the Leng 42 heavy oil block largely improved.

Another field pilot test of N_2-CSS is the Nare heavy oil reservoir in the Mansarovar oil field, Colombia [154]. Its formation depth is 572–732 m, the reservoir thickness is 39–60 m, the horizontal permeability is $700–4000 \times 10^{-3} \mu m^2$, and its oil viscosity is 2965–8890 mPa·s under reservoir conditions. CSS with horizontal wells was the primary recovery process. However, as CSS continued operating, the problems of serious steam breakthrough, low thermal efficiency of steam, and a high water cut became the top priorities. To improve the recovery performance of CSS, from Jun. to Oct. 2013, 23 wells in the Nare heavy oil reservoir performed N_2-CSS. Based on statistical data on 11 wells, the total N_2 volume injected was about 260×10^4 m^3. After its operation, the average oil rate per well increased from 6.6 to 9.0 t/d, the average water cut was reduced from 78% to 73%, and the estimated total increase in oil production volume was over 5000 t.

In addition to its application in an assisted CSS process, N_2 can be used to control bottom/boundary water coning behavior after multiple CSS cycles in heavy oil reservoirs. For some heavy oil reservoirs with a large aquifer, water coning behavior is usually the top concern after CSS. In the Henan oil field, Sinopec, N_2 was applied in the heavy oil blocks of XQ24, XQ25, BQ57, and BQ67 to hinder boundary water coning behavior. Its test effectiveness was about 60%. In some tests, a foaming agent was also injected to implement an antiwater coning process by N_2-foam.

1.8.2.2 Flue gas/multiple thermal fluids—cyclic steam stimulation process

Flue gas is a mixture of N_2 and CO_2. Therefore, a flue gas—based thermal recovery process usually combines the advantages of both N_2 and CO_2. In 2009, a flue gas-CSS process was first applied to improve the performance of CSS wells in the Shengli oil field, China. During the first cycle in well GDN5-604, the injection volume of N_2 was 21×10^4 m^3, the injection volume of CO_2 was 3.7×10^4 m^3, and the injection volume of steam was 480 m^3. After the operation, the average water cut was reduced by about 30%, the oil production rate increased by 7.3 t/d, and the cumulative incremental oil production was 1009 t. Based on the successful operation in this well, more heavy oil blocks in the Shengli oil field were selected as potential test reservoirs for flue gas-CSS. Especially for heavy oil reservoirs after multiple CSS cycles and heavy oil reservoirs with a huge aquifer, this process was attractive.

MTFs stimulation is another proposed hybrid thermal-NCG for heavy oil reservoirs [122,155]. As discussed in Section 1.6.1, MTFs are a mixture of different noncondensable gases. They are generated based on the combustion of fuel, air, and water. Therefore, the main composition of MTFs includes N_2 and CO_2. However, the performance of MTF stimulation is more attractive than flue gas-CSS. It combines the multiple advantages of different hybrid processes. The NB 35—2 reservoir is a complicated offshore heavy oil reservoir with multiple oil—water systems in Bohai Bay, CNOOC. The reservoir depth is 900—1300 m and its permeability is $60-5000 \times 10^{-3}$ μm^2. Considering the limited space in an offshore platform, conventional steam-based recovery is difficult to apply in offshore heavy oil reservoirs. In 2010, MTFs stimulation was performed in four wells in the NB 35-2 heavy oil reservoir, including B14m, B2S, B28h, and B29m. The average oil viscosity under reservoir conditions was 676 mPa·s. During the operation, the maximum liquid production rate was 178 m^3/d and the maximum oil production rate was 127 m^3/d. Compared with conventional steam recovery, the performance of MTF stimulation in offshore heavy oil reservoirs was more attractive.

1.8.3 Field tests of hybrid thermochemical processes

Hybrid thermochemical processes are usually applied in heavy oil reservoirs with serious reservoir heterogeneity and/or widely distributed major zones. Through these processes, a steam injection profile can be significantly improved. Typical hybrid thermochemical processes will be given next.

1.8.31 Noncondensable gas-foam process

An NCG-foam process is one of the most representative hybrid thermochemical processes. Commonly used methods include N_2-foam, CO_2-foam, and flue gas-foam. During field operations, to guarantee their effective performance, a strict evaluation scheme of a foaming agent is usually required, including its static performance (foam stability, foam volume, and foam size) and dynamic performance (its resistance factor and the period of validity). For heavy oil reservoirs with serious reservoir heterogeneity, by performing NCG-foam blocking, a more uniform steam injection profile can be obtained. For heavy oil reservoirs with bottom/boundary aquifers, by performing NCG-foam blocking, water coning behavior can be effectively controlled. In some cases, however, the capacity of conventional NCG-foam is no longer effective and it cannot plug serious channeling path in reservoirs. Therefore, for some EOR projects, a foam stabilizer is proposed. Specifically, polymer is a typical stabilizer in a foam system to enhance foam strength; it is called PEF. Telmadarreie and Trivedi used a micromodel to discuss the performance of CO_2-foam and CO_2-PEF injection in carbonate heavy oil recovery [156].

The LZ27 block in the Henan oil field, Sinopec is a large heterogeneous heavy oil reservoir. Table 1.4 shows the basic parameters of this reservoir. The

TABLE 1.4 Basic reservoir properties of LZ 27 heavy oil block.

Parameter	Value	Parameter	Value
Oil-bearing area (km^2)	0.89	Porosity (%)	34.82
Geological reserves (10^4 t)	21.6	Formation pressure (MPa)	2.58
Reservoir depth (m)	230	Reservoir temperature (°C)	25.2
Thickness (m)	4.5	Oil density (g/cm^3)	0.9698
Permeability (10^{-3} μm^2)	2246	Oil viscosity (mPa·s)	12,749

permeability max-min ratio in this block is above 5.0. In Sep. 2009, because of gradually unsatisfactory recovery performance of CSS, four inverted nine-spot well groups in this heavy oil reservoir were selected for a steam flooding test. However, in Jan. 2010, the oil production rate of these steam flooding well groups was dramatically reduced from 28.8 to 13.1 t/d. Based on a performance analysis, early steam breakthrough was the primary reason. To control steam breakthrough effectively and increase the sweep efficiency of steam, an N_2-foam process was performed in the four well groups. Table 1.5 shows the detailed operation parameters of this N_2-foam test and Table 1.6 gives its

TABLE 1.5 Operation parameters of noncondensable gas-foam operation in LZ27 heavy oil block.

Well group	Date	Volume of foaming agent (t)	N_2 (Nm^3)	Foaming method
1	Jul. 7–12, 2010	5.4	30,000	Surface foaming
2	Jul. 12–17, 2010	5.6	30,000	Surface foaming
3 (first)	Jul. 31-Aug. 3, 2010	4.0	19,200	Surface foaming
4	Aug. 6–9, 2010	6.0	24,000	Surface foaming
3 (second)	Oct. 22–27, 2010	5.5	33,000	Surface foaming

TABLE 1.6 Recovery performance of noncondensable gas-foam operation in LZ27 heavy oil block.

Well group	Date	Incremental oil production (t)	Injection pressure increased (MPa)
1	Jul. 7–12, 2010	844.1	0.60
2	Jul. 12–17, 2010	930.1	0.90
3 (first)	Jul. 31-Aug. 3, 2010	134.3	0.55
4	Aug. 6–9, 2010	118.7	0.49
3 (second)	Oct. 22–27, 2010	37.8	0.90

TABLE 1.7 Recovery performance of multistage noncondensable gas-foam slug process.

Well no.	Date	Volume of foaming agent (t)	Cyclic increase in oil production (t)
1	2015.11	8	276
2	2015.11	6	144
3	2015.7	5	244
4	2015.7	5	173
5	2015.8	6	126

recovery performance. As shown, in 2010, five N_2-foam processes were performed in total. The total foaming agent injected was 26.5 t and the total N_2 volume injected was 136,200 Nm^3. As of Dec. 2010, the cumulative incremental oil production was about 2065 t.

In addition, during a field operation, the performance of a single-stage NCG-foam slug is often ineffective in some tests. One of the most important reasons is serious reservoir heterogeneity. After CSS, it results in more complicated fluid flow in heterogeneous reservoirs. Therefore, a multistage NCG-foam slug process was proposed based on temperature distribution characteristics after CSS. From a reservoir section far from a well to a section around a bottom hole, the temperature gradually increases. Therefore, a combined profile control method of high- to low-temperature NCG-foam processes is desirable. This method can give full play to the advantages of each NCG-foam system. It is more attractive economically. A multistage NCG-foam slug process was applied in the Shengli and Henan oil fields, Sinopec. In 2015, five wells performed this process, as shown in Table 1.7. Total cumulative oil production reached 963 t. In a typical CSS well, a two-NCG-foam slug process was designed. For each slug, the volume of the foaming agent solution was 160 m^3 and the N_2 volume was 25,000 Nm^3. After the operation, the average water cut was reduced by 10% and the cyclic increase in oil production was about 276 t.

1.8.3.2 High-temperature gel process

Compared with NCG-foam, an HTG system is a chemical plugging additive with high viscosity. The plugging ability of a gel system is higher than that of NCG-foam. In field operations, although both NCG-foam and gel systems can effectively plug a steam channeling path and improve the steam injection profile, they have some differences. First, compared with NCG-foam, the major advantage of a gel system is that it is relatively insensitive to some reservoir conditions, such as the presence of oil. For NCG-foam, the presence of oil significantly affects the foam's stability and plugging strength. Second, gels are more sensitive to changes in pressure gradients. In an operation, a critical breakthrough pressure gradient is another important parameter to evaluate the performance of HTG. Although a gel system has been proposed for many years, the application of HTG in thermal recovery processes is limited mainly because of unsatisfactory gelling properties. Because some HTG systems with a higher tolerable temperature have been proposed (>200°C), HTG was applied in poststeam injection heavy oil reservoirs. The Henan and Shengli oil fields, Sinopec are the two representative heavy oil fields [18,139].

From 2015 to 2018, HTG process applied to control a steam injection profile in 12 typical steamed wells in the Shengli oil field. After the operation, cumulative oil production increased by 4203.5 t, average incremental oil production per well was 350 t, and the average water cut was reduced by over 8.0%. During the operation, the gel solution was a liquid with low viscosity before gelling. The optimal gelling temperature was 80–160°C. Fig. 1.13 shows the recovery performance of a typical well among the 12 of them. As shown, before HTG, the average water cut of this well had reached 97.5%. After the operation, the average water cut was reduced from 97.5% to 91.5%, the oil production rate increased from 2.3 to 5.0 t/d, and the total incremental oil production was 209 t.

1.8.3.3 New hybrid thermochemical processes

Except for NCG-foam and HTG, hybrid thermochemical processes are also proposed for poststeamed heavy oil reservoirs. Coal ash, oil sludge,

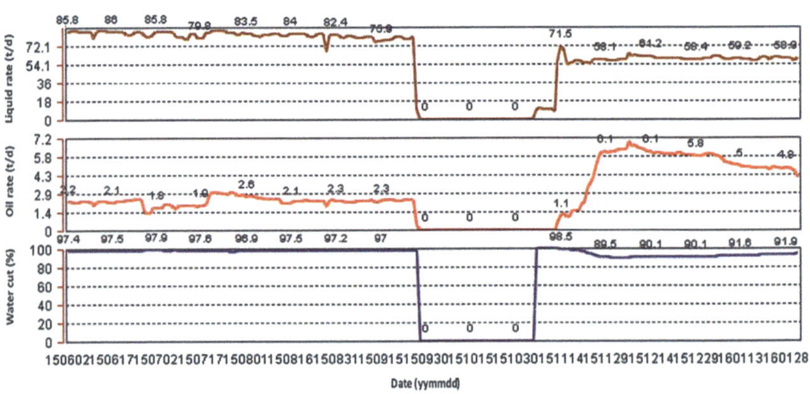

FIGURE 1.13 Recovery performance of high-temperature gel process in a typical well.

thermosetting plugging agents, and low-temperature expandable graphite systems are typical representatives [157,158]. The first three processes have been applied to improve the recovery performance of steamed heavy oil reservoirs.

First, for coal ash (Fig. 1.14A), it was used to improve a steam injection profile in the XQ25 and YQ3 heavy oil blocks in the Henan oil field, Sinopec. In 2010, this process was performed in about 10 well groups; its effectiveness was 88.3%. By the end of Dec. 2010, the cumulative increase in oil production was 1888.4 t. Well X2306 was a typical well in the N3 heavy oil block in the Henan oil field. Until Jun. 2010, this well had performed six CSS cycles. An obvious steam breakthrough phenomenon was observed between this well and five adjacent CSS wells. Therefore, in Jul. 2010, steam injection profile control by coal ash was performed in this well. After testing, five steam breakthrough paths were effectively controlled. By the end of Dec. 2010, the cumulative increase in oil production reached 132.1 t. The recovery performance of this well group was significantly improved.

For oil sludge (Fig. 1.14B), it is another environmentally friendly profile control process. This process has been applied to improve the performance of steam-stimulated wells in the Liaohe oil field, CNPC and the Henan oil fields, Sinopec. In the Liaohe oil field, until 2018, about 45 steamed wells performed an oil sludge test. Before this sludge injection process, all tested wells had experienced seven to 10 CSS cycles, and serious steam breakthrough was observed. In these 45 tests, about 133,200 t of sludge was injected; the average injection volume of sludge for each well was about 800–2000 t. After the operation, a more than 20% increase in cyclic oil production was obtained.

For a thermosetting plugging agent, it is a newly proposed in-depth profile control additive. It is a chemical additive with a high molecular weight and is usually produced from a polycondensation reaction of reactants with a low molecular weight. This plugging agent has a high strength and long effectiveness time in reservoirs. It can resist the conditions of high salinity and high temperature. It has good performance in high permeability layers, fractures,

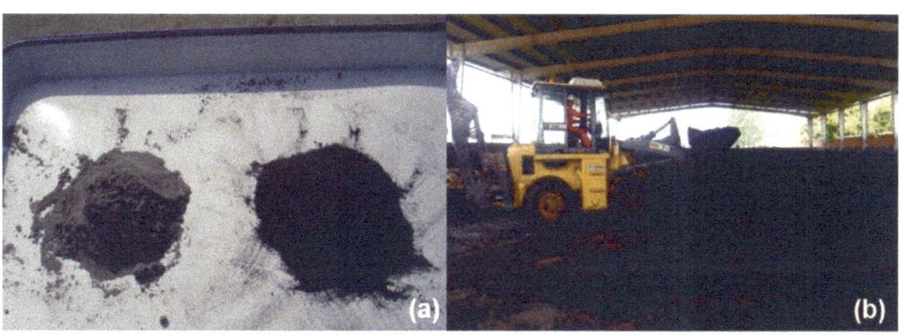

FIGURE 1.14 (A) Coal ash; (B) Oil sludge.

large-scale paths, and cavities. It has been applied to improve the performance of steam flooding in the P612 heavy oil block in the Shengli oil field, Sinopec. After the operation, the average oil production rate increased from 3 to 5 t/d. The water cut and liquid temperature that were produced were also significantly reduced.

Steam injection techniques are the primary exploitation methods for heavy oil resources in the world. However, although heavy oil reservoirs continue to develop, a series of problems after steam injection need to be solved for oil companies. Therefore, when traditional steam-based processes are no longer effective, a newer generation of emerging recovery processes for heavy oil reservoirs is urgently required, particularly for low oil processes. Among many problems, how to increase the sweep efficiency of injected steam is the first issue. The answer to this question is addressed in this book. Hybrid EOR processes are the main technical direction. By using the heat that remains after steam injection, all three hybrid EOR processes presented in this chapter can be effectively applied to improve the recovery performance of heavy oil reservoirs. In the next chapter, the current status of steam-based recovery and some urgent problems will be discussed in detail.

References

[1] https://petrowiki.org/Heavy_oil.

[2] Alberta energy regulator. http://www.eub.gov.ab.ca/.

[3] Liu Z, Wang H, Blackbourn G, et al. Heavy oils and oil sands: global distribution and resource assessment. Acta Geol Sin 2019;93(1):199–212.

[4] Dusseault MB. Comparing Venezuelan and Canadian heavy oil and tar sands. In: Paper 2001-061 presented at the petroleum society's Canadian international petroleum conference, Calgary, Alberta, Canada, June 12–14; 2001.

[5] Meyer RF, Attanasi ED, Freeman PA. Heavy oil and natural bitumen resources in geological basins of the world. Open file-report 2007-1084. U.S. Geological Survey; 2007.

[6] Speight James G. Heavy oil production processes. Gulf Publishing Company; 2013.

[7] https://www.nrcan.gc.ca/our-natural-resources/energy-sources-distribution/10728.

[8] Jimenez J. The field performance of SAGD projects in Canada. In: IPTC 12860 presented at the international petroleum technology conference, Kuala Lumpur, Malaysia, December 3–5; 2008.

[9] Banerjee DK. Oil sands, heavy oil & bitumen: from recovery to refinery. PennWell Corporation; 2012.

[10] Jia C. Status and reserve evaluation method of oilsands resources. Beijing: Petroleum Industry Press; 2006.

[11] Khan J, Parag D. Twenty-five years of oil recovery by steam injection. In: SPE 24198 presented at the SPE/DOE eighth symposium on enhanced oil recovery, Tulsa, Oklahoma, April 22–24; 1992.

[12] Farouq Ali SM. Practical heavy oil recovery. Lecture Notes, University of Calgary; 2007.

[13] Nasr TN, Ayodele OR. Thermal techniques for the recovery of heavy oil and bitumen. In: SPE 97488 presented at the SPE international improved oil recovery conference in Asia Pacific, Kuala Lumpur, Malaysia, December 5–6; 2005.

[14] Dong X, Liu H, Wang Q, Pang Z, Wang C. Non-Newtonian flow characterization of heavy crude oil in porous media. J Pet Explor Prod Technol 2013;3(1):43–53.

[15] Dong X, Liu H, Chen Z, et al. Enhanced oil recovery techniques for heavy oil and oilsands reservoirs after steam injection. Appl Energy 2019;239:1190–211.

[16] Liu HQ. Principle and design of thermal oil recovery processes. Petroleum Industry Press; 2013 [in Chinese].

[17] Alvarez JM, Han S. Current overview of cyclic steam injection process. J Pet Sci Res 2013;2(3):116–27.

[18] Huo GR, Li XM, Zhang GQ. Thermal oil recovery technologies of heavy oil reservoirs in Shengli oilfield. Petroleum Industry Press; 1999.

[19] Zhang FL, Zhao HY. Steam based recovery technologies of heavy oil reservoirs in Liaohe oilfield. Petroleum Industry Press; 2007.

[20] Chang J. Understanding HW-CSS for thin heavy oil reservoir. In: SPE 165386 presented at the SPE heavy oil conference Canada, Calgary, Alberta, Canada, June 11–13; 2013.

[21] Hou J, Zhou K, Zhao H, Kang X, Wang S, Zhang X. Hybrid optimization technique for cyclic steam stimulation by horizontal wells in heavy oil reservoir. Comput Chem Eng 2016;84(4):363–70.

[22] Al-Murayri MT, Maini BB, Harding TG, Oskouei J. Multicomponent solvent co-injection with steam in heavy and extra-heavy oil reservoirs. Energy Fuels 2016;30:2604–16.

[23] Coskuner G. A new process combining cyclic steam stimulation and steam-assisted gravity drainage: hybrid SAGD. J Can Petrol Technol 2009;48(1):8–13.

[24] Xu J, Chen Z, Yu T, Cao J. Numerical thermal simulation and optimization of hybrid CSS/SAGD process in Long Lake with lean zones. In: SPE 170149 presented at the SPE heavy oil conference-Canada, Calgary, Alberta, Canada, June 10–12; 2014.

[25] Escobar MA, Valera CA, Perez RE. A large heavy oil reservoir in Lake Maracaibo basin: cyclic steam injection experiences. In: SPE 37551 presented at the SPE international thermal operations & heavy oil symposium, Bakersield, California, February 10–12; 1997.

[26] Donnelly JK. The best process for Cold Lake: CSS vs. SAGD. J Can Petrol Technol 2000;39(8):5–7.

[27] Nehring R, Hess R, Kamionski M. The heavy oil resources of the United States. Research report R-2946-DOE. The U.S. Department of Energy; 1983.

[28] Farouq Ali SM. CSS - Canada's super strategy for oil sands. J Can Petrol Technol 1994;33(9):16–9.

[29] Batycky J. An assessment of in situ oil sands recovery processes. J Can Petrol Technol 1997;36(9):15–9.

[30] Novak J, Edmunds N, Cimolai M. A history match of CSS recovery in the Grosmont. In: Paper presented at the Canadian international petroleum conference, Calgary, Alberta, June 12–14; 2007.

[31] Butler RM. Thermal recovery of oil and bitumen. Englewood Cliffs, New Jersey: Prentice Hall; 1991.

[32] Holly C, Mader M, Soni S, Toor J. Alberta energy oil sands production profile. Energy Technical Services - Resource Development Policy Division; 2016. January 31.

[33] Edmunds N, Barrett K, Solanki S, Cimolai M. Prospects for commercial bitumen recovery from the Grosmont carbonate, Alberta. J Can Petrol Technol 2009;48(9):26–32.

[34] Alvarez JM, Sawatzky RP, Forster LM, Coates RM. Alberta's bitumen carbonate reservoirs – moving forward with advanced R&D. In: Paper presented at the second world heavy oil congress, Edmonton, AB, March 10–12; 2008.

[35] Mendoza H, Padron A, Portillo F. Steam stimulation in horizontal wells pilot test in Venezuela. In: Paper presented at the annual technical meeting, Calgary, Alberta, June 8–11; 1997.

[36] Shouliang L, Zhang YT, Wu SH, Liu SQ, Li XL, Li SL. Status of heavy oil development in China. In: SPE 97844 presented at the SPE international thermal operations and heavy oil symposium, Calgary, Alberta, Canada, November 1–3; 2005.

[37] Shandrygin Al, Lutfullin A. Current status of enhanced recovery techniques in the fields of Russia. In: SPE 115712 presented at the SPE annual technical conference and exhibition, Denver, Colorado, USA, September 21–24; 2008.

[38] Valbuena OH, Bernal ME, Ramon JC, Tian X. First extra-heavy-oil development in caguan-putumayo basin, Colombia, capella field. In: SPE 171077 presented at the SPE heavy and extra heavy oil conference: Latin America, Medellín, Colombia, September 24–26; 2014.

[39] Patarroyo M, Carvajal G, Milena Q, Urdaneta J, Gomez A, Pacheco J. New art of building horizontal and highly deviated wells help maximize recovery and production from heavy oil fields in Colombia. In: SPE 170009 presented at the SPE heavy oil conference-Canada, Calgary, Alberta, Canada, 10–12 June; 2014.

[40] Aguilar MG, Mares E, Cabra CM, Peralta OO. CSS performance in sands of samaria tertiary field, Mexico. In: SPE 171091 presented at the SPE heavy and extra heavy oil conference: Latin America, Medellín, Colombia, September 24–26; 2014.

[41] Al-Manhali M, Al-Rawahi M, Al-Hinai S, Alwazeer A, Al-Shaqsi K, Brissenden SJ. Application of CSS to develop a thick heavy oil field in south Oman. In: SPE 172855 presented at the SPE international heavy oil conference and exhibition, Mangaf, Kuwait December 8–10; 2014.

[42] Bettembourg S, Holyoak S, Alwazeer A, Manhali M, Rawahi M, Habsi A. Key learnings from first 2 Years of a full field CSS development in Oman. In: SPE 179833 presented at the SPE EOR conference at oil and gas west Asia, Muscat, Oman, March 21–23; 2016.

[43] Blevins TR, Billingsley RH. The ten-pattern steamflood, Kern River field, California. J Petrol Technol 1975;27(12):1505–14.

[44] Williams LL, Fong WS, Mridul K. Effects of discontinuous shales on multizone steamflood performance in the Kern River field. SPE Reservoir Eval Eng 2001;4(5):350–7.

[45] Imperial Oil. Cold Lake approvals 8558 and 4510. 2014 annual performance review. 2014.

[46] Greaser GR, Shore RA. Steamflood performance in the Kern River field. In: SPE 8834 presented at the SPE/DOE enhanced oil recovery symposium, Tulsa, Oklahoma, April 20–23; 1980.

[47] Oglesby KD, Blevins TR, Rogers EE, Johnson WM. Status of the 10-pattern steamflood, Kern River field, California. J Petrol Technol 1982;34(10):2251–7.

[48] Restine JL, Graves Jr WG, Eliasa R. Infill drilling in a steamflood operation: Kern River field. SPE Reservoir Eng 1987;2(2):243–8.

[49] Bousaid IS. Hot-water and steamflood studies using Kern River oil. In: SPE 21543 presented at the SPE international thermal operations symposium, Bakersfield, California, February 7–8; 1991.

[50] An JQ, Li J, Jiang H. Steam-flood trial and research on mid-deep heavy oil reservoir QI40 block in Liaohe oilfield. In: SPE 104403 presented at the international oil & gas conference and exhibition in China, Beijing, China, December 5–7; 2006.

[51] Xiao WQ, Gao XT, Zhang YX, Zhu J, Yan YF. Feasibility and pilot test of heavy oil reservoir steam flooding in henan oilfield. J Oil Gas Technol 2008;30(1):341–3.

[52] Wei RF. Study on Steam flooding supporting technology of extra-heavy oil and external expansion plan of Wa-38 block. Mater Thesis. Northeast Petroleum University; 2015 [in Chinese].

[53] Butler RM. A new approach to the modelling of steam-assisted gravity drainage. J Can Petrol Technol 1985;24(3):42–51.

[54] Butler RM. Rise of interfering steam chambers. J Can Petrol Technol 1987;26(3):70–5.

[55] Al Bahlani AM, Babadagli T. A critical review of the status of SAGD: where are we and what is next?. In: SPE 113283 presented at the SPE western regional and Pacific section AAPG joint meeting, Bakersfield, California, USA, March 29–April 4; 2008.

[56] Dong X, Liu H, Zhang Z, Lu C, Fang X, Zhang G. Feasibility of the steam-assisted-gravity-drainage process in offshore heavy oil reservoirs with bottom water. In: OTC 24763 presented at the offshore technology conference Asia held in Kuala Lumpur, Malaysia, March 25–28; 2014.

[57] Xi C, Qi Z, Jiang Y, Han W, Shi L, Li X, Wang H, Zhou Y, Liu T, Du X. Dual-horizontal wells SAGD start-up technology: from conventional steam circulation to rapid and uniform electric heating technology. In: SPE 189241 presented at the SPE symposium: production enhancement and cost optimisation, Kuala Lumpur, Malaysia, November 7–8; 2017.

[58] Butler RM. Thermal recovery of oil and bitumen. GravDrain's Blackbook; 1997.

[59] Al Bahlani AM, Babadagli T. SAGD laboratory experimental and numerical simulation studies: a review of current status and future issues. J Petrol Sci Eng 2009;68:135–50.

[60] Polikar M, Cyr TJ, Coates RM. Fast-SAGD: half the wells and 30% less steam. In: SPE 65509 presented at the SPE/petroleum society international conference on horizontal well technology, Calgary, Alberta, Canada. November 6–8; 2000.

[61] Gotawala DR, Gates ID. Stability of the edge of a SAGD steam chamber in a bitumen reservoir. Chem Eng Sci 2011;66:1802–9.

[62] Mojarad M, Dehghanpour H. Analytical modeling of emulsion flow at the edge of a steam chamber during a steam-assisted-gravity-drainage process. SPE J 2016;21(2):353–63.

[63] Akin S. Mathematical modeling of steam Assisted gravity drainage. SPE Reservoir Eval Eng 2005;8(5):372–6.

[64] Sharma J, Gates I. Multiphase flow at the edge of a steam chamber. Can J Chem Eng 2010;88(3):312–21.

[65] Reis JCA. Steam-assisted gravity drainage model for tar sand: radial geometry. J Can Petrol Technol 1993;32(8):43–8.

[66] Irani M, Ghannadi S. Understanding the heat-transfer mechanism in the steam assisted gravity-drainage (SAGD) process and comparing the conduction and convection flux in bitumen reservoirs. SPE J 2013;18(1):134–45.

[67] Zhang Z, Liu H, Dong X, Qi P. Unified model of heat transfer in the multiphase flow in Steam Assisted Gravity Drainage process. J Petrol Sci Eng 2017;157:875–83.

[68] Keshavarz M, Harding T, Chen Z. A new approach to analytical treatment of steam-assisted gravity drainage: a prescribed interface model. SPE J 2019;24(1):492–510.

[69] Noik C, Delmazzone C, Goulay C, Glenat P. Characterisation and emulsion behaviour of Athabasca extra heavy oil produced by SAGD. In: SPE 97748 presented at the international thermal operations and heavy oil symposium, Calgary, Alberta, Canada, November 1–3; 2005.

[70] Ezeuko CC, Wang J, Gates ID. Investigation of emulsion flow in steam-assisted gravity drainage. SPE J 2013;18(3):440–7.

[71] Hascakir B. How to select the right solvent for solvent-aided steam injection processes. J Petrol Sci Eng 2016;146:746–51.

[72] Mohebati MH. Experimental and numerical investigation of hydrocarbon Co-injection with steam in the steam Assisted gravity drainage process for in-situ heavy oil and bitumen recovery. Ph.D dissertation. University of Calgary; 2014.

[73] Tavallali M. Physical and numerical modeling of SAGD under new well configurations. Ph.D dissertation. University of Calgary; 2013.

[74] Ni J, Zhou X, Yuan Q, Lu X, Zeng F, Wu K. Numerical simulation study on steam-assisted gravity drainage performance in a heavy oil reservoir with a bottom water zone. Energies 2017;10(12):1999.

[75] Dong X, Liu H, Hou J, Zhang T, Chen Z. An empirical correlation to predict the SAGD recovery performance. In: SPE 176410 presented at the SPE/IATMI Asia Pacific oil & gas conference and exhibition, Nusa Dua, Bali, Indonesia, October 20–22; 2015.

[76] Gates ID, Chakrabarty N. Optimization of steam Assisted gravity drainage in McMurray reservoir. J Can Petrol Technol 2006;45(9):54–62.

[77] Yang L. Field test of SAGD as follow-up process to CSS in Liaohe oil field of China. J Can Petrol Technol 2007;46(4):12–5.

[78] Mendoza HA, Finol JJ, Bulter RM. SAGD, pilot test in Venezuela. In: SPE 53687 presented at the SPE Latin American and Caribbean petroleum engineering conference, Caracas, Venezuela, April 21–23; 1999.

[79] Nguyen KN, Doan LT, Kato K. Detailed history matching of a SAGD well pair using discretized wellbore modeling. In: SPE 174502 presented at the SPE Canada heavy oil technical conference, Calgary, Alberta, June 9–11; 2015.

[80] Wu Y, Li X, Zhao R, Li J, Liu X, Zhou Y, Du X. Case study of solvent-assisted start-up in Fengcheng SAGD project. In: SPE 174440 presented at the Canada heavy oil technical conference, Calgary, Alberta, Canada, June 9–11; 2015.

[81] Nasr TN, Ayodele OR. New hybrid steam-solvent processes for the recovery of heavy oil and bitumen. In: SPE 101717 presented at the Abu Dhabi international petroleum exhibition and conference, Abu Dhabi, U.A.E. November 5–8; 2006.

[82] Nesse T. Experimental comparison of hot water propane injection to steam propane injection for recovery of heavy oil. Master thesis. Texas A&M University; 2004.

[83] Behzad R, Peyman P, Alireza F, Mahmood RY, Kamran H, Maryam K, Mohammad M, Ahmad D. A new approach to characterize the performance of heavy oil recovery due to various gas injections. Int J Multiphas Flow 2018;99:273–83.

[84] Nasr TN, Pierce GE. Steam-CO_2 recovery processes for bottom water oil reservoirs. J Can Petrol Technol 1995;34(7):42–9.

[85] Shedid SA, Abbas AA. Comparison of chemical steam floods through vertical and horizontal wells. In: SPE 65482 presented at the SPE/CIM international conference on horizontal well technology, Calgary, Alberta, Canada, November 6–8; 2000.

[86] Saboorian-Jooybari H, Dejam M, Chen Z. Heavy oil polymer flooding from laboratory core floods to pilot tests and field applications: half-century studies. J Petrol Sci Eng 2016;142:85–100.

[87] Richardson WC, Kibodeaux KR. Chemically assisted thermal flood process. US Patent. US6305472B2, October 23. 2001.

[88] Li Z, Lu T, Tao L, Li B, Zhang J, Li J. CO_2 and viscosity breaker assisted steam huff and puff technology for horizontal wells in a super-heavy oil reservoir. Petrol Explor Dev 2011;38(5):600–5.

[89] Wang C. Research on enhanced recovery of super-heavy oil reservoir by HDCS flooding technology. PhD dissertation. China University of Petroleum (East China); 2015.

[90] Dong L. Effect of vapor–liquid phase behavior of steam-light hydrocarbon systems on steam assisted gravity drainage process for bitumen recovery. Fuel 2012;95:159–68.

[91] Ardali M, Barrufet M, Mamora DD, Qiu F. A critical review of hybrid steam/solvent processes for the recovery of heavy oil and bitumen. In: SPE 159257 presented at the SPE annual technical conference and exhibition, San Antonio, Texas, USA, October 8–10; 2012.

[92] Li H, Zheng S, Yang D. Enhanced swelling effect and viscosity reduction of solvent(s)/ CO_2/Heavy-oil systems. SPE J 2013;18(4):695–707.

[93] Arshad M, Li H. Multiphase equilibria of solvent-steam-bitumen system within SAGD steam-chamber boundary. In: SPE 174444 presented at the SPE Canada heavy oil technical conference, Calgary, Alberta, Canada, June 9–11; 2015.

[94] Kar T, Williamson M, Hascakir B. The role of asphaltenes in emulsion formation for steam Assisted gravity drainage (SAGD) and expanding solvent - SAGD (ES-SAGD). In: SPE 171076 presented at the SPE heavy and extra heavy oil conference: Latin America, Medellín, Colombia, September 24–26; 2014.

[95] Leaute RP. Liquid addition to steam for enhancing recovery of bitumen with CSS: evolution of technology from research concept to a field pilot at Cold Lake. In: SPE 79011 presented at the SPE international thermal operations and heavy oil symposium and international horizontal well technology conference, Calgary, Alberta, Canada, November 4–7; 2002.

[96] Leaute RP, Carey BS. Liquid addition to steam for enhancing recovery (LASER) of bitumen with CSS: results from the first pilot cycle. J Can Petrol Technol 2007;46(9):22–30.

[97] Hedden R, Verlaan M, Lastovka V. Solvent enhanced steam drive. In: SPE 169070 presented at the SPE improved oil recovery symposium, Tulsa, Oklahoma, USA, April 12–16; 2014.

[98] Zhao DW, Wang J, Gates ID. Solvent-aided steam-flooding strategy optimization in thin heavy oil reservoirs. In: IPTC 16793 presented at the international petroleum technology conference, Beijing, China, March 26–28; 2013.

[99] Huang S, Liu H, Xue Y, Xiao P, Xiong H. Performance prediction of solvent enhanced steam flooding for recovery of thin heavy oil reservoirs. In: SPE 184962 presented at the SPE Canada heavy oil technical conference, Calgary, Alberta, Canada, February 15–16; 2017.

[100] Gupta SC, Gittins SD, Picherack P. Field implementation of solvent aided process. In: PETSOC-2002-299 presented at the Canadian international petroleum conference, Calgary, Alberta, June 11–13; 2002.

[101] Orr B. ES-SAGD; past, present and future. In: SPE-129518-STU presented at the SPE annual technical conference and exhibition, New Orleans, Lousiana, USA. October 4–7; 2009.

[102] Ardali M. Investigation of hybrid steam/solvent injection to improve the efficiency of the SAGD process. PhD Dissertation. Texas A&M University; 2013.

[103] Jiang H, Deng X, Huang H, Beaulieu G, Heck G, Akinlade O, Nasr TN. Study of solvent injection strategy in ES-SAGD process. In: SPE 157838 presented at the SPE heavy oil conference Canada, Calgary, Alberta, Canada, June 12–14; 2012.

[104] Zhao L. Steam Alternating solvent process. SPE Reservoir Eval Eng 2007;10(2):185–90.

[105] Zhao L, Nasr TN, Huang H, Beaulieu G, Heck G, Golbeck H. Steam Alternating solvent process lab test and simulation. J Can Petrol Technol 2005;44(9):37–43.

[106] Stone T, Malcolm JD. Simulation of a large steam–CO_2 co-injection experiment. J Can Petrol Technol 1985;24(6):51–9.
[107] Srivastava RK, Huang S, Dong M. Comparative effectiveness of CO_2 produced gas, and flue gas for enhanced heavy-oil recovery. SPE Reservoir Eval Eng 1999;2(3):238–47.
[108] Liu Y, Yang H, Zhao L, Sun Y, Cui G, Zhao M, Hu S, Zhong L. Improve offshore heavy oil recovery by compound stimulation technology involved thermal, gas and chemical methods. In: OTC 20907 presented at the offshore technology conference, Houston, Texas, USA, May 3–6; 2010.
[109] Dong X, Liu H, Zhang Z, Wang C. The flow and heat transfer characteristics of multi-thermal fluid in horizontal wellbore coupled with flow in heavy oil reservoirs. J Petrol Sci Eng 2014b;122:56–68.
[110] Liu Y, Zou J, Meng X, Zhong L, Wang Q, Zhang W, Zhou F, Han X. Progress in Bohai offshore heavy oil thermal recovery. In: OTC 26354 presented at the offshore heavy oil conference Asia, Kuala Lumpur, Malaysia, March 22–25; 2016.
[111] Tang XX, Ma Y, Sun YT. Research and field test of complex thermal fluid huff and puff technology for offshore viscous oil recovery. China Offshore Oil Gas 2011;23:185–8 [in Chinese].
[112] Alnoaimi KR. Addition of condensable or non-condensable gas to steam flood processes for improved heavy oil recovery by gravity drainage. Master thesis. Stanford University; 2010.
[113] Wang Y, Liu H, Chen Z, Wu Z, Pang Z, Dong X, Chen F. A visualized investigation on the mechanisms of anti-water coning process using nitrogen injection in horizontal wells. J Petrol Sci Eng 2018;166:636–49.
[114] Butler RM. Steam and gas push (SAGP). In: Paper 97-137 presented at the 48th annual technical meeting of the petroleum society, Calgary, Alberta, Canada, June 8–11; 1997.
[115] Butler RM, Jiang Q, Yee C. Steam and gas push (SAGP)-3; recent theoretical developments and laboratory results. J Can Petrol Technol 2000;39(8):51–60.
[116] Jiang Q, Butler RM, Yee C. The steam and gas push (SAGP)—2, mechanism analysis and physical model testing. In: Paper 98-43 presented at the 49th annual technical meeting of the petroleum society, Calgary, AB, June 8–10; 1998.
[117] Canbolat S, Akin S, Kovscek AR. Noncondensable gas steam-assisted gravity drainage. J Petrol Sci Eng 2004;45:83–96.
[118] Yuan J-Y, Chen J, Pierce G, Wiwchar B, Golbeck H, Wang X, Beaulieu G, Cameron S. Noncondensable gas distribution in SAGD chambers. J Can Petrol Technol 2011;50(3):11–20.
[119] Al-Murayri MT, Harding TG, Maini BB. Impact of noncondensable gas on performance of steam-assisted gravity drainage. J Can Petrol Technol 2011;50(7/8):46–54.
[120] Jonasson HP, Kerr RK. SAGDOX-steam assisted gravity drainage with the addition of oxygen injection. In: SPE 165509 presented at the SPE heavy oil conference Canada, Calgary, Alberta, Canada, June 11–13; 2013.
[121] Rahnema H, Mamora D. Combustion assisted gravity drainage (CAGD) appears promising. In: SPE 135821 presented at the Canadian unconventional resources & international petroleum conference, Calgary, Alberta, Canada, October 19–21; 2010.
[122] Dong X, Liu H, Hou J, Zhang Z, Chen Z. Multi-thermal fluid assisted gravity drainage process: a new improved-oil-recovery technique for thick heavy oil reservoir. J Petrol Sci Eng 2015;133:1–11.
[123] Dong X, Liu H, Zhang Z, Wang L, Chen Z. Performance of multiple thermal fluids assisted gravity drainage process in post SAGD reservoirs. J Petrol Sci Eng 2017;154:528–36.

[124] Okoye CU, Tiab D. Enhanced recovery of oil by alkaline steam flooding. In: SPE 11076 presented at the SPE annual technical conference and exhibition, New Orleans, Louisiana, September 26–29; 1982.

[125] Tiab D, Okoye CU, Osman MM. Caustic steam flooding. J Petrol Technol 1982;34(8): 1817–27.

[126] Wu Z, Liu H, Wang X, Zhang Z. Emulsification and improved oil recovery with viscosity reducer during steam injection process for heavy oil. J Ind Eng Chem 2018;61:348–55.

[127] Gupta SC, Zeidani K. Surfactant-steam process: an innovative enhanced heavy oil recovery method for thermal applications. In: SPE 165545 presented at the SPE heavy oil conference-Canada, Calgary, Alberta, Canada, June 11–13; 2013.

[128] Babadagli T, Er V, Naderi K, Burkus Z, Ozum B. Use of biodiesel as an additive in thermal recovery of heavy oil and bitumen. J Can Petrol Technol 2010;49(11):43–8.

[129] Srivastava P, Castro L. Successful field application of surfactant additives to enhance thermal recovery of heavy oil. In: SPE 140180 presented at the SPE Middle East oil and gas show and conference, Manama, Bahrain, September 25–28; 2011.

[130] Jayasekera AJ, Goodyear SG. The development of heavy oil fields in the United Kingdom continental shelf: past, present, and future. SPE Reservoir Eval Eng 2000;3(5):371–9.

[131] Delamaide E, Moreno WP. Enhanced oil recovery of heavy oil in reservoirs with bottom aquifer. In: SPE 174050 presented at the SPE western regional meeting, Garden Grove, California, USA, April 27–30; 2015.

[132] Han M, Xiang W, Zhang J, Jiang W, Sun F. Application of EOR technology by means of polymer flooding in Bohai oilfields. In: SPE 104432 presented at the international oil & gas conference and exhibition in China, Beijing, China, December 5–7; 2006.

[133] Zhou X, Zeng F. Feasibility study of using polymer to improve SAGD performance in oil sands with top water. In: SPE 170164 presented at the SPE heavy oil conference-Canada, Calgary, Alberta, Canada, June 10–12; 2014.

[134] Fortenberry R. Experimental demonstration and improvement of chemical EOR techniques in heavy oils. University of Texas at Austin; 2013.

[135] Taghavifar M, Fortenberry R, Rouffignac ED, Sepehrnoori K, Pope GA. Feasibility study of using polymer to improve SAGD performance in oil sands with top water. In: SPE 170164 presented at the Canada heavy oil technical conference, Calgary, Alberta, Canada, June 10–12; 2014.

[136] Lau HC. Alkaline steam foam: concepts and experimental results. In: SPE 144968 presented at the SPE enhanced oil recovery conference, Kuala Lumpur, Malaysia, July 19–21; 2011.

[137] Dong X, Liu H, Wang C, Lu C, Yan F. Polymer-enhanced foam injection process: an improved-oil- recovery technique for light oil reservoirs after polymer flooding. Energy Sources, Part A Recovery, Util Environ Eff 2016;38(3):354–61.

[138] Altunina L. Improved cyclic-steam well treatment with employing thermoreversible polymer gels. In: SPE 104330 presented at the SPE Russian oil and gas technical conference and exhibition, Moscow, Russia, October 3–6; 2006.

[139] Wang C, Liu H, Zheng Q, Liu Y, Dong X, Hong C. A new high-temperature gel for profile control in heavy oil reservoirs. ASME J Energy Res Technol 2016;138(2):022901.

[140] Moradi-Araghi A. A review of thermally stable gels for fluid diversion in petroleum production. J Petrol Sci Eng 2000;26(1–4):1–10.

[141] He H, Wang Y, Zhao M, Cheng L, Liu P. Laboratory evaluation of thermoreversible gel for in-depth conformance control in steam-stimulated wells. In: SPE 157871 presented at the SPE heavy oil conference Canada, Calgary, Alberta, Canada, June 12–14; 2012.

[142] Li H. Study on the viscosity reducer assisted SAGD technique. Master thesis. Beijing: China University of Petroleum; 2017.
[143] Chen Q, Geertrui M, Kovscek AR. Improving steam-assisted gravity drainage using mobility control foams: foam assisted-SAGD (FA-SAGD). In: SPE 129847 presented at the SPE improved oil recovery symposium, Tulsa, Oklahoma, USA, April 24–28; 2010.
[144] Li R. Chemical additives and foam to enhance SAGD performance. Master thesis. University of Calgary; 2016.
[145] Abdul M, Ayache SV, Batôt G, Gasser-Dorado J, Delamaide E. Improvement of the SAGD process by use of steam-foam: design and assessment of a pilot through reservoir simulation. In: SPE 196676 presented at the SPE reservoir characterisation and simulation conference and exhibition, Abu Dhabi, UAE, September 17–19; 2019.
[146] Adetunji LA, Ben-Zvi A, Filstein A. Foam formulation for high temperature SAGD applications. In: SPE 198919 presented at the SPE thermal well integrity and design symposium, Banff, Alberta, Canada, November 19–21; 2019.
[147] CMG manual, computer modeling group. 2015.
[148] Pang Z. Seepage mechanisms & applications of steam (nitrogen) foam compound flooding. Doctoral thesis. Beijing: China University of Petroleum; 2008.
[149] Khajehpour M, Etminan SR, Goldman J, Wassmuth F, Bryant S. Nanoparticles as foam stabilizer for steam-foam process. SPE J 2018;23(6):2232–42.
[150] Bayestehparvin B, Farouq Ali SM, Abedi J. Use of solvents with steam - state-of-the-art and limitations. In: SPE 179829 presented at the SPE EOR conference at oil and gas west Asia, Muscat, Oman, March 21–23; 2016.
[151] Zhang X. Applied research of steam-carbon dioxide-auxiliary agent huff and puff technology. Acta Perolei Sinica 2006;27(2):80–4 [in Chinese].
[152] Wang G. Field experiment of CO_2 assisted steam stimulation technology. Sino-Global Energy 2015;20(7):68–71.
[153] Xu Z, Liu P, Zhang S, Yuan Z, Li X, Hao M, Liu L. Physical experiment and numerical simulation study for start-up of ES-SAGD in heavy oil reservoir. Pet Geol Recovery Effic 2017;24(3):110–5 [in Chinese].
[154] Li X, Zhao K, Zhu M. Simulation technique for MECL heavy oil reservoirs in Colombia. Pet Drill Tech 2015;43(1):100–5.
[155] Sun Y, Zhao L, Lin T, Zhong L, Yu D, Lin H. Enhance offshore heavy oil recovery by cyclic steam–gas-chemical co-stimulation. In: Paper SPE 149831 presented at the SPE heavy oil conference and exhibition, Kuwait City, Kuwait, December 12–14; 2011.
[156] Telmadarreie A, Trivedi JJ. New insight on carbonate-heavy-oil recovery: pore-scale mechanisms of post-solvent carbon dioxide foam/polymer-enhanced-foam flooding. SPE J 2016;21(5):1655–68.
[157] Dai C, Gu C, Liu B, Lyu Y, Yao X, He H, Fang J, Zhao G. Preparation of low-temperature expandable graphite as a novel steam plugging agent in heavy oil reservoirs. J Mol Liq 2019;293:111535.
[158] Zhong L, Liu J, Yuan X, Wang C, Teng L, Zhang S, Wu F, Shen W, Jiang C. Subsurface sludge sequestration in cyclic steam stimulated heavy-oil reservoir in Liaohe oil field. SPE J 2020;25. SPE-195415-PA. (online).

Chapter 2

Existing problems for steam-based enhanced oil recovery processes in heavy oil reservoirs

2.1 Current status of steam-based enhanced oil recovery processes

High oil viscosity is the key indicator for the effective development of heavy oil reservoirs. In addition, reducing oil viscosity by a thermal process for these reservoirs is the primary mature technology. Cyclic steam stimulation (CSS), steam flooding, and steam-assisted gravity drainage (SAGD) are three classical steam-based enhanced oil recovery (EOR) processes for heavy oil reservoirs. CSS is usually considered the first recovery process. Next, steam flooding is an important follow-up process of CSS. However, a successful operating transition from CSS to steam flooding remains scarce. Compared with other processes, CSS is more easily operated and reliable. Through the scheme of injection, soaking, and production in a single well, this process has a rational recovery rate. It is attractive for heavy oil reservoirs with an oil viscosity lower than 10,000 mPa s [1,2]. Horizontal well-based CSS has also been applied in heavy oil reservoirs. There are problems in a poststeam injection stage:

(1) Because of a large density difference between steam and heavy crude oil, steam is more prone to moving to a reservoir top under the effect of gravitational differentiation (steam overlap). Therefore, the displacement efficiency in the upper reservoir part is much higher, and it is lower in the lower part. An obvious nonuniform degree of steaming can be observed in the vertical direction. This nonuniform degree can further increase vertical reservoir heterogeneity and hinder the recovery performance of heavy oil reservoirs. This problem is more significant in thick, heavy oil reservoirs.
(2) A large viscosity difference between steam and heavy oil can also induce a serious fingering problem. Hence, widely spread unswept remaining oil

Hybrid Enhanced Oil Recovery Processes for Heavy Oil Reservoirs.
https://doi.org/10.1016/B978-0-12-823954-4.00005-9

resources outside flow paths of steam and condensate can be observed. Similarly, this problem increases horizontal reservoir heterogeneity and reduces the unlocking degree of heavy oil reservoirs. On the other hand, it also reduces the expansion area of steam. Thus, thermal efficiency is significantly decreased.

(3) After multiple CSS cycles, a large volume of water can be observed around a bottom hole. It increases the heat loss rate in subsequent steam; most injected steam is used to reheat the water around the bottom hole. Thus, thermal efficiency is reduced. Moreover, the duration of the water drainage stage in CSS is increased.

(4) Serious wellbore heat loss also affects the recovery performance of steam injection processes for heavy oil reservoirs. During pure steam injection, a large heat loss rate results in lower steam quality under bottom hole conditions.

In China, steam injection is still the main exploitation method for heavy oil reservoirs. For a typical heavy oil field in eastern China, its production has gradually increased since 2001, and most of the heavy oil production has been from CSS. Total CSS wells were 4928; the average CSS cycle for each well was 6. Until the end of 2018, annual heavy oil production from CSS reached 380.4×10^4 t, accounting for 84.3% of total heavy production for this field. Another typical heavy oil field is in northern China, as shown in Table 2.1. In 2018, total thermal wells were more than 11,000; more than 60% of them are still in production. In 2018, total heavy oil production in this oil field was about 602×10^4 t, and about 90.0% was from thermal recovery processes. CSS accounted for 61.4% of thermal heavy oil production for this field. The annual steam–oil ratio of CSS was 3.57 and the average recovery factor of CSS was 25.9%.

As steam injection continues, most heavy oil reservoirs have entered into the later stages of thermal recovery. A series of problems have emerged and hindered the continuous normal development of heavy oil reservoirs. How to enhance heavy oil recovery is the highest concern for heavy oil corporations around the world. Among many problems, low steam efficiency and high steam–oil ratios are the top two problems to be addressed for the next stage of development of heavy oil reservoirs. On the other hand, after lengthy steam stimulation, the fluid properties of heavy oil in porous media have also changed. The oil composition, density, and viscosity are no longer the same as their initial values, which increases the difficulty of effective development. In addition, some marginal heavy oil reservoirs whose physical properties are complicated have become important sources to increase heavy oil production. Compared with conventional heavy oil reservoirs, more requirements will arise for the effective development of marginal heavy oil reservoirs.

TABLE 2.1 Heavy oil production of typical heavy oil field in Northern China in 2018.

Exploitation method	Recovery process	Annual heavy oil production ($\times 10^4$ t)	Production proportion (%)
Thermal heavy oil production	Steam-assisted gravity drainage	105.7	17.6
	Steam flooding	72	12.0
	In situ combustion	31.3	5.2
	Cyclic steam stimulation	333	55.3
	Sum	542	90.0
Cold heavy oil production		60	10.0

2.2 Steam overlap

2.2.1 Characteristics of steam overlap

Steam overlap refers to the phenomenon of a vertical velocity difference between steam and a liquid. It mainly results from the gravitational differentiation of steam and condensate during steam injection [2,3]. During steam injection, viscosity, gravity, and capillary forces are the three main forces in reservoirs. Steam overlap is an important result of complicated interactions between reservoir fluids and these different forces. Viscous force is caused by displacement; and it has an important role in horizontal fluid flow. Gravity is caused by the density difference of fluids; it is the dominating force of steam overlap. Capillary force is caused by interfacial tension; it usually determines the distribution of initial fluid saturation and residual oil saturation. In unconsolidated sandstone heavy oil reservoirs, capillary force is usually small.

2.1.1.1 Linear displacement process of steam injection

As shown in Fig. 2.1, during steam injection, once a steam−liquid interface reaches a stable status, the fluid potential at various locations perpendicular to the horizontal direction is consistent. Thus, the fluid potential relative to plane $z = h$ can be expressed as:

$$\Phi_{s1} = p_{s1} - \rho_s g h_{s1} \tag{2.1}$$

$$\Phi_{s2} = p_{s2} - \rho_s g h_{s2} \tag{2.2}$$

$$\Phi_{o1} = p_{o1} - \rho_o g h_{s1} \tag{2.3}$$

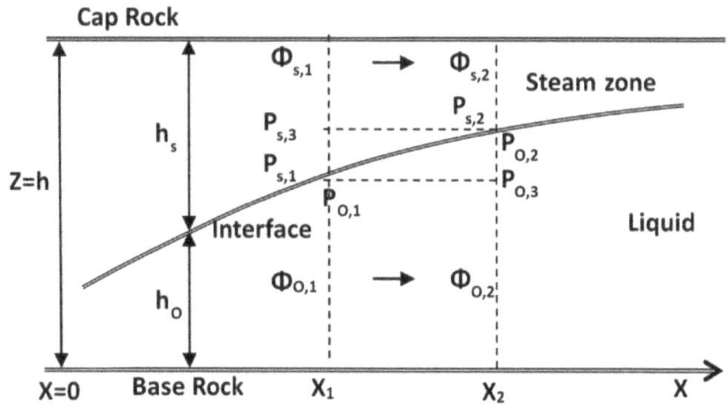

FIGURE 2.1 Schematic of steam–liquid interface in linear displacement process.

$$\Phi_{o2} = p_{o2} - \rho_o g h_{s2} \tag{2.4}$$

At the steam–liquid interface:

$$p_{s1} - p_{o1} = p_{s2} - p_{o2} \tag{2.5}$$

As $\Delta x \rightarrow 0$:

$$\frac{\partial \Phi_s}{\partial x} - \frac{\partial \Phi_o}{\partial x} = (\rho_o - \rho_s) g \frac{\partial h_s}{\partial x} \tag{2.6}$$

Based on Darcy's law:

$$\frac{\partial \Phi_s}{\partial x} = -\frac{\mu_s i_s}{B h_s K_s \rho_s} \tag{2.7}$$

$$\frac{\partial \Phi_o}{\partial x} = -\frac{\mu_o i_o}{B(h - h_s) K_o \rho_o} \tag{2.8}$$

The relationship between the mass flow rates of steam and oil in a reservoir and the mass flow rates of steam and oil at a steam–liquid interface is:

$$i_s = i_{sb} \frac{h_s}{h} \tag{2.9}$$

$$i_o = i_{oe} \left(1 - \frac{h_s}{h}\right) \tag{2.10}$$

From Eqs. (2.8)–(2.10):

$$\frac{\partial h_s}{\partial x} = -\frac{\mu_s i_{sb}}{B(\rho_o - \rho_s) g K_s \rho_s h} \left(1 - \frac{\mu_o}{\mu_s} \frac{K_s}{K_o} \frac{i_{oe}}{i_{sb}} \frac{\rho_s}{\rho_o}\right) \tag{2.11}$$

Eq. (2.11) indicates the shape of a steam–liquid interface in linear displacement. It can be used to manifest the degree of steam overlap.

We set:

$$A_{ld} = \frac{\mu_s i_{sb}}{B(\rho_o - \rho_s) g K_s \rho_s h} \tag{2.12}$$

$$M^*_{eq} = \frac{\mu^*_o}{\mu_s} \frac{K_s}{K_o} \frac{i_{oe}}{i_{sb}} \frac{\rho_s}{\rho_o} \tag{2.13}$$

where A_{ld} is a shape factor of steam overlay in linear displacement and M^*_{eq} is the equivalent mobility ratio. Thus, Eq. (2.11) can be expressed as:

$$\frac{\partial h_s}{\partial x} = -A_{ld}\left(1 - M^*_{eq}\right) \tag{2.14}$$

Integrating the left-hand side:

$$\int_0^{h_s} dh_s = -A_{ld}\left(1 - M^*_{eq}\right)\int_{x_e}^{x} dx \tag{2.15}$$

$$h_s = A_{ld}\left(1 - M^*_{eq}\right)(x_e - x) \tag{2.16}$$

$$x_b(t) = x_e(t) - \frac{h}{A_{ld}\left(1 - M^*_{eq}\right)} \tag{2.17}$$

where P is pressure, Pa; Φ is the fluid potential, Pa; h_s is the height of steam overlap, m; h is the reservoir thickness, m; B is the reservoir width, m; i_s is the steam mass flow rate in the reservoir, kg/s; i_o is the oil mass flow rate in the reservoir, kg/s; i_{sb} is the steam mass flow rate below the steam–liquid interface, kg/s; i_{oe} is the oil flow rate above the steam–liquid interface, kg/s; μ_s is the steam viscosity, Pa·s; μ^*_o is the condensate viscosity under the steam temperature, Pa·s; ρ_o is the oil density, kg/m^3; ρ_s is the steam density, kg/m^3; K_o and K_s are the effective permeabilities of oil and steam, m^2, respectively; and g is the gravitational acceleration, 9.8 m/s^2.

For the steady heat transfer of a steam–liquid interface, the condensate fluids below the interface include heated heavy oil and condensate water. The viscosity of condensate is usually close to the viscosity of water. Under a predefined displacement pressure, the viscosity ratio and effective permeability difference between steam and condensate are not significant. However, the density of steam is significantly lower than that of condensate. Therefore, M^*_{eq} is usually below 1.0. If $M^*_{eq} << 1$, the shape of the steam–liquid interface mainly depends on the shape factor, A_{ld}. In this parameter, the effects of a steam injection rate, steam density, and steam permeability are included. As the oil viscosity increases, the viscosity of condensate below the steam–liquid interface increases accordingly. Thus, M^*_{eq} also increases. Once $M^*_{eq} \geq 1$, the displacement interface is no longer stable.

To calculate a shape factor for steam overlay in a linear displacement process (A_{ld}), because the value of the steam mass flow rate (i_{sb}) at x_b is difficult to obtain, the steam mass flow rate around an injector (i_{si}) is often chosen. However, considering the existence of heat loss at x_b, i_{sb} is usually lower than i_{si}. Therefore, the value of A_{ld} calculated is higher than its actual value.

2.1.1.2 Radial displacement process of steam injection

A schematic of a steam–liquid interface in a radial displacement process is shown in Fig. 2.2. Similar to the process of linear displacement, as $\Delta r \rightarrow 0$, we have:

$$\frac{\partial \Phi_s}{\partial r} - \frac{\partial \Phi_o}{\partial r} = (\rho_o - \rho_s) g \frac{\partial h_s}{\partial r} \tag{2.18}$$

Based on Darcy's law:

$$\frac{\partial \Phi_s}{\partial r} = -\frac{\mu_s i_s}{2\pi r h_s K_s \rho_s} \tag{2.19}$$

$$\frac{\partial \Phi_o}{\partial r} = -\frac{\mu_o i_o}{2\pi r (h - h_s) K_o \rho_o} \tag{2.20}$$

From Eqs. (2.18)–(2.20), we obtain:

$$\frac{\partial h_s}{\partial r} = -\frac{\mu_s i_s}{2\pi r (\rho_o - \rho_s) g K_s \rho_s h_s} \left[1 - M_{eq}^* \frac{h_s}{h - h_s} \frac{i_o}{i_s} \frac{i_{sb}}{i_{oe}} \right] \tag{2.21}$$

Eq. (2.21) indicates the shape of a steam–liquid interface in radial displacement. It can be used to manifest the degree of steam overlap in radial displacement.

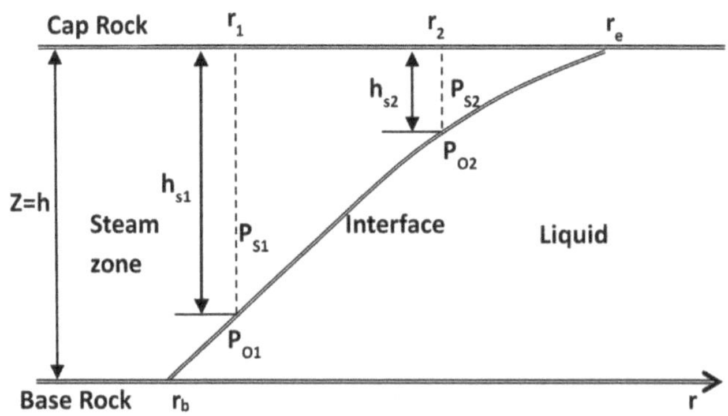

FIGURE 2.2 Schematic of steam–liquid interface in radial displacement process.

We set:

$$A_{rd}^2 = \frac{\mu_s i_{si}}{\pi(\rho_o - \rho_s) g h^2 K_s \rho_s} \tag{2.22}$$

where i_{si} is the steam mass flow rate at an injector, kg/s. Analogously, the steam mass flow rate around an injector (i_{si}) is often chosen as the steam mass flow rate below the steam–liquid interface (i_{sb}). A_{rd} is called the shape factor of steam overlay in radial displacement. Generally,

$$\frac{h_s}{h - h_s} \frac{i_o}{i_s} \frac{i_{sb}}{i_{oe}} \approx 1 \tag{2.23}$$

Because $M_{eq}^* << 1$, Eq. (2.21) can be written as:

$$\frac{\partial h_s}{\partial r} = -A_{rd}^2 \frac{h^2}{2 r h_s} \frac{i_s}{i_{si}} \tag{2.24}$$

The shape factor of a steam–liquid interface is a function of the steam injection rate. Under a predefined steam injection rate, if the slope of a steam–liquid interface is high, we can have a small degree of steam overlap and a high shape factor of the steam–liquid interface.

Under a constant steam quality at a bottom hole, as the steam injection rate (steam mass flow rate) is reduced, the possibility of steam overlap in a reservoir increases. Therefore, for CSS, the steam injection rate is usually high and the effect of steam overlap can be neglected. On the other hand, steam overlap is highly related to reservoir thickness. Under a given steam injection rate, the higher the reservoir thickness, the smaller the linear flow rate of steam. Thus, the possibility of steam overlap occurring in a reservoir is high. For a typical heavy oil reservoir with serious steam overlap, by intermittently adjusting an injection/production work system, redistribution of reservoir fluids in porous media can be induced, which is helpful for increasing the macroscopic heating range of a reservoir.

2.2.2 Experimental test of steam overlap

Steam overlap is one of the main challenges for vertical sweep efficiency in steam injection. To simulate the real behavior of steam overlap experimentally, an experimental method was proposed to study the characteristics of steam overlap under different reservoir conditions [4].

2.2.2.1 Experimental method

Fig. 2.3 shows the schematic of an experimental setup, and a wellbore model was proposed. The latter is a three-dimensional (3D) wellbore model. This model can be used to investigate steam conformance experimentally along a vertical/horizontal wellbore under different conditions [5]. In this section,

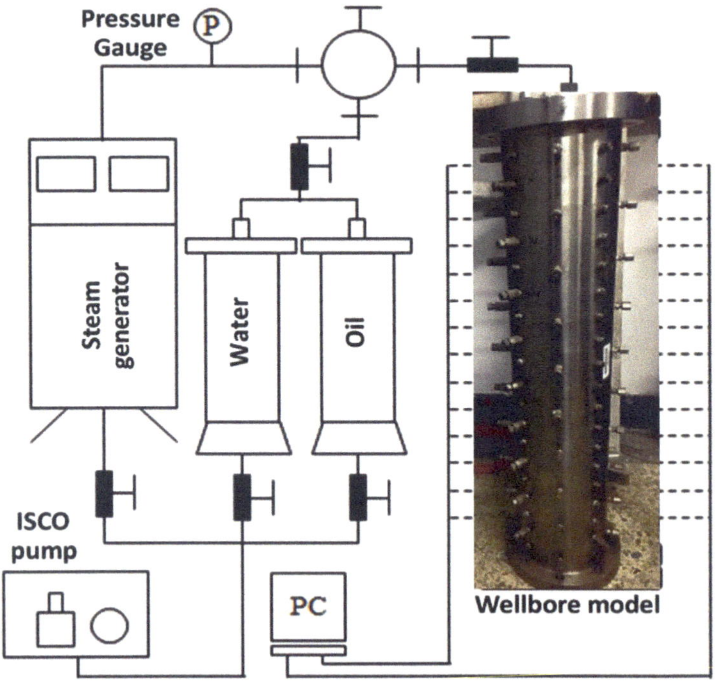

FIGURE 2.3 Experimental setup and wellbore model. *ISCO*, Teledyne ISCO pump; *PC*, Personal computer.

based on this model, the distribution of a vertical steam injection profile and the characteristics of steam overlap are discussed. The length of this wellbore model is 78 cm, the inner radius is 20 cm, and the inner radius of the simulated well is 4 cm. During experiments, 30 temperature measuring points and 10 pressure measuring points are designed within this wellbore model.

In the procedure, three different experiments are performed to investigate the effect of reservoir features on steam overlap. The features include a homogeneous reservoir, a positive reservoir (permeability increasing from the upper to lower sections), and a negative reservoir (permeability decreasing from the upper to lower sections). The heavy oil sample used in the experiments is a wellhead dead oil; its viscosity at 50°C is about 12,600 mPa s. The water used is distilled. Quartz sand with 40–80 mesh is used to simulate the porous medium environment.

The detailed experimental procedure is:

➢ sand filling to simulate permeability conditions,
➢ connecting experimental apparatuses,
➢ gas tightness testing to check for leaks,
➢ distilled water injection to test the pore volume and porosity,

➢ oil sample injection to establish the initial oil saturation condition,
➢ opening the valve and starting steam injection, and
➢ data acquisition for temperature distribution along the wellbore.

2.2.2.2 Experimental results

Fig. 2.4 shows the experimental results. For all three experiments, obvious steam overlap can be observed. In the early stage ($t < 40$ min), the temperature in the upper section of this model is higher, which indicates that most of the injected steam is used to heat the upper section. Comparatively, the heating range in the negative case is more concentrated at the top of the

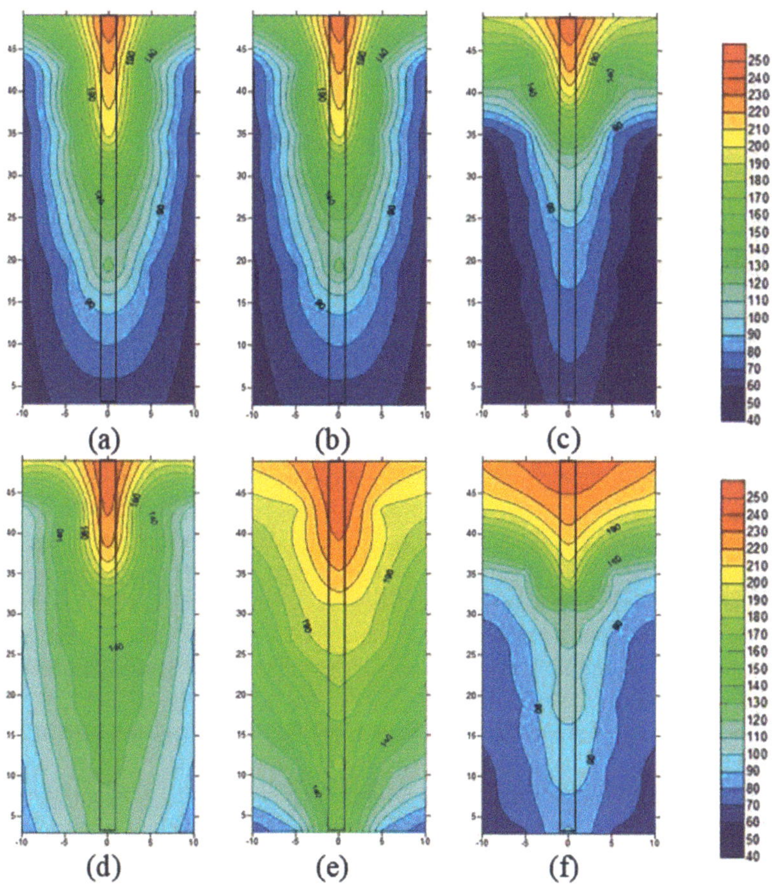

FIGURE 2.4 Characteristics of steam overlap for three different models. (A) 40 min for positive rhythm; (B) 40 min for homogeneous model; (C) 40 min for negative rhythm; (D) 90 min for positive rhythm; (E) 90 min for homogeneous model; and (F) 90 min for negative rhythm.

reservoir, and the heating range in the positive and homogeneous reservoirs is larger. As steam injection continues, a steam front gradually expands downward and the temperature in the lower section rises. Because steam injection lasts for 90 min, the whole reservoir in the positive case is effectively heated for the three experiments. This is because permeability distribution in the positive case can offset the effect of steam overlap. In the negative case, after the 90-min operation, some bottom sections within the model are not effectively heated and most of the injected steam is used to heat the upper and middle sections. The heating range in the negative case is limited because the permeability distribution in this case can aggravate the effect of steam overlap. Under the same condition of steam injection, the sweep efficiency of the positive reservoir is the highest, and that of the negative reservoir is the lowest. The heating performance of the homogeneous reservoir is between them.

2.3 Steam breakthrough

2.3.1 Characteristics of steam breakthrough

Steam injection is usually the first exploitation method used for heavy oil reservoirs. Once steam injection is conducted, an oriented fluid flow can be formed [2,6]. During steam injection, steam will first flow along high permeability paths, and thus nonuniform steam distribution in the reservoir can be formed. Then, the injected steam and condensate are produced from nearby wells, especially for seriously heterogeneous reservoirs. The production of injected steam indicates heat waste and reduced thermal efficiency. Also, during steam injection, the liquid production rate gradually increases. Moreover, after lengthy steam injection, both the water cut and temperature of the produced liquid increase. Once the water cut approaches 100%, it indicates the production of steam (i.e., the occurrence of steam breakthrough). At that time, the heat sweep efficiency and the range of mobilized heavy crude oil are still low in a reservoir. This phenomenon further affects the performance of thermal recovery and disturbs the normal production of thermal wells. The occurrence of steam breakthrough between an injector and a producer usually represent pressure interference behavior between wells, a rise in the working fluid level, and a rise in the production rate.

Fig. 2.5 shows a vertical steam injection profile of a typical well. Because of the large vertical permeability heterogeneity in this well, layer I has the highest ratio of injected steam distribution and layer II is the lowest. As steam injection continues, this nonuniform steam injection profile expands. Finally, the injected steam will first be produced from layer I, and steam breakthrough occurs.

Fluid channeling behavior between wells often leads to steam breakthrough. It is the extreme stage of heat connection in thermal recovery. There are usually two types of steam breakthrough: steam and hot water. Reasons for steam

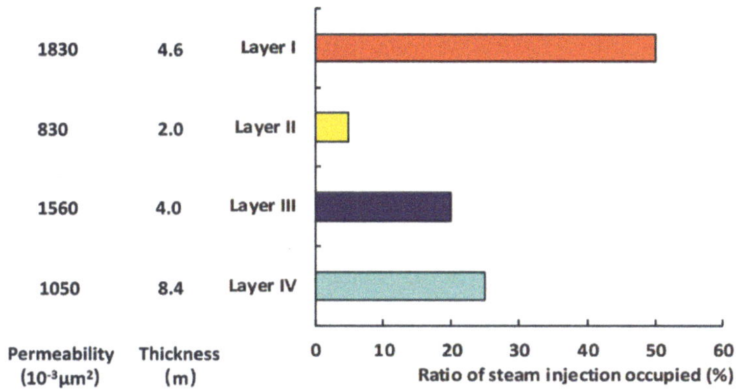

FIGURE 2.5 Vertical steam injection profile of a typical well.

breakthrough include serious intraformation/interformation permeability heterogeneity, serious steam overlap in thick, heavy oil reservoirs, and unintentional fractures caused by unreasonable steam injection operation. Reasons for hot water breakthrough include fluid fingering along a main stream line and a heat–pressure connection between wells because of continuous steam injection process [2,7]. For a steam injection process in heavy oil reservoirs, steam breakthrough can generate high water cut paths between wells. In this situation, condensate fingering can also be observed. For hot water breakthrough, if the temperature of the produced liquid is higher than 100°C, the steam status caused by flash evaporation can be observed under well head conditions. However, in reality, it is hot water breakthrough, not a real steam breakthrough.

According to the degree of interference of steam breakthrough between wells, the wells are usually classified into three categories. The first one is a steam injection well that must be closed because of the serious steam breakthrough. The second one is a well whose liquid production rate, liquid temperature, and water cut dramatically increase. To maintain normal production, measures of adjusting injection/production strategies should be performed. The last is a well whose liquid production rate, liquid temperature, and water cut change slightly. No additional measures may be required.

2.3.2 Mechanisms of steam breakthrough

As presented in Section 1.1.3, a steady displacement front in a tilted formation meets the requirement [8]:

$$(M_0 - 1)v_c < \frac{K_1}{\mu_1}(\rho_1 - \rho_2)g \sin \alpha \tag{2.25}$$

However, steam injection is different from conventional water flooding. Injected steam condenses in a displacement front, and a phase change behavior at a steam–liquid interface can be observed. Therefore, the fluid flow rate before the interface is different from that after the interface. Conventional mobility analysis for water flooding is no longer applicable for the characteristics of a steam–liquid interface in steam injection. Based on the definition of the equivalent mobility ratio, Eq. (2.13) can be rewritten as:

$$M_{eq}^{*} = \frac{\mu_o^*}{\mu_s}\frac{K_s}{K_o}\frac{v_{oe}}{v_{sb}} = \frac{\nabla p_o}{\nabla p_s} \tag{2.26}$$

where v_{oe} is the flow velocity of the heated oil and condensate, m/s; v_{sb} is the flow velocity of steam, m/s; ρ_s is the steam density, kg/m^3; ρ_o is the oil density, kg/m^3; ∇p_o is the pressure gradient of the fluids after the steam–liquid interface, MPa/m; and ∇p_s is the pressure gradient of steam, MPa/m.

Eq. (2.26) indicates that a ratio of equivalent mobilities is actually a ratio of pressure gradients. It is related to the viscosity ratio, permeability ratio, and velocity ratio. At a steam–liquid interface, the velocity ratio (v_{oe}/v_{sb}) is relatively low because of the condensation effect of steam. Although the steam viscosity is much lower than the initial viscosity of heavy crude oil in a reservoir, the oil viscosity at a steam–liquid interface can be reduced by a large proportion because of lateral heat transfer between the steam condensate and the interface. The viscosity of the displaced fluid (μ_o^*) is actually oil viscosity under steam temperature conditions ($\mu_o(T_s)$). Therefore, the viscosity ratio at the steam–liquid interface (μ_o^*/μ_s) can be dramatically reduced. Because of the pressure dependence behavior of steam properties, the characteristics of an equivalent mobility ratio and a steam–liquid interface are also related to pressure and temperature.

If we set the velocity of a displacement interface by v_f, the displaced oil within unit time is:

$$i_{oe} = v_f B h \phi \frac{S_{oi} - S_{or}}{B_o} \rho_o \tag{2.27}$$

Considering the thermal efficiency in a reservoir, within unit time, the steam volume required for the same volume reservoir is:

$$i_{sb} = \frac{v_f B h M_R (T_h - T_i)}{H_m E_{hs}} \tag{2.28}$$

From Eqs. (2.27) and (2.28), the ratio of the mass flow rates of the displaced oil and steam (displaced oil–steam ratio) is:

$$\frac{i_{oe}}{i_{sb}} = F_{OSR} = \frac{\phi(S_{oi} - S_{or})\rho_o H_m E_{hs}}{B_o M_R (T_h - T_i)} \tag{2.29}$$

where i_{sb} is the steam mass flow rate, kg/s; i_{so} is the oil mass flow rate, kg/s; v_f is the interface velocity, m/s; B is the reservoir width, m; h is the reservoir thickness, m; S_{oi} is the initial oil saturation, fraction; S_{or} is the residual oil saturation, fraction; B_o is the oil formation volume factor, m^3/m^3; ϕ is porosity,

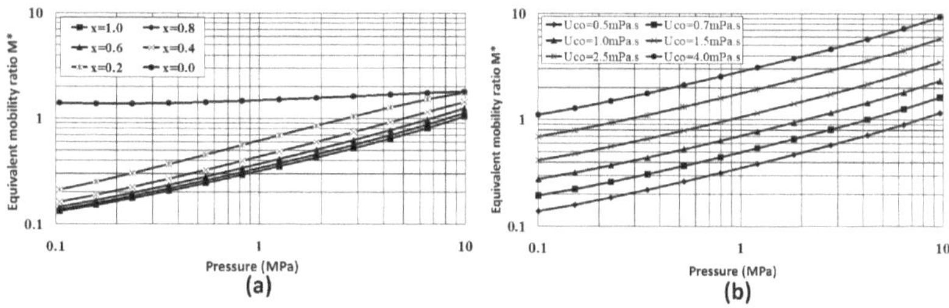

FIGURE 2.6 Equivalent mobility ratio versus pressure. (A) Different steam qualities; (B) Different condensate viscosities.

fraction; ρ_o is the oil density, kg/m^3; M_R is the reservoir thermal capacity, kJ/(m^3·K); T_h is the temperature in a steam chamber, K; T_i is the initial reservoir temperature, K; H_m is steam enthalpy, kJ/kg; and E_{hs} is the reservoir thermal efficiency, fraction.

Therefore, Eq. (2.26) can be rewritten as:

$$M_{eq}^{*} = \frac{\mu_o^{*}}{\mu_s}\frac{K_s}{K_o}\frac{\rho_s}{\rho_o}F_{OSR} \tag{2.30}$$

From Eq. (2.30), the equivalent mobility ratio under different pressure and temperature conditions can be obtained, as shown in Fig. 2.6. Under certain pressure conditions, a low equivalent mobility ratio can be observed. In Fig. 2.6A, for a hot water flooding process ($X = 0$), the equivalent mobility ratio is always above 1.0. This indicates that hot water flooding is unstable. However, for steam flooding, a steam–liquid interface can be stable under low steam injection pressure. The variation in behavior of an equivalent mobility ratio on pressure provides a theoretical foundation for a transition from CSS to steam flooding.

On the other hand, considering the unsteady heat transfer behavior of a steam–liquid interface, the real viscosity of the displaced oil μ_o^{*} is usually lower than that under steam temperature conditions ($\mu_o(T_s)$). As the oil viscosity increases, the condensate viscosity increases. From the results in Fig. 2.6B, for a heavy oil reservoir with a high oil viscosity, the equivalent mobility ratio is higher than 1.0. Thus, steam flooding can be an unstable process. As the oil viscosity increases, the possibility of steam breakthrough increases.

2.3.3 Volume and strength of steam breakthrough

2.3.3.1 Volume of steam breakthrough

For heavy crude oil, as the temperature increases, the non-Newtonian fluid behavior of shear thinning gradually disappears and a Newtonian fluid behavior emerges [9]. The volume of steam breakthrough is defined as the volume of a porous medium whose temperature is higher than the transition temperature from a non-Newtonian to a Newtonian fluid. On a reservoir surface, the volume of steam breakthrough involves a lateral sweep volume along a main stream line. From the performance of steam breakthrough in field operations, the fluids in steam breakthrough paths include condensate water and remaining oil after steam flooding. In most cases, the remaining oil in steam breakthrough paths is in a residual state. Therefore, after steam breakthrough occurs, fluid flow in steam breakthrough paths is a two-phase flow process of steam and water. It is different from a conventional two-phase flow; during steam injection, heat transfer behavior between steam and surrounding rocks can lead to the condensate of steam. A non-piston steam displacement model in steam breakthrough paths can be developed as:

$$\int_0^{t_{bj}} \left[C_j i_s v_{ws} - \overline{G}_j V_{tj} \right] dt = \phi \frac{V_{tj}}{f_s'(S_{gf})} \tag{2.31}$$

$$\overline{G}_j = \frac{2\lambda(T_h - T_i)}{\sqrt{\pi\alpha[t - \delta(x)]} L_v \rho_s \overline{B}} \tag{2.32}$$

where C_j is a dividing coefficient of the steam injection volume in the j-direction, nondimensional; i_s is the steam injection rate, g/s; v_{ws} is the specific volume of steam, m^3/g; $\overline{G}_j$ is the condensate volume per rock volume within unit time in the j-direction, m^3/(m$^3 \cdot$d); $f_s'(S_{gf})$ is the derivative of the steam fraction in a displacement front, nondimensional; S_{gf} is the steam saturation in the displacement front, nondimensional; V_{tj} is the rock volume in the j-direction, m^3; t_{bj} is the time of steam breakthrough in the j-direction, s; λ is the reservoir thermal conductivity, W/(m$\cdot$K); α is the reservoir thermal diffusivity, m^2/s; $\overline{B}$ is a feature size of the lateral sweep of a steam breakthrough zone, m; L_v is the latent heat of vaporization, J/g; ρ_s is the steam density, g/m^3; $\delta(x)$ is the moving time of the steam front, s; T_h is temperature in the stream zone, K; and T_i is the initial reservoir temperature, K.

During a steam injection process in heavy oil reservoirs, a multidirectional steam breakthrough path can be generated, as shown in Fig. 2.7A. In the same producing layer of an injector/producer group, considering the shielding effect between different wells, the angle of steam breakthrough (β) can be derived. On the other hand, because of the difference in reservoir properties, the degree of connection between an injector and a producer can be determined, as shown in Fig. 2.7B.

Thus, the dividing coefficient of a steam injection volume can be calculated based on the angle of steam breakthrough and formation coefficients:

$$C_j = \frac{\beta_j \left(\overline{K}\overline{h}\right)_j}{\sum_{j=1}^{M} \beta_j \left(\overline{K}\overline{h}\right)_j} \tag{2.33}$$

where M is the number of the directions of steam breakthrough and β_j is the angle of steam breakthrough in the j-direction.

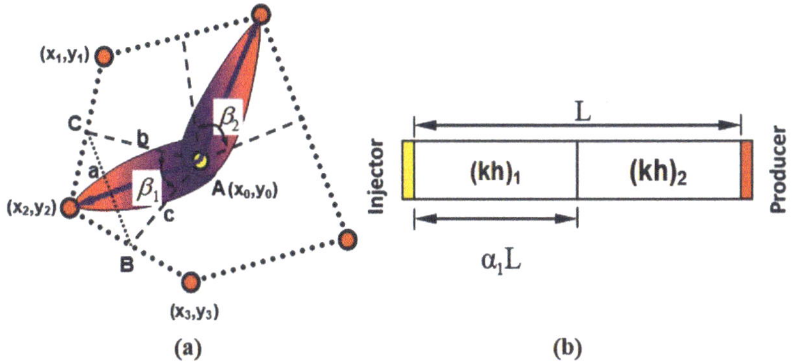

FIGURE 2.7 Schematic of steam breakthrough between injector and producer. (A) Angle of steam breakthrough; (B) heterogeneous property distribution.

The average formation coefficient can be derived by the method of equivalent flow resistance:

$$\left(\overline{Kh}\right)_j = \frac{(Kh)_{1j}(Kh)_{2j}}{\left(1-\alpha_{1j}\right)(Kh)_{1j} + \alpha_{1j}(Kh)_{2j}} \tag{2.34}$$

where $\overline{(Kh)}_j$ is the average formation coefficient between an injector and a producer; $(Kh)_{1j}$ is the formation coefficient of the injector in the j-direction; $(Kh)_{2j}$ is the formation coefficient of the producer in the j-direction; and α_{1j} is a length factor for the injector (default at 0.5).

The derivative of the steam fraction in a displacement front in Eq. (2.31) can be obtained by a graphical method. The relationship for steam–water relative permeability is:

$$K_{rw} = S_{wn}^{1.5} S_w^3 \tag{2.35}$$

$$K_{rs} = (1 - S_{wn})\sqrt{1 - S_{wn}^{0.25} S_w^{0.5}} \tag{2.36}$$

$$S_{wn} = \frac{S_w - S_{lr}}{1 - S_{lr}} \tag{2.37}$$

$$S_{lr} = S_{wc} + S_{ors} \tag{2.38}$$

Thus, the equation for the steam fraction is:

$$f_s = \frac{K_{rs}/\mu_s}{K_{rs}/\mu_s + K_{rw}/\mu_w} \tag{2.39}$$

where K_{rw} is the water relative permeability; K_{rs} is the steam relative permeability; S_w is water saturation; S_{wn} is normalized water saturation; S_{lr} is the minimum liquid saturation; S_{wc} is irreducible water saturation; S_{ors} is residual oil saturation by steam flooding; f_s is the steam fraction; μ_s is steam viscosity, mPa.s; and μ_w is water viscosity, mPa.s.

2.3.3.2 Permeability of steam breakthrough path

In most cases, fluid flow in porous media obeys Darcy's law. However, as permeability and pressure drop dramatically increase, the flow state in porous media shows the characteristics of nonlinear flow. In addition, inertial resistance dominates fluid flow. Thus, the drop in pressure and the flow rate in porous media no longer have a linear relationship. It is a nonlinear flow process. Darcy's law is not applicable. A binomial expression between the fluid flow rate and drop in pressure is [10]:

$$-\frac{dp}{dx} = \frac{\mu}{K}v + \xi \rho v^2 \tag{2.40}$$

From this equation, if the fluid flow rate is small, the contribution of the second part in the right-hand side can be neglected and this equation will transit to the conventional Darcy's law. Therefore, for Eq. (2.40), the first part on the right-hand side is the drop in pressure caused by viscous resistance, and the second part on the right-hand side is the drop in pressure caused by inertial resistance. Once the fluid flow rate is small, the first part dominates the flow. Otherwise, the second part dominates the flow in porous media:

$$v = \frac{q}{\overline{B}\overline{h}} \tag{2.41}$$

$$\xi = \frac{c}{\sqrt{K}} \tag{2.42}$$

Substituting Eqs. (2.41) and (2.42) into Eq. (2.40),

$$\frac{\Delta p}{L} = \frac{\mu_{ws}}{K_s}\frac{q_{ws}}{\overline{B}\overline{h}} + \rho_{ws}\frac{c}{\sqrt{K_s}}\frac{q_{ws}^2}{\left(\overline{B}\overline{h}\right)^2} \tag{2.43}$$

where $\overline{B}\overline{h}$ is the average flow area in the *j*-direction, m^2; Δp is the drop in pressure in the *j*-direction, Pa; ξ is a velocity factor; c is a constant; μ_{ws} is steam viscosity, Pa.s; q_{ws} is the volume flow rate of steam, m^3/s; K_s is the effective permeability of steam, m^2; and L is well spacing in the *j*-direction, m.

From Eq. (2.43), the effective permeability of steam can be derived. It is actually permeability for the status of a residual fluid. Then, the absolute permeability of a reservoir can be obtained by:

$$K = K_s exp\left(\frac{S_{lr}}{1 - S_{lr}}\right) \tag{2.44}$$

where S_{lr} is residual fluid saturation ($S_{or} + S_{wc}$).

2.4 Fine migration

2.4.1 Introduction of fine migration in steam injection process

Heavy oil resources are generally deposited in an unconsolidated porous medium environment. During steam injection processes for heavy oil reservoirs, on account of the long-term steam stimulation, some solid particles or fine grains come off from formation rocks and begin to migrate with reservoir fluids in porous media [11,12]. This process can tremendously affect the physical properties and recovery performance of reservoirs. Once the solid particles move to a pore throat, a bridging effect can be formed to plug the pore space and reduce the reservoir permeability. On the other hand, under a fluid flow rate, the solid particles can simultaneously move in reservoirs. Once the flow rate is reduced, the fine grains will deposit at rock surfaces and narrow the pore space. In addition, for some hybrid thermal recovery processes, solid particles and steam are co-injected into reservoirs. The dissolution of rock skeleton particles and clay minerals also produces a large number of solid particles or fine gains.

2.4.1.1 Source of solid particles

Under the condition of high-temperature or high-alkali concentration, the dissolution and transformation of rock minerals are accelerated compared with conventional recovery [13,14]. This process can provide a large number of solid particles. The dissolution of rock skeleton particles diminishes the size of rock grains and increases the number of fine grains; it also breaks the rock grains and results in chemical reactions. In addition, a series of changes in mineral transformation and generation as well as grain deposition can be observed in reservoirs. Therefore, from this point of view, mineral dissolution is the basis of formation damage and the change in a fluid flow condition. In heavy oil reservoirs, the dissolution of quartz and feldspar is more significant. It will further damage the rock skeleton and lead to serious problems of sand production and steam breakthrough.

2.4.1.2 Fine migration by mechanical interaction

Because of the mechanical interaction of steam and condensate, unconsolidated rock grains can move in porous media. Thus, reservoir permeability is damaged from the mechanical movement of fine grains [13,15]. If a large number of fine grains are created in porous media, they can accumulate in pore throats and reduce reservoir permeability. The movement of rock grains and damage to reservoir permeability are caused by velocity sensitivity. Steam with a high temperature and high velocity dissolve a rock skeleton and migrate rock grains by mechanical interaction. Most heavy oil reservoirs in the world are classified as unconsolidated sandstone formations. They are more sensitive to velocity damage from injection rates.

2.4.1.3 Fine migration by chemical reactions

As steam is injected into heavy oil reservoirs, a series of chemical reactions are activated owing to the release of rock grains. The swelling, scattering, and movement of clay minerals and pore blockages significantly damage formation properties.

First, the condensate of steam can reduce ion concentration in reservoir fluids. If critical salinity is reached, the repulsion of a double layer in clay minerals is increased. It can prompt the scattering and movement of clay minerals. Finally, clay grains can accumulate in pore throats. When the salinity is higher than critical salinity, fine clay grains adsorb on rock surfaces. When it is lower than critical salinity, repulsion of the double layer dominates the interaction. The fine grains will be released.

Second, based on field operations, the pH value of the liquid phase of steam from a steam boiler usually reaches 102.12. It is strongly alkaline [16]. In general, low pH fluids can lead to the release of small grains, and high pH fluids can benefit from the release of large grains. A high pH environment can prompt the same electric charge environment for rock grains, clay particles, and clay surfaces. In such a situation, clay grains can be released and accumulate in pore throats, damaging reservoir properties.

Third, a high temperature condition is the most significant characteristic of thermal recovery processes. The condensate of steam usually has a low salinity and a high alkalinity. Under a high temperature condition, a high alkaline environment can accelerate the dissolution of silicon and induce a mineral reaction. As a silicon-contained fluid moves to a pore space with a low temperature, a cementing reaction is activated. Then it will block pores. As the condensate contacts with formation minerals, due to a mineral transformation, the dissolution reaction of dolomite and kaolinite and the formation reaction of analcite, calcite, montmorillonite and chlorite will be activated. The accumulated analcite, montmorillonite and chlorite can tremendously reduce the reservoir permeability.

Moreover, during thermal recovery processes, various operation parameters and fluid parameters usually vary with operation time. Rapid variations of flow rates, salinity and temperature may damage the reservoir permeability. But, if the variation process is slow, the damage reaction can be neglected. Because of the characteristics of high temperature, high alkalinity and low salinity of steam, after a rapid steam injection, they can sharply reduce the reservoir salinity. Thus, a large number of clay grains are released, and the grain concentration in the reservoir fluid is rapidly increased. As the grain concentration increases, grains will accumulate in pore throats by bridging and/or storage. Therefore, the formation permeability is reduced.

2.4.2 Experimental tests of fine migration

As steam injection continues, a series of reactions is activated between the injected steam and the reservoir rocks, called hydrothermal reactions. It is a complicated process. These include physical reactions, chemical reactions, and a mechanical process. One of the most important results from hydrothermal reactions is fine migration. It damages the reservoir pore structure and reduces reservoir porosity and permeability. Even the normal development of a heavy oil reservoir can be greatly affected.

2.4.2.1 Experimental method

A set of flow experiments were conducted to test the behavior of reduced permeability during steam injection [14,16]. Flow experiments are an easily way to test changes in reservoir permeability by positive flow and reverse flow. Because of the complicated mineral composition and pore structure of a heavy oil reservoir, changing behavior in reservoir properties in a steam injection process is also complicated. Fine migration, mineral dissolution, and clay swelling are three of the most common factors for reservoir damage. How to recognize the dominating factor of damage is the highest concern. The process of positive and reverse flow tests is an effective method for analyzing reservoir damage caused by fine migration. In this experiment, a fluid is first injected into a porous medium with a constant injection rate for a certain volume. Afterward, we can change the injection direction of the fluid and retest reservoir permeability in reverse flow. During reverse flow, if reservoir permeability begins to rise again and then decrease after a lengthy injection, that represents the occurrence of fine migration. On the other hand, if reservoir permeability remains unchanged or continues to decrease, that indicates that a reduction in reservoir permeability is not caused by fine migration. Because of a rapid change in fluid injection direction, the mobilized fine grains also flow back based on mechanical movement. Therefore, some plugged pore throats will be reopened and permeability will rise back. However, as reverse injection continues, the mobilized fine grains will accumulate again and plug the pore throats. Thus, another reduction in permeability can be observed.

In the experiments, a sand pack model is applied to test permeability in positive flow and reverse flow. The experimental setup is shown in Fig. 2.8. The sand pack model is 30 cm long and 3.8 cm in diameter. During the experiments, it is filled with unconsolidated rock grains. The rock grain samples are from the Liaohe oil field. The mineral composition of these rock grains is 33.5% quartz, 32% feldspar, and 6.75% clay. The water sample we use is formation water with a salinity of 2000–8000 mg/L. During the experiments, two different fluids are applied to test the characteristics of permeability damage. The first is brine water (formation water); the other uses an alkali solution (NaOH). The concentration is 0.4% by weight. The pH value is about 13.

The detailed experimental procedure is:

- sand pack preparation using unconsolidated rock grains,
- connecting experimental apparatuses,
- gas tightness testing to check for leaks,

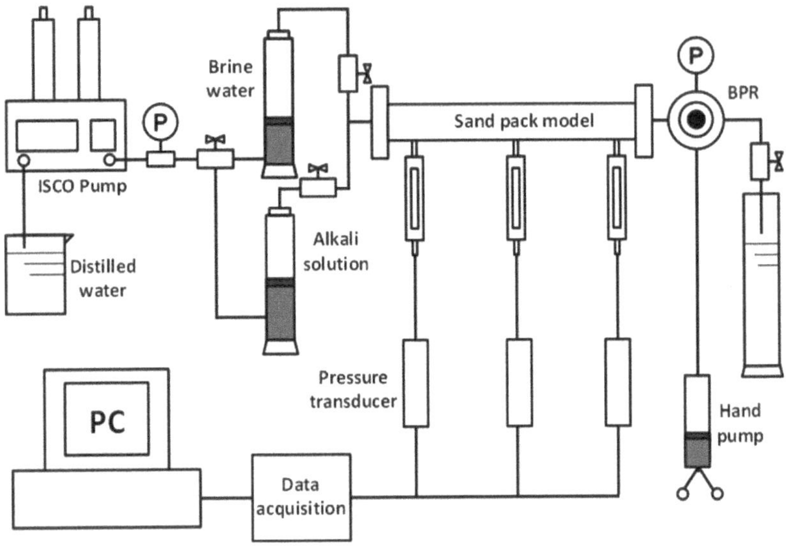

FIGURE 2.8 Schematic of experimental setup for test of fine migration. *BPR*, Back pressure regulator.

- ➢ fluid injection for a certain volume,
- ➢ permeability testing,
- ➢ repeating the steps of fluid injection and permeability testing until the total fluid volume has reached 10 PV,
- ➢ changing the fluid injection direction for reverse flow,
- ➢ fluid injection for a certain volume,
- ➢ permeability testing,
- ➢ continuing fluid injection and testing permeability until the total volume has reached 10 PV.

2.4.2.2 Experimental results

Table 2.2 shows the permeability after the positive and reverse flow experiments. For the positive flow test, reservoir permeability is reduced as brine water and alkali solution injection continues. For the reverse flow test, reservoir permeability can rise back, but it is still below the initial permeability. This indicates the migration behavior of fine grains in the porous medium. For comparison, the effect of the alkali solution on reservoir permeability is more significant; for the alkali solution injection, both fine migration and mineral dissolution can have an important effect on reservoir permeability. Also, as the temperature increases, the degree of permeability damage increases.

TABLE 2.2 Reservoir permeability after flow experiments [16].

Parameter	50°C		100°C		150°C		200°C	
	Brine water	Alkali solution	Brine water	Alkali solution	Brine water	Alkali solution	Brine water	Alkali solution
Initial permeability (10^{-3} μm^2)	2200	2205	2121	2102	2065	2047	2037	2075
Permeability after positive flow experiment (10^{-3} μm^2)	1880	1428	1713	1344	1590	1275	1535	1172
Permeability after reverse flow experiment (10^{-3} μm^2)	2051	1725	1869	1586	1769	1468	1697	1376

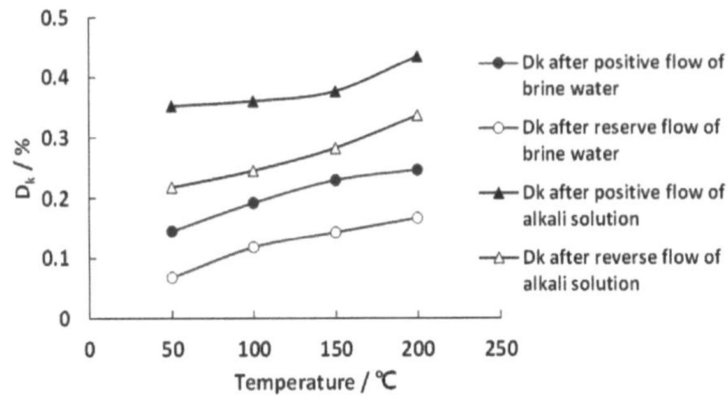

FIGURE 2.9 Ratios of permeability damage at different temperatures.

To evaluate the effect of fine migration on permeability damage effectively, the ratio of permeability damage is introduced:

$$D_K = \frac{K_i - K_p}{K_i} \tag{2.45}$$

where D_k is the ratio of permeability damage, K_i is the initial permeability, 10^{-3} μm^2, and K_p is permeability after a flow experiment, 10^{-3} μm^2. D_k can be used to evaluate the changing percentage of reservoir permeability.

The results for the ratio of permeability damage at different temperatures are shown in Fig. 2.9. As the temperature increases, the ratio of permeability damage gradually increases. This indicates that reservoir permeability after a flow experiment is reduced as the temperature increases. The higher the temperature, the larger the difference between the initial and final permeability after flow experiments. Compared with the flow experiments of brine water, the ratio of permeability damage for the alkali solution experiment is higher because an alkaline environment can benefit the behavior of mineral dissolution and transformation, which further damages reservoir permeability.

2.5 Mineral dissolution and transformation

A hydrothermal reaction is one of the most important causes of permeability damage during steam injection. Mineral dissolution is a major reaction. It can result in a serious change in pore structure and a great reduction in reservoir permeability. In steam injection, under the effect of saturated steam at a high temperature, high pressure, and high pH, many different types of mineral dissolution and transformation will be activated.

2.5.1 Characteristics of mineral dissolution and transformation

2.5.1.1 Mechanisms of mineral transformation

During steam injection in heavy oil reservoirs, mineral transformation behavior can usually be observed in clay minerals of kaolinite, montmorillonite, illite, and chlorite. This behavior can also be found in non-clay minerals of quartz and feldspar. Because of mineral transformation, the formation sensitivity is significantly increased (e.g., water, acid, and velocity sensitivity). The precipitation of montmorillonite and zeolite and the dissolution of opal, kaolinite, calcite, and some other clay minerals are two of the most significant reactions. The specific manifestation of formation damage caused by mineral dissolution has two aspects. First, during development, as the reservoir temperature or fluid temperature is reduced, the precipitation of dissolved minerals can deposit in a formation and induce damage. Second, some dissolvable minerals involve undissolved fine grains. Once the minerals are dissolved in reservoir fluids by high-temperature condensate, the undissolved fine grains are released. Then, these fine grains can migrate with reservoir fluids and plug pores and throats by grain bridging.

2.5.1.2 Mechanisms of rock–condensate reactions

In an initial reservoir condition, non-clay minerals of a rock skeleton (i.e., quartz and feldspar) are usually more stable than clay minerals. They cannot react physically or chemically with the injected fluids. However, for steam injection, because of the high pH values of steam and its corresponding condensate, the possibility of interactions between fluids and rock minerals is tremendously increased. A high alkaline environment can accelerate the occurrence of dissolution, decomposition, and synthetization reactions for different minerals in a formation. The strength of reactions is related to the rock minerals, fluid composition, temperature, and pH value.

The mineral dissolution and transformation of several different clay and non-clay minerals during steam injection are presented next [11–15].

2.5.1.2.1 Kaolinite

In the alkaline condition, kaolinite is unstable. Temperature and pH value have an important effect on the mechanism and products of a kaolinite transformation reaction.

If a pH value is about 9 and the reservoir temperature is 250°C, kaolinite is translated into montmorillonite:

$$Me + \text{kaolinite} + H_2S_iO_4 \rightarrow \text{montmorillonite} + H_2O + H^+ \quad (2.46)$$

where *Me* represents the metal ion of Ca^+ or Na^+.

Otherwise, if the pH value is above 10 (strongly alkaline) and the reservoir temperature is 250°C, kaolinite is translated into analcite:

$$Me + \text{kaolinite} + H_2S_iO_4 \rightarrow \text{analcite} + H_2O + H^+ \quad (2.47)$$

If the reservoir temperature is between 200°C and 250°C, the formation of montmorillonite and the dissolution of dolomite occur simultaneously. The chemical reaction is shown as:

$$\text{kaolinite} + \text{dolomite} + \text{quartz} \rightarrow \text{montmorillonite} + \text{calcite} + H_2O + CO_2 \quad (2.48)$$

In the neutral solution of NaCl, a chemical reaction can be activated in the minerals kaolinite, illite, and calcite:

$$\text{Kaolinite} + \text{calcite} + \text{quartz} \rightarrow \text{montmorillonite} + H_2O + CO_2 \quad (2.49)$$

Montmorillonite is a water-sensitive clay mineral. Hydration swelling is the most distinguishable feature of montmorillonite in a reservoir. Under the effect of a hydrodynamic force, montmorillonite will decompose and migrate into the pore space in the reservoir. Finally, the migrated grains can plug pore throat paths and reduce reservoir permeability. Moreover, under high temperature, besides the products of montmorillonite translated from kaolinite, the chemical reactions of kaolinite in a formation generate a small tubular structure of halloysite, entangled shreds of enlarged plates of dickite, and separated, broken book stacks of kaolinite [17,18]. They also reduce reservoir permeability.

2.5.1.2.2 Montmorillonite

In a solution environment of NaOH and $NaHCO_3$, if the temperature is higher than 350°C, montmorillonite can translate into an illite–montmorillonite mixed layer, and then illite. Once the pH is higher than 11, montmorillonite can translate into illite if the metal ion of K^+ can be provided from the dissolution of potash feldspar. If temperature equals 350°C, a new mineral of boehmite can be translated. In an Na^+ contained environment, with the dissolution of a large number of quartz minerals, montmorillonite can translate into analcite. In a solution environment of NaOH and KH_2PO_4, montmorillonite mainly translates into illite.

2.5.1.2.3 Carbonate minerals

During steam injection, the dissolution of carbonate rocks and the transformation of new minerals usually occur simultaneously in a reservoir. As shown in chemical reactions (2.48) and (2.49), the decomposition of kaolinite is related to the decomposition of carbonate minerals. The presence of carbonate minerals in a reservoir can further induce a translation from kaolinite to montmorillonite. Also, the calcium ion produced from calcite decomposition is mainly applied in the synthesis of montmorillonite. In heavy oil reservoirs, as kaolinite contacts with the neutral or alkaline steam condensate, kaolinite can decompose into illite (montmorillonite) or calcite. In addition, an acid environment is released during this reaction, as shown in Eqs. (2.46) and (2.47). Thus, a chemical reaction can be activated between the carbonate rocks and acid fluid to release CO_2:

$$\text{calcite} + 2H^+ \rightarrow H_2O + Ca^{2+} + CO_2 \quad (2.50)$$

Although an acid environment can be also produced from a reaction between some other minerals (e.g., feldspar) and kaolinite, the chemical reaction kinetics condition is beneficial to the reaction of carbonate minerals. If carbonate minerals are present in a reservoir, they will dominate the assumption of an acid fluid and the production of CO_2. Another chemical reaction can be activated between carbonate minerals and aluminosilicate to produce precipitation:

$$\text{carbonate} + \text{aluminosilicate} \rightarrow \text{clay} + CO_2 \tag{2.51}$$

Aluminosilicate usually includes the minerals of feldspar and kaolinite, and carbonate includes the minerals of siderite, limestone, and dolomite. The clay minerals produced refer to montmorillonite. Thus, serious formation damage of water sensitivity can occur. Another type of formation damage comes from the precipitation of calcium carbonate caused by CO_2 produced in chemical reactions.

2.5.1.2.4 Fine quartz

Quartz usually refers to silicon dioxide. Its chemical properties are stable at room temperature. The solubility of silicon dioxide in water is slow. However, as the temperature increases, its solubility increases. For steam injection, because of the alkaline environment of steam condensate, the solubility of silicon dioxide dramatically increases. Quartz has higher solubility in an alkaline fluid, which increases as the pH value and temperature increase. When the pH is lower than 10, the solubility of quartz is relatively low. When the pH equals 13, its solubility dramatically increases. Around a steam injector, the injected fluids may have the saturation state of quartz and other minerals. When the injected fluids move into formation, the temperature decreases. Silica can precipitate into amorphous silica, quartz, and some other new minerals. The precipitation of silica minerals can damage the formation and lead to a reversible change in reservoir permeability.

2.5.1.3 Mechanisms of formation damage caused by mineral transformation

2.5.1.3.1 Permeability reduction caused by mineral dissolution and precipitation

For thermal recovery, the pore geometry in a reservoir can be changed because of changes in local temperature and effective stress [12]. A reduction in reservoir permeability will further increase the injection pressure and operation costs. The recovery factor will also be reduced. As the temperature increases, the solubility of silicon dioxide increases. Therefore, the concentration of silicon dioxide in fluids is increased. Also, a high-viscosity silicon-dioxide cement can be formed in a pore space. It also precipitates on the surfaces of fine grains. Thus, the radii of pore throat paths are reduced. The mechanisms of the effects of temperature and confined pressure on absolute reservoir permeability include clay swelling, the generation of fine grains, thermal expansion, changes in the elastic constant, aggravation of reservoir heterogeneity, and chemical interactions between the reservoir matrix and pore fluids.

2.5.1.3.2 Serious fine migration caused by mineral transformation

During steam injection, because of a series of chemical reactions, feldspar, quartz, and carbonate can translate into different clay minerals. Therefore, the number of fine grains in thermal recovery is usually higher than that in conventional recovery. The migration of clay minerals will damage reservoir performance. Because of the laminar structure of kaolinite, although kaolinite grains are adsorbed on the surface of quartz grains, they are still movable in a reservoir under the effect of hydrodynamic tension on kaolinite grains. Therefore, reservoir permeability is significantly reduced.

2.5.2 Experimental tests of mineral dissolution and transformation

Using a high-temperature high-pressure reactor, a systemic experimental investigation was performed on the characteristics of mineral dissolution and transformation [16]. The target heavy oil reservoir was Qi 40 block in the Liaohe oil field, China National Petroleum Corporation, China. The reservoir depth is 924 m. It is an unconsolidated heavy oil reservoir. The average percentages of quartz, feldspar, rock debris, and clay minerals in the formation are respectively 33.5%, 32.0%, 23.7%, and 6.75%. For clay minerals in this heavy oil reservoir, montmorillonite is the main component; its relative percentage is about 64.4%. The relative percentages of kaolinite and illite are 18.3% and 17.3%, respectively.

For the effect of mineral dissolution and transformation on reservoir properties, mineral dissolution can take place around a steam injector. Because of the conditions of high temperature and a high pH value, reservoir rocks will be more unconsolidated from mineral dissolution. Also, the dissolved fine grains will migrate in a region far from the wellbore. Because of a reduction in temperature, the crystallization of minerals and the synthesis of some new minerals can plug pore throats and reduce reservoir permeability.

2.5.2.1 Dissolution of quartz grains

Table 2.3 shows experimental results for the dissolution of quartz grains in a high-temperature high-pressure (HTHP) reactor. The test pressure is 6 MPa and the pH of the fluid environment is controlled by $NaHCO_3$ and NaOH. The duration of each test is 72 days. The solubility of quartz is relatively lower for pH 8−10. As the pH increases to 13, the solubility dramatically increases. It is about 10−20 times higher than that at a low pH. The temperature is also an

TABLE 2.3 Experimental results of dissolution of quartz grain.

No.	Temperature (°C)	pH value	Duration time (h)	Solubility of quartz (mg/L)
1	100	8	72	33
2	200	8	72	152
3	300	8	72	175
4	100	10	72	178
5	200	10	72	367
6	300	10	72	440
7	100	13	72	478
8	200	13	72	6250
9	300	13	72	7250

important parameter. For case 7 in Table 2.3, although the pH increases to 13, because the temperature remains at 100°C, solubility does not increase significantly. For field steam injection, the pH of the steam condensate is an important parameter for controlling the dissolution behavior of quartz.

2.5.2.2 Dissolution of clay minerals

The clay minerals used in this test are collected by settling separation on the oil sand samples from the Qi 40 heavy oil block. Table 2.4 shows experimental results for the dissolution of clay minerals in the HTHP reactor. The test pressure is 6 MPa. As the temperature and pH increase, the solubility of clay minerals increases. Because of the highest temperature and pH, case 9 has the maximum solubility. Moreover, the pH after the test is lower than the initial pH because of the effect of hydrothermal reactions in the reservoir.

2.5.2.3 Dissolution of mixed fine grains

In this section, collected oil sand samples from the Qi 40 block are used directly to test the characteristics of mineral dissolution. The real oil sands represent the actual behavior of mineral dissolution in heavy oil reservoirs. Table 2.5 shows the experimental results of the dissolution of mixed fine grains in the HTHP reactor. The solubility of mixed fine grains at a high temperature and high pH (case 9) is about 12.7 times higher than that at a low temperature and low pH (case 1). Therefore, the pH of injection fluids in field operations should be low. It is usually recommended to have a value of 9.0.

TABLE 2.4 Experimental results of dissolution of clay minerals.

No.	Weight of clay samples (mg)	Temperature (°C)	Initial pH	Duration time (h)	Solubility of clay minerals (mg/L)	pH after test
1	1000	150	8	48	1873	6.5
2	1000	150	10	48	1971	7.5
3	1000	150	13	48	2619	12
4	1000	200	8	48	2574	7.5
5	1000	200	10	48	2353	8
6	1000	200	13	48	3217	12
7	1000	300	8	48	3466	7.2
8	1000	300	10	48	3302	7.5
9	1000	300	13	48	5387	10

TABLE 2.5 Experimental results of dissolution of mixed fine grains.

No.	Temperature (°C)	Initial pH	Duration time (h)	Solubility (mg/L)	pH after test
1	100	8	48	913	7.5
2	100	10	48	823	9.5
3	100	13	48	1333	12.0
4	200	8	48	2157	7.5
5	200	10	48	1270	9.5
6	200	13	48	12,660	12.0
7	300	8	48	1893	7.5
8	300	10	48	880	9.5
9	300	13	48	11,580	12.0

2.6 Clay swelling

2.6.1 Effect of clay minerals

Clay is a finely grained natural rock or soil material that combines one or more clay minerals with possible traces of quartz, metal oxides, and organic matter

[19]. The main chemical compositions of clay include silicon oxide, aluminum oxide, water, iron, alkali metal, and alkaline earth metal oxide.

Heavy oil reservoirs are unconsolidated shallow media with a low initial temperature. Montmorillonite and kaolinite are the main clay minerals in these reservoirs. In steam injection, as high-temperature steam contacts with reservoirs, reservoir permeability is reduced by the swelling, separation, and the migration of clay grains [20,21]. Because of the heterogeneity of the composition and distribution of clay minerals in the reservoirs, the effects of swelling and migration of clay grains on permeability reduction are also heterogeneously distributed.

For formations with a higher concentration of clay minerals, the effect of clay swelling is more significant. However, for formations with a lower concentration, it has less of an effect. Therefore, to some extent, the swelling behavior of clay grains can further aggravate reservoir heterogeneity. On the other hand, the behavior of clay swelling and separation in reservoirs also damages the pore structure and even results in the occurrence of sand production. In reservoirs, because of their pore structure and grain filling in rock, injected fluids usually first contact clay minerals. Because of the small grain size and large specific area of clay grains, the chemical reaction between fluids and clay minerals will be easily activated. For steam injection operations in heavy oil reservoirs, controlling the behavior of clay swelling effectively is important.

2.6.2 Mechanisms and sensitivity of clay swelling

2.6.2.1 Mechanisms of clay swelling

The hydration swelling behavior of different clay minerals varies. It is dominated by three forces: surface hydration, osmotic hydration, and capillary force.

2.6.2.1.1 Surface hydration force

Surface hydration force is caused by the adsorption behavior of water molecules on the surface of clay minerals. The hydration force at the rock surface refers to surface hydration energy. A hydrogen chain connection may be formed between the first layer of water molecules and the oxygen molecules in a hexagon grid of the surface of clay minerals. Therefore, through a hydrogen chain, water molecules will be connected by a hexagonal ring. The next layer is similar to the first layer with water molecules. A structure of surface hydration has a crystalline nature. The viscosity of surface water is higher than that of free water. For kaolinite, the adsorption of water molecules happens only on outside surfaces and grain edge locations. Because the adsorption enthalpy of this type of adsorption water is small, a small increase in temperature can induce evaporation. For montmorillonite, hydration in both outside and inside surfaces is important. For low humidity, surface hydration happens only on an outside surface. For an inside surface, an interlaminar exchangeable cation exists only between different layers. As the humidity increases, the continuous adsorption of water molecules can result in multi-layer adsorption and even fill the micropores outside.

2.6.2.1.2 Osmotic hydration force

When the cation concentration between different crystal layers is higher than that in reservoir fluids, water molecules can easily enter the inner crystal layers and increase the distance between adjacent crystal layers. A diffusing electric double layer can also be developed (osmotic hydration). Theoretically, the volume increased by osmotic expansion is higher than that by lattice expansion. In addition, the high expansibility of montmorillonite is caused by a large inner surface area and a plenty of exchangeable cations. However, for kaolinite, chlorite, and illite, their expansibility is significantly lower than that for montmorillonite because of a lower specific area and freedom from exchangeable cations.

2.6.2.1.3 Capillary force

During steam injection, once the swelling of clay grains is observed, the water volume in the pore space of clay minerals increases. Thus, the attractive force between clay particles is reduced and the pore radius is increased. As the water volume increases, the contact angle between water and a clay grain surface decreases. Therefore, both the capillary pressure and surface tension are reduced. From the macroscopic view, the rock strength is reduced and is softened.

2.6.2.2 Sensitive factors for clay swelling

2.6.2.2.1 Effect of crystal location on a hydration film

Most of negative charges in clay grains are concentrated on a layer surface; there are relatively fewer on a side surface. Therefore, the cations adsorbed by clay grains are also concentrated on the layer surface. The hydration of clay minerals mainly refers to cation hydration on a layer surface. For different distributions of adsorbed cations in clay minerals, the thickness of a hydration film of surface hydration is also different. A hydration film is thicker on a layer surface and thinner on a side surface.

2.6.2.2.2 Effect of clay species on hydration behavior

Different clay minerals have varying hydration behavior. Montmorillonite has a large exchangeable capacity of cation and high dispersion. Its hydration behavior is obvious. For kaolinite, it has a low exchangeable capacity of cation and nonsignificant hydration behavior. For illite, because of the special structure of interlaminar cation (K^+), it is a nonexpansive mineral and its hydration behavior is low.

2.6.2.2.3 Effect of exchangeable cation on hydration behavior

In clay minerals, two different forces can be observed in crystal layers. The first is the swelling force caused by the hydration of interlaminar cation and the repulsive force between negatively charged crystal layers. The second is the electrostatic attractive force between a crystal layer of clay grains and interlaminar cation, and that between different crystal layers. These two forces result in the difference of hydration behavior for various exchangeable cations. The percentage of these two forces dominates the degree of swelling and dispersion of clay minerals. If the electrostatic attractive force is higher than the repulsive force between crystal layers, lattice swelling can be activated in clay minerals. On the contrary, if the repulsive force between crystal layers is high enough to resist the electrostatic attractive force, osmotic swelling can be activated in clay minerals, even forming a diffusing electric double layer.

2.6.3 Migration of clay grains

The migration of clay grains is important for reducing reservoir permeability. It can be classified into two categories: chemically induced fine migration and mechanically induced fine migration. Both variations in flow rates and physicochemical conditions can result in fine migration and permeability reduction.

2.6.3.1 Critical salinity

For the development of heavy oil reservoirs, because of the injection of external fluids, the original salinity of a reservoir fluid will be diluted. As steam injection continues, the salinity is reduced. If the salinity of the reservoir fluid is lower than the critical salinity, the reservoir permeability remains constant. However, during the reduction of salinity to the critical value, reservoir permeability is rapidly reduced. The critical salinity is related to the fluid temperature and the cation species in the fluid.

2.6.3.2 Critical flow rate

Kaolinite and illite are typical velocity-sensitive clay minerals. When a fluid flow rate in porous media is higher than the critical flow rate, serious migration of clay grains can tremendously reduce reservoir permeability; it is a relationship of linear decrease. On the contrary, if the fluid flow rate is lower than the critical flow rate, a reduction in reservoir permeability may not be observed.

The migrated clay grains are from external fluids (e.g., drilling, completion, and injection fluids) or existing fine grains in reservoirs. For the latter, they are caused by compatibility between steam and formation rocks/fluids.

For external fluids, the three mechanisms of pore plugging are gradual plugging, single-pore plugging, and filter cake plugging. Plugging caused by fine migration is highly related to the relationship between the size of migrated grains and the throat size. When the size of migrated grains is far less than the throat size, precipitation and de-precipitation will coincide in porous media. When the grain size is higher than the throat size, single-pore plugging will be activated. When the grain size is greatly higher than the throat size, a permeable filter cake may be formed in throats because of accumulated fine grains. In a later stage of a fluid flow, as solid grains are stripped from the filter cake, permeability gradually recovers.

2.7 Water coning

For heavy oil reservoirs with a bottom/boundary aquifer, water coning is a major problem hindering thermal recovery [22]. Reasons for this behavior include low cementing quality, aquifer perforation, reservoir heterogeneity, unreasonable operation parameters, unsuitable stimulation treatment, geological reasons, and some operation processes. Once a connection is made between the aquifer and thermal wells, the average water cut dramatically increases, and then normal development is affected. In field operations, to recognize water sources accurately during water coning, the water exploration methods of liquid-producing profile measurements, well logging, and mechanical devices have been widely applied. For watering coning, a lengthy water washing from an aquifer can induce sand production and damage wellbore integrity; thermal wells may even shut down. On the other hand, corrosion can also be observed in casings and production tubings because of polar and nonpolar ions from the aquifer. For different thermal wells, water coning is also different. It is caused by differences in reservoir properties, wellbore characteristics, and production performance. The most important issues for the water coning behavior of thermal recovery in heavy oil reservoirs are to evaluate the water coning effectively and inhibit water coning.

2.7.1 Evaluation methods of water coning behavior

In this section, two different methods are introduced to evaluate water coning in aquifer-contained heavy oil reservoirs.

2.7.1.1 Evaluation method of recovery performance

This method is proposed based on real production data. In the field, we can obtain cumulative oil and water production volumes (Q_O, Q_W) and a cumulative steam injection volume (I_S). A steam injection volume is usually replaced with the cold water volume equivalent. Also, for steam-based recovery processes, the difference between a cumulative water production volume and a cumulative steam injection volume is the water influx. To evaluate water coning behavior accurately, two evaluation indicators are proposed: the ratio of cumulative water to cumulative oil (RWO) production volumes and a water intrusion volume (WIV). The slope of the curve of RWO versus Q_o can be used to evaluate the speed of water coning effectively and thus to screen the optimal method to control water coning. For a serious water coning well, a high-strength gel can be applied to prevent water coning. For a slight water coning well, noncondensable gas (NCG) or NCG-foam can be used as a chemical agent to prevent water coning. On the other hand, the slope of the curve of WIV versus Q_o can be used to evaluate the strength of water coning, and thus to design the suitable injection volume of a plugging agent.

2.7.1.2 Evaluation method of Hall's curve

Because of its certain geologic condition, a heavy oil reservoir and bottom water can come into direct contact through a low permeable layer or a non-fully covered baffle. All three contact relationships can result in serious water coning. A direct contact relationship can generate conventional water coning behavior. For the other two, after a lengthy steam stimulation, the properties of a low permeable layer or baffle have changed compared with their initial properties. After that, an obvious response to steam injection pressure can be observed for these two types of water coning. For these complicated water coning behaviors, the method of Hall's curve can be applied to evaluate them.

Hall's curve was initially proposed by Hall in 1963. In 1989, it was introduced to evaluate the recovery performance of different EOR processes (e.g., polymer flooding). The slope of Hall's curve represents fluid flow resistance during different recovery processes [23].

2.7.1.2.1 Theory of Hall's curve in vertical wells

Hall's curve is proposed based on a radial steady-state single-phase flow theory for Newtonian fluids. For a vertical well, the relationship of the Hall integral term and the cumulative fluid injection term is linear in Cartesian coordinates:

$$W_i = \frac{0.535626K \cdot h}{\mu_w B_w [\ln(R_e/R_w) + S]} \int (p_{inwf} - p_e) dt \tag{2.52}$$

$$\int (p_{inwf} - p_e) dt = \frac{1.867\mu_w B_w [\ln(R_e/R_w) + S]}{K \cdot h} W_i \tag{2.53}$$

$$m = \frac{1.867\mu_w B_w [\ln(R_e/R_w) + S]}{K \cdot h} \tag{2.54}$$

where P_{inwf} is the bottom hole pressure of an injector, MPa; P_e is the reservoir pressure, MPa; W_i is a cumulative fluid injection volume within a defined time,

m^3; K is the reservoir permeability, 10^{-3} μm^2; h is the reservoir effective thickness, m; R_e is an oil drainage radius, m; and S is the skin factor.

2.7.1.2.1.1 Theory of Hall's curve in horizontal wells For a horizontal well, based on Hall's curve, the relationship between the integral term of injection pressure and the integral term of a fluid injection volume still obeys a linear relationship in Cartesian coordinates. Therefore, the slope of this curve can represent the rising behavior of fluid injection pressure, as shown in Eq. (2.55):

$$m = \frac{\int \Delta p dt}{\int Q dt} \tag{2.55}$$

For a horizontal well, the classic Joshi's productivity model can be applied [24]:

$$q = \frac{542.8 K h \Delta p}{\mu B \left(\ln \frac{a + \sqrt{a^2 - L^2/4}}{L/2} + \frac{h}{L} \ln \frac{h}{2\pi r_w} \right)} \tag{2.56}$$

where q is the flow rate, m^3/d; Δp is a pressure differential, MPa; K is the reservoir permeability, μm^2; h is the thickness, m; r_w is the wellbore radius, m; L is the horizontal wellbore length, m; B is the fluid volumetric factor, m^3/m^3; and a is an oil drainage radius, m.

During a predefined time, the pressure drops can be a constant, which indicates that a steady flow rate is achieved. Therefore, the slope equation of Hall's curve can be simplified to:

$$m = \frac{\Delta p}{Q} \tag{2.57}$$

Based on Eqs. (2.56) and (2.57):

$$m = \frac{\mu_w B_w}{542.8 K h} \left(\ln \frac{a + \sqrt{a^2 - L^2/4}}{L/2} + \frac{h}{L} \ln \frac{h}{2\pi r_w} \right) \tag{2.58}$$

Fig. 2.10 shows a Hall curve chart. Obvious changing behavior on Hall's curve slope indicates an increase or reduction in fluid flow resistance around a wellbore. The curve slope will increase under the two different conditions: the injected fluids connect a low permeability path to a wellbore; the fluid flow paths are blocked around a wellbore. On the other hand, the curve slope will be reduced in the other two different conditions: the injected fluids connect a high permeability path to a wellbore; secondary fractures are formed around a wellbore.

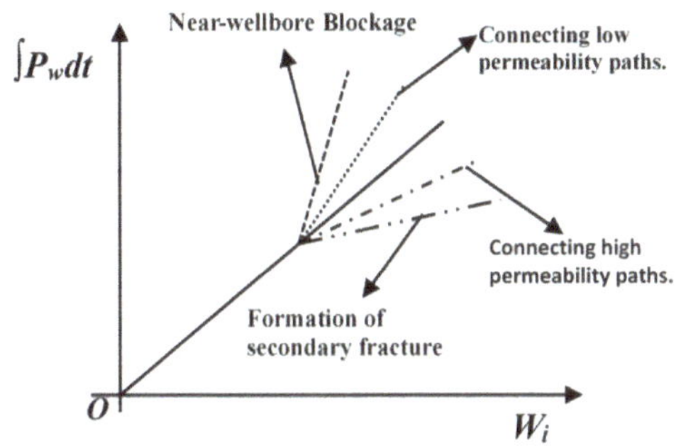

FIGURE 2.10 Hall curve chart.

2.7.1.3 Numerical simulation of water coning behavior for different heavy oil reservoirs

As mentioned, a heavy oil reservoir and bottom aquifer water can make direct contact by a low permeable layer or via a non-fully covered baffle. All three different contact relationships can result in serious water coning. Because of the difference of the type of contact between the bottom aquifer and heavy oil reservoir, their water coning characteristics are significantly different. In this section, using a commercial simulator (CMG STARS), the water coning behavior of three different heavy oil reservoirs will be discussed. Three different numerical models are developed, as shown in Fig. 2.11. The basic data of these three numerical simulation models comes from a typical bottom aquifer contained heavy oil reservoir in Chun-Guang (CG) oil field. Its average permeability is $3000 \times 10^{-3}\ \mu m^2$ and its oil viscosity under reservoir conditions is above 10,000 mPa s. It is a typical extraheavy oil field with a large volume of bottom water in western China. Because of water coning, many CSS horizontal wells in this heavy oil field have been temporarily closed to save operation costs. The ratio of water coning thermal wells to all wells in this oil field reached above 50%. Furthermore, to describe water coning accurately, a new water component is defined in the numerical models. The simulation results are shown in Fig. 2.12.

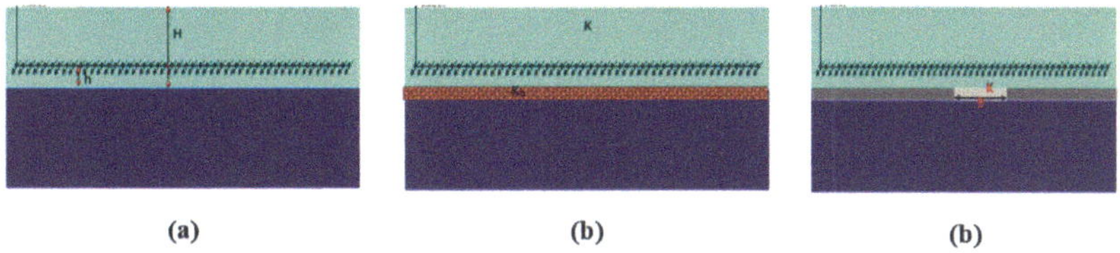

FIGURE 2.11 Three different bottom-aquifer containing heavy oil reservoirs. (A) Direct contact; (B) Contact by a low permeable layer; and (C) contact by a non-fully covered baffle.

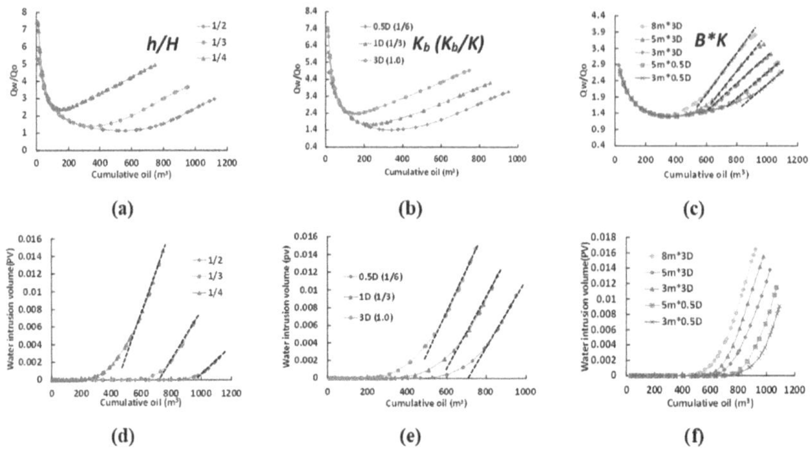

FIGURE 2.12 Simulation results of water coning behavior for different heavy oil reservoirs. (A) Ratio of cumulative water to cumulative oil (RWO) for direct contact relationship; (B) RWO for contact relationship by a low permeable layer; (C) RWO for contact relationship by non-fully covered baffle; (D) water intrusion volume (WIV) for direct contact relationship; (E) WIV for contact relationship by a low permeable layer; (F) WIV for contact relationship by a non-fully covered baffle.

First, for a direct contact relationship, the most sensitive factor is the distance between the bottom aquifer and the heavy oil zone (*h*). Therefore, a new parameter for the nondimensional distance between the bottom aquifer and the heavy oil reservoir is proposed (*h*/*H*). A smaller nondimensional distance indicates that the horizontal well is closer to the bottom aquifer. Fig. 2.12A shows the RWO of different nondimensional distances for the direct contact relationship. When the nondimensional distance decreases, the time of water coning is earlier and the cumulative oil production volume decreases. For a given cumulative oil volume, a smaller nondimensional distance indicates a higher water intrusion volume. Fig. 2.12D shows the WIVs of different nondimensional distances for the direct contact relationship. As the nondimensional distance decreases, the slope of this curve grows higher and the strength of water coning grows higher. In some field cases, water coning can be observed even in the first CSS cycle. The most likely reason is a small distance between a horizontal well and a bottom aquifer.

Second, for a contact relationship by a low permeable layer, the most sensitive factor for water coning is the permeability of this layer. As the permeability of this layer (K_b) approaches the reservoir permeability (K), water coning approaches the direct contact relationship. Fig. 2.12B and E show the RWOs and WIVs of different permeabilities in this case. As the permeability of this layer increases, the time of water coning grows earlier and the water intrusion volume becomes higher. Higher permeability of a low permeable layer also results in a higher slope of the WIV curve and a higher degree of water coning. Compared with the direct contact relationship, the presence of a low permeable layer delays the occurrence of water coning.

Third, for a contact relationship by a non-fully covered baffle, the most sensitive factor for water coning is the fluid flowing ability in this uncovered baffle (permeable gap). Thus, a parameter called a flow coefficient of a permeable gap is proposed. It is the product of the width and permeability of this gap ($B \times K$). This parameter can be used to evaluate the incompleteness of a baffle effectively. Fig. 2.12C and F show the RWOs and WIVs of different flow coefficients of the permeable gap in this case. As the flow coefficient of the permeable gap increases, the slope of an RWO curve significantly increases. It indicates a higher speed of water coning. Fig. 2.12F shows that a higher flow coefficient of the permeable gap also indicates a higher water intrusion volume. In field operations, because of an interaction between steam and a baffle, a part of the impermeable baffle can regain permeable capacity. In this case, to prohibit the occurrence of water coning, we must effectively address the continuous expansion of a permeable gap.

2.7.2 Prohibition methods of water coning

Once water coning is observed, prohibiting continuous coning from an aquifer using a suitable method is the only option for regaining the normal development of a reservoir. If water coning continues to expand, this thermal well will reach a high water cut level within a short time. Fig. 2.13 shows a schematic for the prohibition of water coning in thermal horizontal wells. Once a suitable agent for an antiwater coning process is injected, it can enter water coning paths preferentially because of their low flow resistance. Then, under the performance of the chemical agent, water will return to the bottom water zone. Therefore, for the next CSS cycle, because of the high flow resistance in previous water coning paths, the other unlocked reservoir sections will dominate production, as shown in Fig. 2.13C.

The most commonly used agents for an antiwater coning process include NCG, foam, high-temperature gels, and solid particle. In Chapter 1, their mechanisms and some field operation tests were discussed. In field operations, a suitable agent may be selected according to specific reservoir properties and

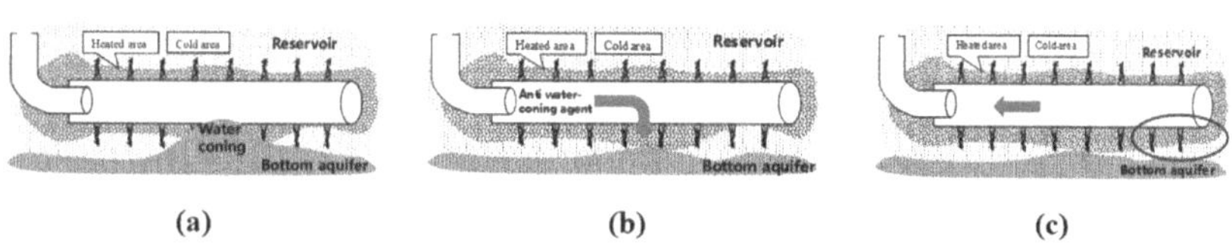

FIGURE 2.13 Schematic for prohibition of water coning in thermal horizontal wells. (A) Occurrence of water coning; (B) Injection of chemical agent; (C) normal production of thermal horizontal well.

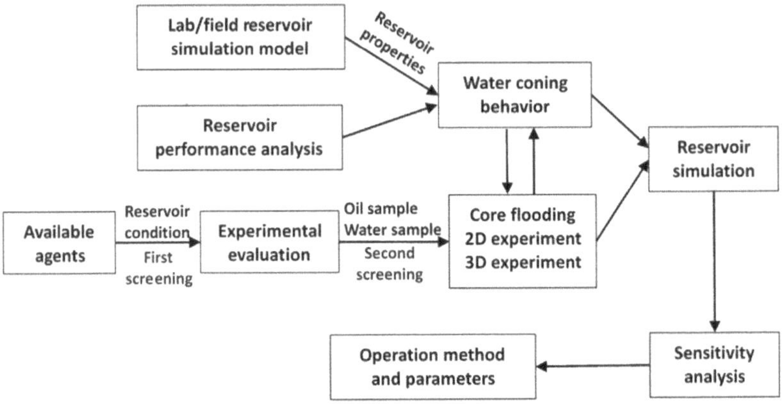

FIGURE 2.14 ramework for screening of antiwater coning methods. *2D*, two-dimensional (etc.).

characteristics of water coning. Fig. 2.14 shows a framework for the screening process of antiwater coning methods. To screen a suitable agent, a series of experimental and numerical simulation studies are required. First, according to specific reservoir conditions and recovery performance, water coning behavior can be discussed. After that, combining available agents for antiwater coning, a strict experimental evaluation process for these agents may be performed. Based on the evaluation results, suitable agents can be initially screened. Then, using the oil and water samples from the oil field, the different scale physical simulation runs can be conducted to confirm or verify the feasibility of the screened agents. For the experimental models, these include core flooding experiments (e.g., single core flooding, or parallel displacement), 2D visualization experiments, and 3D experiments. For a reservoir model, because of the characteristics of high porosity, high permeability, and an unconsolidation pore structure, a sand pack model is the most commonly used method to simulate a real porous medium environment for a heavy oil reservoir. Based on the experimental results from the different scale sand pack models, the performance of an agent in a porous medium environment can be checked. In addition, a suitable agent for antiwater coning can be determined. Finally, for the specific operation parameters of an antiwater coning process, the method of reservoir numerical simulation can be applied.

2.8 Other steam–rock interactions

Besides these problems, there are other steam–rock interactions in heavy oil reservoirs. The recovery performance of some EOR processes also has an important effect, including for asphaltene deposition, wettability alternation, and emulsification.

2.8.1 Asphaltene deposition

For different components in crude oil, asphaltene is the heaviest and most polar component. For heavy oil reservoirs, asphaltene deposition can be usually observed in steam flooding with a low steam injection pressure and solvent coinjected thermal recovery. It refers to the separation of asphaltene from heavy oil. It remains in a suspension state in heavy crude oil after separation. Then, as the particle size of suspended asphaltene grows, an asphaltene flock can be formed. Finally, the retention of these flocks on pore wall surfaces can cause asphaltene deposition [25]. Because of deposition, asphaltene is usually called the cholesterol of petroleum [26]. Primary reasons for asphaltene deposition include the mechanisms of steam distillation and solvent extraction [2,27]. After thermal recovery, three configurations of asphaltene deposition can be observed in heavy oil reservoirs:

(1) Star and dotted configuration. Deposited asphaltene is distributed in a reservoir pore space in a star and dotted configuration. The characteristics of this configuration are isolated and scattered. Because the proportion of this configuration in a reservoir is small, its damage to reservoir properties is usually negligible. The size of this configuration is about 1–10 μm. It is mainly observed in a thermal recovery process at 50–100°C.

(2) Clustered aggregate configuration. Deposited asphaltene is distributed in a reservoir pore space by the configuration of a large cluster or an isolated island. This configuration usually coexists with the star and dotted configuration. It is mainly observed in thermal recovery at 100–200°C. Because of the effect of a large shape, it can block a pore throat space. Obvious damage to reservoir properties can be observed.

(3) Package and network configuration. Deposited asphaltene is distributed in a reservoir pore space by a local connection. Some can form a large network and accumulate in the pore space. Even rock fine grains are fully covered by asphaltene. Thus, the wettability of rock will alter from water wet to oil wet. The size of this configuration is about 10–100 μm. This configuration is mainly observed in thermal recovery at 200–300°C. For this asphaltene configuration, it plugs a pore throat space and increases fluid flow resistance by wettability alternation. Therefore, the damage from this asphaltene configuration on reservoir properties is significant.

Reservoir damage caused by asphaltene deposition has three patterns. The first bridges blockage accumulated in a narrow pore and throat space. The second is wettability alternation of a rock surface from water wet to oil wet. The last one is a water–oil emulsion. It can further increase oil viscosity and reduce flowability.

2.8.2 Wettability alteration

During steam injection for heavy oil reservoirs, an important feature is the variation in temperature in the reservoir. A rise in rising results in changes in capillary pressure, rock wettability, and relative permeability curves [16,20]. Because of heterogeneous distributions of reservoir rock properties (mineral composition, a specific surface area, and surface roughness), fluid properties (fluid composition and active material fractions) and reservoir conditions (temperature, pH, and pressure), the rock wettability in a reservoir is also heterogeneous in space.

In oil—water two-phase flow, as the temperature increases, the oil—-water relative permeability curve gradually moves to the right side. It indicates an increase in irreducible water saturation and a reduction in residual oil saturation. The two-phase flow region in the relative permeability curves also shrinks for two reasons: the first is an increase in rock hydrophilicity and the second is the thermal expansion of reservoir rock. An important consequence of thermal expansion is a reduction in pore space and the two-phase coexistence region. On the other hand, for steam flooding in heavy oil reservoirs, the liquid phase refers to oil and the gas phase refers to steam. As steam injection continues, residual oil gradually attaches to pore wall surfaces and steam mainly flows in pore centers, especially after steam breakthrough occurs. Such fluid distribution is an important result of wettability alternation.

2.8.3 Emulsification

A number of surface-active materials can be found in heavy crude oil (e.g., asphaltene, resin, and naphthenic acid). They are an important emulsifier in a reservoir. In particular, asphaltene can effectively adsorb in a water—oil interface and form a stable interface membrane to enhance the stability of an emulsion. Therefore, during steam injection for heavy oil reservoirs, in situ emulsification is one of the most common phenomena. Emulsion is a dispersed system in reservoir fluids. It is distributed in porous media by spherical droplets. The behavior of transformation, trapping, retention, and blocking is commonly observed in the movement of emulsion in porous media. Therefore, an important result of emulsification in thermal recovery is a reduction in sweep efficiency, and recovery performance is affected. Compared with a conventional fluid system, an emulsion system is a complicated fluid system. The physical properties of emulsion (e.g., viscosity, density, flow behavior, and thermal properties) are tremendously different from those of a single fluid system. This difference results in more complicated flow behavior in heavy oil reservoirs. In addition, emulsification can be observed in hybrid thermochemical processes [28]. It is an important EOR mechanism of these hybrid processes.

2.9 Remaining oil saturation distribution

The remaining oil saturation distribution after steam-based recovery is an important basis for follow-up hybrid EOR processes. Because of problems in heavy oil reservoirs, the remaining oil saturation distribution after steam injection operations is more complicated than that after conventional water flooding. In this section, based on a case study, macroscopic distribution and microscopic distribution of remaining oil saturation will be discussed.

2.9.1 Macroscopic distribution of remaining oil saturation

The macroscopic distribution of remaining oil saturation is dominated by reservoir geologic conditions and operation schemes. These include five categories: remaining oil between adjacent wells, vertical remaining oil between layers, remaining oil by an incomplete well pattern, remaining oil by a fault barrier, and remaining oil by an incomplete perforation scheme. In this section, based on a field-scale numerical simulation in a typical heavy oil reservoir from the Henan oil field, Sinopec, the macroscopic distribution of remaining oil saturation is discussed. This heavy oil reservoir was discovered in 1997. CSS remains the primary recovery method. There are 53 wells in this reservoir, about 19 of which are still active. The average water cut has reached 77.78%. After more than 20 years of CSS recovery, the oil saturation distribution in this heavy oil reservoir differs significantly from the initial distribution. Based on a field-scale numerical simulation, the remaining oil saturation distribution was obtained. It is helpful for the next development of this heavy oil reservoir. Fig. 2.15 shows the numerical simulation results for remaining oil saturation in this heavy oil reservoir.

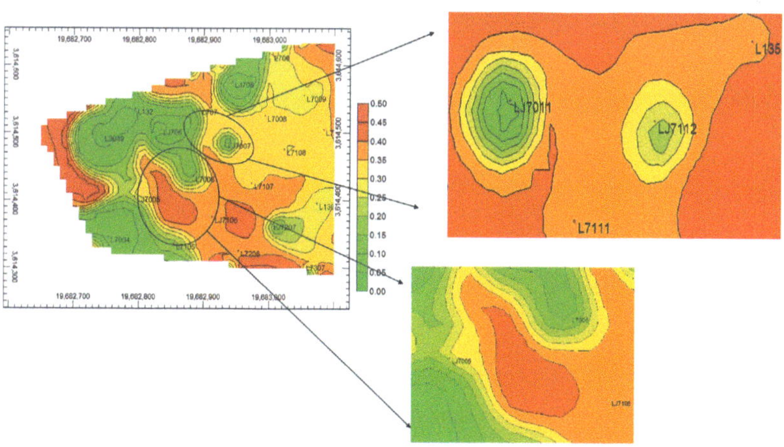

(a)

FIGURE 2.15 Numerical simulation results for the macroscopic distribution of remaining oil saturation. (A) Remaining oil between adjacent wells; (B) Vertical remaining oil between layers; (C) Remaining oil by incomplete well pattern; (D) Remaining oil by a fault barrier; and (E) remaining oil by incomplete perforation scheme.

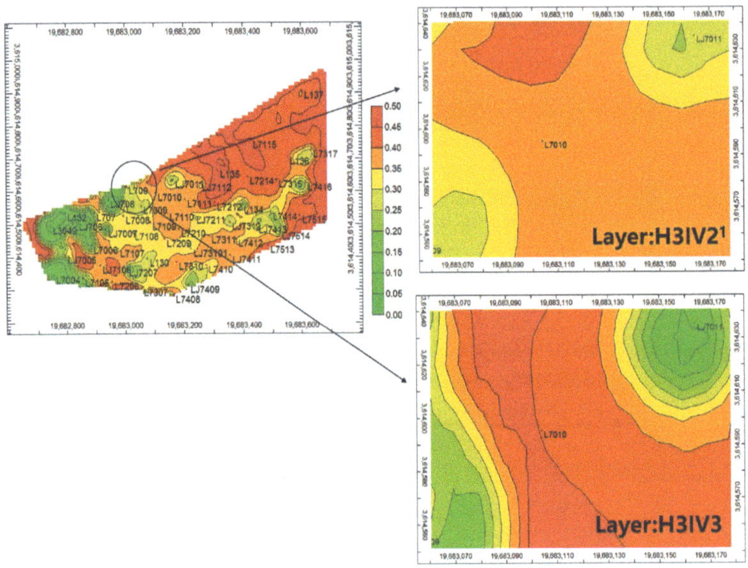

(b)

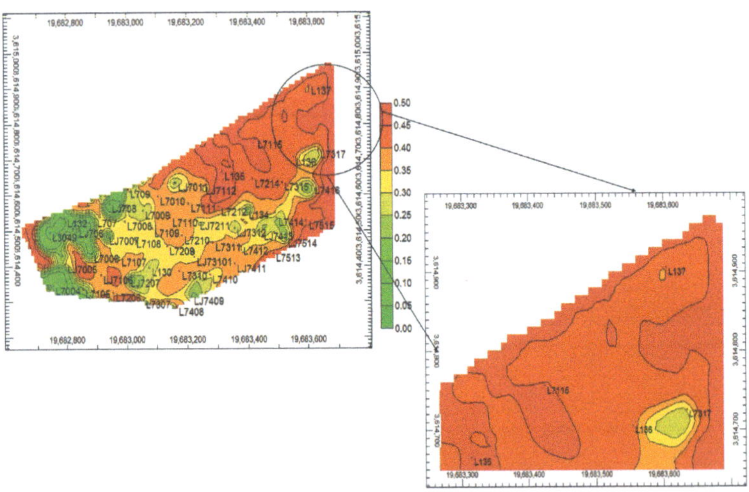

(c)

FIGURE 2.15 (*Cont'd.*)

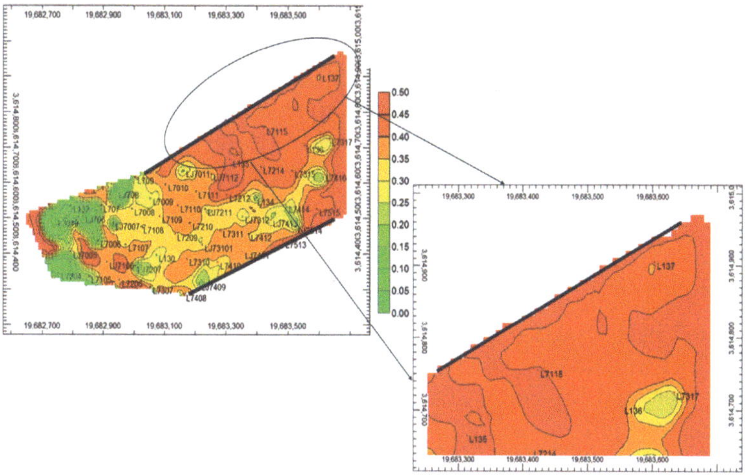

(d)

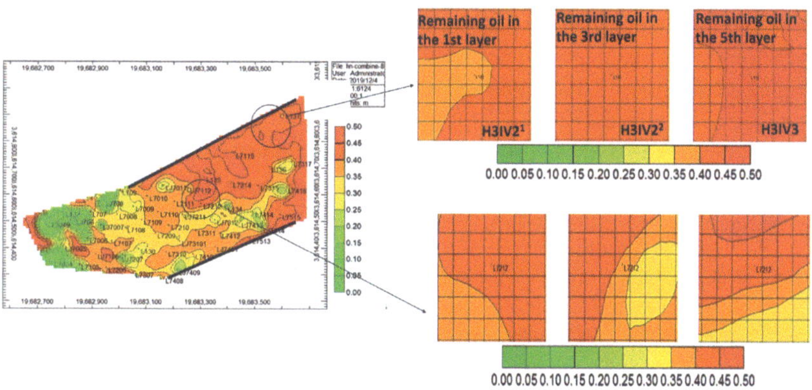

(e)

FIGURE 2.15 *(Cont'd.)*

The type of remaining oil between adjacent wells can be observed (Fig. 2.15A). After lengthy steam stimulation, remaining oil saturation within the reservoir zone that was swept by steam is low and the oil recovery factor is relatively higher. However, large amounts of heavy oil resources can be observed between adjacent wells. Simultaneously, under the effects of steam breakthrough and a high steam injection rate, the thermal efficiency of steam is lower, indicating that the reservoir zone between adjacent wells cannot be effectively heated through thermal energy carried by steam. Therefore, an accumulation of heavy crude oil between adjacent wells was formed.

Second, the vertical remaining oil between layers is an important type of remaining oil (Fig. 2.15B). It is mainly caused by the permeability difference between layers. The recovery performance of a high permeability layer is much better than that for a low permeability layer. A high permeability layer is a chief flow path. Therefore, the remaining oil can be established in a low permeability layer.

In this reservoir, a well pattern infilling process was performed in 2006. After infilling, a five-spot well pattern with a spacing of 70 × 100 m was applied. However, because of the reservoir's properties, a low well density was observed in some reservoir sections (Fig. 2.15C). Around the well groups of L137 and L7115, because of large well spacing and the incomplete well pattern, large amounts of heavy oil resources in the northeast part of this reservoir were not effectively unlocked and remaining oil caused by an incomplete well pattern formed.

Fourth, this reservoir has two important faults with an approximate east–west strike. The first fault dominates the northern oil-bearing boundary, and the fault displacement changes from 13 to 312 m. The second fault extends to the middle section of this reservoir, and the fault displacement changes from 7 to 143 m. For reservoir sections around the faults, because of the blocking effect of faults, the oil drainage radius is small. Therefore, the effect of a well pattern on oil reserves around faults is limited and oil recovery is low. In addition, a remaining oil accumulation zone formed around the faults (Fig. 2.15D). An obvious remaining oil accumulation zone may be observed in the well groups of LJ708, LJ7011, and L7513.

On the other hand, among the 53 wells in this heavy oil reservoir, a full perforation state on all three layers was performed in only 14 of them. Most of the wells perforate only part of the reservoir. For the unperforated oil-bearing layers, high oil saturation was observed (Fig. 2.15E). For well L137, the oil layer in H3IV3 was not perforated. For well L7212, the oil layers of $H3IV2^1$ and H3IV3 were not perforated. Therefore, high remaining oil saturation caused by an incomplete perforation scheme may be observed.

Based on these five remaining oil types, improving the sweep efficiency vertically and horizontally should be the main EOR directions after steam injection in the future. For remaining oil reserves between adjacent wells, follow-up methods of steam flooding and hybrid steam-additive processes are commonly applied EOR processes. For hybrid EOR processes, both cyclic operation methods and continuous injection methods may be considered (e.g., cyclic steam-N_2/CO_2 coinjection, cyclic steam-solvent coinjection, and steam-chemical flooding). For vertical remaining oil between layers, a follow-up EOR method to improve the vertical steam injection profile is recommended (e.g., NCG-foam, gels, and solid particles). For remaining oil caused by an incomplete well pattern and a fault barrier, some infill wells can be planned to improve the lateral sweep efficiency. For remaining oil caused by an incomplete well perforation scheme, reperforating can be applied to improve the injection–production relationship.

Furthermore, performing steam flooding through complete well patterns by adjacent perforated wells is another method.

2.9.2 Microscopic distribution of remaining oil saturation

From the microscopic point of view, the distribution of remaining oil saturation depends on the pore structure, rock wettability, and fluid properties. In this section, based on a visualized experimental investigation at the pore scale, remaining oil distribution after steam injection is discussed. During the process, to guarantee the accurate identification of remaining oil, the whole picture of a reservoir will be classified into a set of small views. Based on the observation of these small views, the types of remaining oil at the pore scale and their proportions are obtained. Table 2.6 shows the experimental results of the microscopic distribution of remaining oil saturation, as well as the reasons.

TABLE 2.6 Different microscopic remaining oil types.

Remaining oil	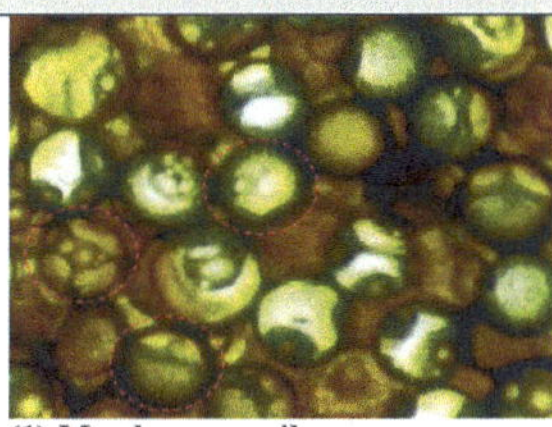 (1) Membranous oil	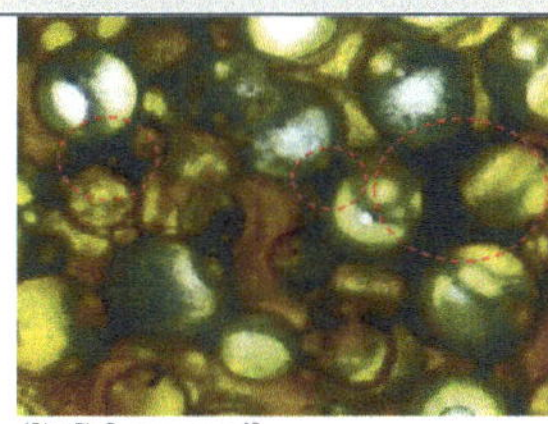 (2) Columnar oil
Causes	Steam displacement, Rock wettability.	Pore structure, Capillary pressure.
Remaining oil	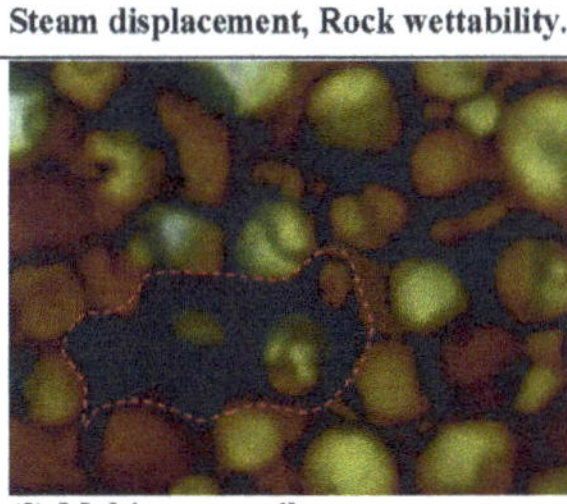 (3) Multi-porous oil	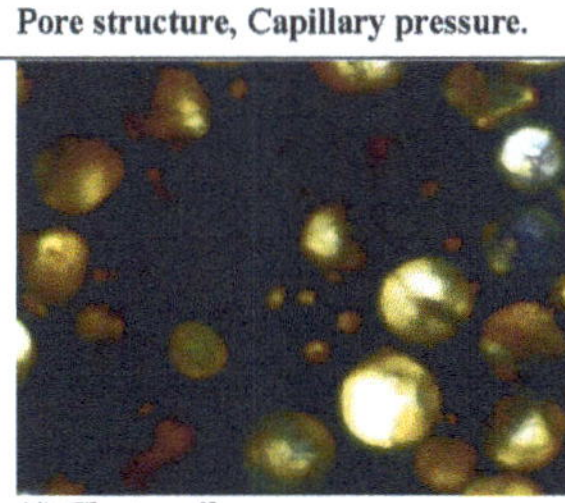 (4) Cluster oil
Causes	Steam unswept, Microscopic heterogeneity.	Microscopic heterogeneity, Steam breakthrough.
Remaining oil	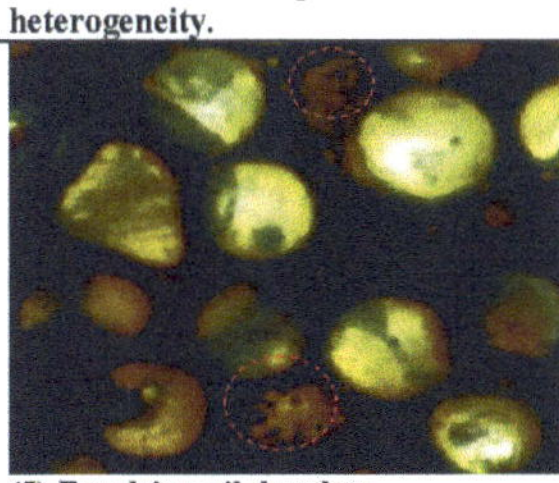 (5) Emulsion oil droplets	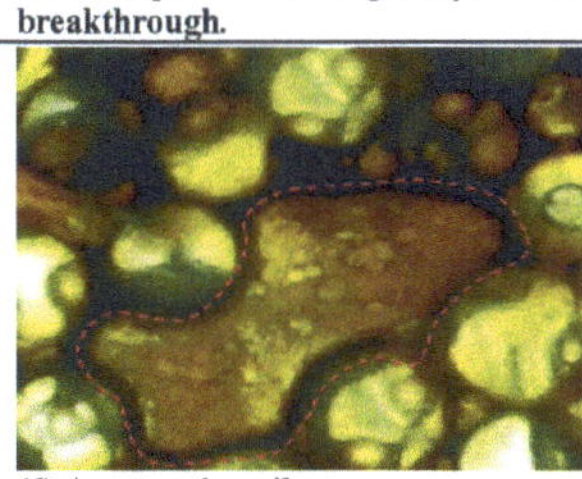 (6) Aggregation oil
Causes	Emulsification, Interfacial tension.	Rock wettability; External additive.

The remaining oil distributions in porous media have six different types: membranous oil, columnar oil, multiporous oil, cluster oil, emulsion oil droplets, and aggregation oil.

Membranous remaining oil refers to an oil film attached to a pore wall surface. It is mainly caused by oil–wet rock wettability and steam displacement. The oil phase in pore centers is effectively unlocked. Based on statistical data, this type of remaining oil represents only a small proportion of all types of remaining oil. Columnar oil refers to heavy oil reserves within a blind pore space. For this type of pore space, the radius of a connecting throat is usually small. The pore structure and capillary pressure are the primary reasons for this type of remaining oil. The third type is multiporous remaining oil. It is heavy oil resources within a multiple pore space, usually caused by microscopic pore heterogeneity. Thus, this part of remaining oil is not swept by steam. For cluster remaining oil, an oil-bearing area is larger than that in multiporous remaining oil. The mechanisms of this type of remaining oil are same as those for multiporous remaining oil. It is mainly caused by microscopic heterogeneity and steam breakthrough. Emulsion oil droplets refer to oil–water emulsions in a pore space. They are caused by emulsification and interfacial tension. Aggregation oil refers to a widely expanded oil film in a pore space; it is usually observed in large-scale pores and in external additive coinjection. In these special pores, the large porous space has been replaced by steam condensate or an external additive, and oil films on pore wall surfaces remain. Rock wettability and an external additive are the primary reasons for this type of remaining oil.

Based on statistical work on the distributions of remaining oil types in a whole heavy oil reservoir, the proportions of multiporous oil and cluster oil are the highest. The proportions of membranous oil and emulsion oil droplets are the lowest. Therefore, to enhance heavy oil recovery performance at the pore scale after steam injection, increasing microscopic sweep efficiency and improving the degree of activation of multiporous oil and cluster oil should be the main research directions.

2.10 Discussion of enhanced oil recovery research directions

Based on this discussion, after a lengthy steam injection process, serious problems hinder the further effective development of heavy oil reservoirs. In this section, EOR research directions after steam injection will be discussed.

2.10.1 Enhanced oil recovery research directions after cyclic steam stimulation process

After multiple cycles of CSS operation, the thermal efficiency of steam is dramatically reduced. Especially in some heavy oil reservoirs, serious problems of steam breakthrough and steam overlap are observed. To continue CSS is no longer economical and may even cause the ineffective circulation of steam around a well bottom hole. To enhance recovery after multiple cycles of CSS operation, improving sweep efficiency should be the top concern. The specific research directions are represented in the following two aspects.

2.10.1.1 Improving the performance of reservoir heating

For CSS, because of the effects of steam breakthrough and steam overlap, the normal expansion of a reservoir heating zone will be significantly affected for heavy oil reservoirs with multiple cycles of CSS operation. The maximum heating radius of CSS for a vertical well is only about 30 m. This indicates that the optimal well spacing is only about 60 m. It is unacceptable for most heavy oil fields. Therefore, to enhance heavy oil recovery after CSS, the first issue we must address is the problem of reservoir heating, which includes vertical heating and horizontal heating. For the vertical perspective, for heavy oil reservoirs after multiple cycles of CSS, a steam front has reached a reservoir top. Because of steam overlap, the heating radius of an upper formation is usually smaller than that of a lower formation. An approximate funnel-shaped heating configuration can be achieved. For the horizontal perspective, because of reservoir heterogeneity, the occurrence of steam fingering and steam breakthrough will be the key obstacles to a horizontal heating radius. Therefore, the reservoir heating radii along different horizontal directions vary significantly.

Compared with hybrid thermal solvent and chemical processes, hybrid thermal-NCG processes are more attractive for improving heating vertically and horizontally. Specifically, a hybrid steam-N_2 process was applied in heavy oil reservoirs with multiple cycles of CSS. During this process, N_2 can preferably enter the steamed paths to move vertically and finally accumulate in a reservoir top. The accumulation of N_2 in a reservoir top can tremendously improve the heat loss of formation overburden and hinder the process of steam overlap. Also, the heating ranges of middle and lower layers are improved. The vertical steam injection profile is more uniform. Based on this, injecting N_2 first and then steam can yield a higher oil production rate, and the mechanism of auxiliary cleanup of N_2 in this injection sequence is more significant. Furthermore, because of the low solubility of N_2 in heavy crude oil and condensate, N_2 can move to a reservoir section far from a wellbore. It is also beneficial for recovering reservoir pressure. Thus, the reservoir can have high expansibility after N_2 injection. After soaking, as the production stage is achieved, heavy oil reserves far from a wellbore can be unlocked by the movement of hot N_2. This indicates that a horizontal heating radius is increased. In some hybrid steam-N_2 tests, the behavior of foamy oil can be also observed because of the cutoff mechanism of N_2 bubble slugs.

2.10.1.2 Improving the performance of oil viscosity reduction

On the other hand, for heavy crude oil with a high viscosity ($\mu_o > 10{,}000$ mPa s), how to reduce the oil viscosity based on of steam injection is another important research direction. Its viscosity depends highly on the fractions of asphaltene and resin in heavy crude oil (Fig. 2.16). The higher the fractions of asphaltene and resin, the higher the oil viscosity, and an approximate exponential relationship between them can be observed. On a molecule scale, a colloidal nucleus may form between asphaltene and resin. Also, resin molecules with a small polarity can be adsorbed on a surface of colloidal nucleus to form a solvation layer. It can protect the colloidal nucleus. As the fractions of asphaltene and resin increase, the volume fraction of a dispersion phase in crude oil increases, and thus, the oil viscosity increases. As more fractions of resin are included, the oil viscosity can increase significantly.

The methods of oil viscosity reduction include three techniques: heating, dilution, and the use of a surfactant.

Oil viscosity reduction by heating is the most economical process for heavy oil. The purpose of heating is to reduce cohesion energy between oil molecules. As the temperature increases, the inner structure of the heavy oil becomes unconsolidated.

For dilution, it is based on the dissolution behavior of a light liquid diluent. Thus, this method can reduce oil viscosity by reducing the concentration of asphaltene and resin in crude oil. It can improve the oil quality in situ. Advantages of this method are that it is easily operated and it has high efficiency, high flowability, and low density. The method of CO_2 injection to reduce oil viscosity is an important dilution process.

To use a surfactant, three viscosity reduction mechanisms are involved. The first is viscosity reduction by an emulsification effect. Under the performance of surfactant, the water—oil emulsion can transfer to an oil—water emulsion to reduce oil viscosity. The second is demulsification. Surfactant is beneficial for

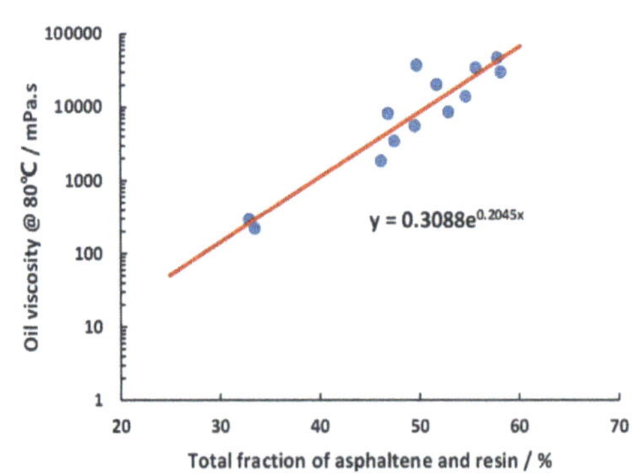

FIGURE 2.16 Oil viscosity dependence on fractions of asphaltene and resin.

the demulsification process of a water–oil emulsion to form free water. According to the volume and flow rate, different oil–water flow statuses can be developed to reduce viscosity. The third is adsorption behavior. Surfactant molecules will adsorb on reservoir rocks to reduce friction loss. These three mechanisms work simultaneously. Considering different operation conditions, the dominant mechanism is also different in various cases. The surfactants used to reduce viscosity (viscosity reducer [VR]) are oil-soluble and water-soluble. The mechanism of oil-soluble VR is to destroy a network configuration of asphaltene molecules. It can form a small-grain structure of heavy crude oil. Thus, resin molecules will dominate the configuration basis and cannot develop a large aggregation. For water-soluble VR, the oil-wet group in VR molecules can form a stable hydrogen bond connection with oil molecules. Water-wet chains can further destroy intermolecular forces between aggregations. Thus, the water–oil emulsion can transfer to the oil–water emulsion to reduce viscosity.

For post-CSS heavy oil reservoirs, how to increase a heating radius and reduce oil viscosity are always major concerns for most reservoirs. Based on this discussion, hybrid EOR processes can address these concerns. Both hybrid steam-NCG and hybrid steam-chemical processes have been applied. In some cases, both NCGs and surfactants have been used in the same cycle to improve the recovery of CSS.

2.10.2 Enhanced oil recovery research directions after steam flooding process

Steam flooding is usually the first follow-up EOR process after CSS. For heavy oil reservoirs with low CSS efficiency, steam flooding is an important transition EOR process. However, once steam flooding has reached a later stage, serious steam channeling behavior and heterogeneous remaining oil saturation distribution may be observed. They hinder the further development of heavy oil reservoirs. Considering these serious problems, how to improve sweep efficiency horizontally and vertically becomes the most important research direction after steam flooding.

Steam fingering is one of the most important reasons for low horizontal sweep efficiency in steam flooding. Steam fingering refers to the phenomenon of unstable displacement fronts caused by the low viscosity and high mobility of steam. It does not help steam flooding. For steam flooding, the flow rate of steam is significantly higher than crude oil. Also, considering inherent reservoir heterogeneity, steam breakthrough can easily be achieved in steam flooding. Once steam breakthrough paths are formed in a reservoir, reservoir heterogeneity will be aggravated. Thus, most subsequent steam injected will reenter these paths. An important result of this behavior is the reduction of the thermal efficiency of steam and horizontal sweep efficiency. However, in such paths, the displacement efficiency is usually low because of multiple interactions of steam and oil. Therefore, improving the horizontal sweep efficiency is the first important step after steam flooding. Among the three hybrid EOR processes discussed in Chapter 1, a hybrid thermochemical process is more significant to improve the horizontal sweep efficiency compared with the other two. Specifically, the systems of NCG-foam, high temperature gels, and solid particles were applied to plug steam breakthrough paths and improve horizontal sweep efficiency.

Similarly, vertical sweep efficiency is an important guarantee for the recovery of heavy oil reservoirs after steam flooding. Steam overlap is a fundamental reason for low vertical sweep efficiency in steam flooding. As presented in Section 2.2, once serious steam overlap is observed, heavy oil reserves in a lower formation section will be difficult to unlock. Also, the continuous expansion of steam overlay will further aggregate vertical or interlayer reservoir heterogeneity. It can significantly damage normal recovery from steam flooding. To improve vertical sweep efficiency, the first issue to be addressed is steam overlap. Therefore, for a thick, heavy oil reservoir, the uniform vertical distribution of a steam injection profile should be an important EOR research direction after steam flooding.

2.10.3 Enhanced oil recovery research directions after steam-assisted gravity drainage process

For an SAGD process, steam chamber expansion has an important effect on the successful operation of SAGD. To improve the expansion of a steam chamber in the entire SAGD process, a hybrid thermal-additive SAGD process was proposed and tested in many heavy oil fields. The methods include expanding solvent SAGD (ES-SAGD), NCG-SAGD, and chemical SAGD. These different hybrid EOR processes can improve the configuration of a steam chamber according to different aspects. In addition, these hybrid SAGD processes reduce steam consumption and the steam−oil ratio. First, for a solvent-based hybrid SAGD process (ES-SAGD), these coinjected light solvents (dilution) can improve steam chamber expansion by altering the fluid phase behavior at a steam chamber front. Then, for an NCG-based hybrid SAGD process (N_2/O_2/CH_4-SAGD), NCGs can accumulate in a reservoir top to assist the lateral expansion of a steam chamber. In this hybrid SAGD process, a steam chamber is usually shaped like a liquid drop instead of a conventional inverted triangle from SAGD. Reducing the heat loss rate between a steam chamber and overburden is an important aspect of this process. Third, for chemical additive-based hybrid SAGD (surfactant-assisted SAGD or chemical additive and foam-assisted SAGD), it can alter the physical properties of emulsion at a steam front. For foam-assisted SAGD, it can also increase steam conformance along a horizontal wellbore to improve steam chamber expansion. For this hybrid SAGD, the most important issues are to develop a thermal stable chemical agent. A hybrid SAGD process will be the most important research direction after SAGD. It has been tested in heavy oil and oil sands reservoirs in the world.

On the other hand, considering the effect of complicated reservoir conditions (lean zones, shale barriers, and permeability heterogeneity) on SAGD recovery, improvements in wellbore configurations and well patterns is another important research direction. The first process improves the wellbore configuration by flow control devices or dual pipes. For a horizontal well pair with a long wellbore length, this measure can significantly improve the distribution of a steam injection profile along the horizontal wellbore. More uniform steam conformance can yield a larger steam chamber volume. The other direction is to perform infill wells or multilateral wells. For infill well-assisted SAGD, some new vertical wells or another horizontal well may be placed between adjacent horizontal well pairs. Also, infill wells can operate by CSS to improve steam chamber expansion. For SAGD with multilateral wells, a multilateral well pair may be designed to replace a conventional horizontal well pair. This process was tested in oil sands reservoirs in Alberta, Canada. The key points of this process are the wellbore angle and the number and locations of lateral wells.

References

[1] Liu WZ. The development modes of heavy oil reservoirs by thermal recovery. Beijing: Petroleum Industry Press; 1998 [in Chinese].

[2] Liu HQ. Principle and design of thermal oil recovery processes. Beijing: Petroleum Industry Press; 2013 [in Chinese].

[3] Farouq Ali SM. Practical heavy oil recovery. Lecture Notes, University of Calgary; 2007.

[4] Wu Z. Stimulation mechanisms and application of air assisted steam for heavy oil reservoirs [Ph.D. dissertation]. Beijing: China University of Petroleum; 2018.

[5] Dong X, Liu H, Wang C, Chen Z. Experimental investigation on the steam injection profile along horizontal wellbore. Energy Rep 2020;6:264−71.

[6] Dong X-H, Liu H-Q. Investigation of the features about steam breakthrough in heavy oil reservoirs during steam injection. Open Petrol Eng J 2012;5:1−6.

[7] Zhang H, Liu H, Wang H, Wang S, Bao S. Optimization design of profile control parameters for steam stimulation wells. Acta Pet Sin 2007;28(2):105−8.

[8] Dong X. The development mechanism and method screening for offshore heavy oil reservoirs with multi-thermal fluid [Ph.D. dissertation]. Beijing: China University of Petroleum; 2014.

[9] Dong X, Liu H, Wang Q, Pang Z, Wang C. Non-Newtonian flow characterization of heavy crude oil in porous media. J Pet Explor Prod Technol 2013;3(1):43−53.

[10] Cheng L. Advanced fluid flow mechanism in porous media. Beijing: Petroleum Industry Press; 2011.

[11] Hajdo LE. formation damage of an unconsolidated reservoir during steamflooding. In: SPE 27385 presented at the SPE formation damage control symposium, Lafayette, Louisiana, USA, 7−10 February, 1994; 1994.

[12] Schembre JM, Kovscek AR. Thermally induced fines mobilization: its relationship to wettability and formation damage. In: SPE 86937 presented at the SPE international thermal

operations and heavy oil symposium and western regional meeting, Bakersfield, California, USA, 16–18 March, 2004; 2004.

[13] Okoye CU, Onuba NL, Ghalambor A, Hayatdavoudi A. formation damage in heavy-oil formation during steamflooding. In: SPE 22980 presented at the SPE asia-pacific conference, Perth. Australia, 4–7 November, 1991; 1991.

[14] Pang Z, Liu H. The study on permeability reduction during steam injection in unconsolidated porous media. J Petrol Sci Eng 2013;106:77–84.

[15] McCorriston LL, Demby RA, Pease EC. Study of reservoir damage produced in heavy oil formations due to steam injection. In: SPE 10077 presented at the SPE annual technical conference and exhibition, San Antonio, Texas, USA, 4–7 October, 1981; 1981.

[16] Liu H. Altering behavior of reservoir properties and formation damage mechanism of heavy oil reservoirs in Liaohe oilfield. Technical Report. Beijing: China University of Petroleum; 2009.

[17] Hayatdavoudi A, Ghalambor A. Controlling formation damage caused by kaolinite clay minerals: part I. In: SPE 31118 presented at the SPE formation damage control symposium, Lafayette, Louisiana, USA, 14–15 February, 1996; 1996.

[18] Hayatdavoudi A, Ghalambor A. Controlling formation damage caused by kaolinite clay minerals: part II. In: SPE 39464 presented at the SPE formation damage control conference, Lafayette, Louisiana, USA, 18–19 February, 1998; 1998.

[19] https://en.wikipedia.org/wiki/Clay.

[20] Bennion DB, Thomas FB, Sheppard DA. Formation damage due to mineral alteration and wettability changes during hot water and steam injection in clay-bearing sandstone reservoirs. In: SPE 23783 presented at the SPE formation damage control symposium, Lafayette, Louisiana, USA, 26–27 February, 1992; 1992.

[21] Fauré M, Sardin M, Vitorge P. Release of clay particles from an unconsolidated clay-sand core: experiments and modelling. J Contam Hydrol 1997;26(1–4):169–78.

[22] Delamaide E, Moreno WP. Enhanced oil recovery of heavy oil in reservoirs with bottom aquifer. In: SPE 174050 presented at the SPE western regional meeting, garden grove, California, USA, 27–30 April, 2015; 2015.

[23] Liu Y, Li Y, Jiang S, Li S, Zhang L, Zhang C, Liu D. An application of Hall curve to thermal recovery wells in Bohai oilfields. China Offshore Oil Gas 2013;25(5):43–5.

[24] Joshi SD. Augmentation of well productivity with slant and horizontal wells. J Petrol Technol 1988;40:729–39.

[25] Gharbi K, Benyounes K, Khodja M. Removal and prevention of asphaltene deposition during oil production: a literature review. J Petrol Sci Eng 2017;158:351–60.

[26] Boek ES, Wilson AD, Padding JT, Headen TF, Crawshaw JP. Multiscale simulation and experimental studies of asphaltene aggregation and deposition in capillary flow. Energy Fuel 2010;24(4):2361–9.

[27] Liu P, Yuan Z, Zhang S, Xu Z, Li X. Experimental study of the steam distillation mechanism during the steam injection process for heavy oil recovery. J Petrol Sci Eng 2018;166:561–7.

[28] Dong X, Liu H, Chen Z, Lu N, Zhang Q. Enhanced oil recovery techniques for heavy oil and oil sands reservoirs after steam injection. Appl Energy 2019;239:1190–211.

Chapter 3

Calculations of wellbore heat loss

3.1 Introduction to wellbore heat loss

Wellbore heat loss refers to heat loss caused by a heat transmission process between a wellbore and its surrounding formation rock. For a hot fluid injection process in heavy oil reservoirs, this loss usually determines the amount of thermal energy carried by a hot fluid under bottom hole conditions. It usually refers to the successful operation of a thermal recovery process. According to different wellbore configurations, wellbore heat loss can be classified into two categories: a single-pipe wellbore configuration and a dual-pipe wellbore configuration.

3.1.1 Wellbore heat loss in a single-pipe wellbore configuration

In 1962, Ramey first proposed heat transfer between a wellbore and a formation; he developed a mathematical model to calculate wellbore temperature [1]. This model is a classical model for wellbore heat loss. Based on Ramey's model, many different attempts have been made to extend his model to different applications. Satter [2] and Horne and Shinohara [3] considered the effects of steam condensation and variable fluid properties along a wellbore on wellbore heat loss. Meyer developed a heat balance equation in fracture-containing reservoirs and discussed heat conduction behavior in a matrix-fracture dual medium [4]. Hansan and Kabir reexamined the effect of nondimensional time on wellbore heat transfer and proposed a new time function to modify the temperature error in Ramey's model [5]. Wu and Pruess modified the assumptions of Ramey's model and proposed a revised model for wellbore heat transmission in layered formations [6]. Shiu and Beggs predicted wellbore temperature in flowing oil wells by modifying Ramey's model by a newly proposed coefficient [7]. According to the basic thermodynamic principles, Sagar et al. improved the correlation proposed by Shiu and Beggs; this newly modified model was closer to a real situation [8]. Hagoort proposed

Hybrid Enhanced Oil Recovery Processes for Heavy Oil Reservoirs.
https://doi.org/10.1016/B978-0-12-823954-4.00006-0

a simple graphical correlation to correct the temperature error in the early stage of Ramey's model [9]. Furthermore, some commercial software packages have been applied to heat transfer behavior between a wellbore and a formation [10]. Shoushtari et al. used the commercial simulators ANSYS and FLUENT to investigate wellbore heat transmission behavior in a high-temperature and high-pressure formation and discussed the effect of annulus pressure on the serving life of tubing and casing [11].

For wellbore heat loss in a horizontal well with a single pipe configuration, Gao et al. developed a radial heat transfer model and discussed the heat transmission behavior of horizontal wellbores [12]. Nuong proposed the concepts of cooling time and heating rings of thermal horizontal wells [13]. Through a combination of cooling time and formation thermal diffusivity, the effectiveness of heat injection during various steam heating processes can be evaluated. A heating ring can be applied to measure heat storage in a heated reservoir volume. Based on these two parameters, a thermal transient analysis is applied in thermal horizontal wells [13]. Arthur et al. also discussed wellbore temperature distributions in steam injection and production processes [14]. Commonly used methods for single pipe wellbore heat loss include:

(1) Using a node analysis method to develop a heat transfer model between a wellbore and a formation. In this method, only a radial heat transfer process is considered to compute wellbore heat loss.
(2) Using commercial software (e.g., Computer Modeling Group's FlexWell, ANSYS, and FLUENT) to simulate fluid thermodynamic properties along a wellbore (e.g., steam quality, pressure, and temperature).
(3) For the wellbore heat loss of a special hot fluid injection process, experimental investigations can be performed to simulate heat transfer, and mathematical models can be developed to obtain thermodynamic properties along a wellbore.

On the other hand, considering the effect of reservoir heterogeneity and a variable mass flow process along a horizontal well, a conventional single pipe well configuration is no longer valid.

3.1.2 Wellbore heat loss in a dual-pipe wellbore configuration

Besides a single-pipe well configuration, a dual-pipe well configuration and the method of flow control devices (FCDs) have been applied [15]. For thermal recovery processes in horizontal wells, because of their long horizontal wellbore and reservoir heterogeneity, a steam injection profile distribution along the wellbore is usually nonuniform. It can significantly reduce the heating performance of a steam injection process in horizontal wells. The application of a dual-pipe well configuration and FCDs in horizontal wells can dramatically improve a steam injection profile along the horizontal wellbore.

They are proposed based on a multipoint steam injection method along horizontal wells. In some cases, they can also be applied in an alternate steam injection process along two different horizontal intervals.

A dual-pipe well configuration includes a parallel dual-pipe configuration and a concentric dual-pipe configuration. Compared with a single-pipe well configuration, wellbore heat loss in a dual-pipe well configuration is more complicated. The thermal coupling effect between two different tubings should be considered. Based on Ramey's model, Griston et al. developed a mathematical model for wellbore heat loss and pressure drops based on energy conservation and momentum conservation [16]. Barua proposed a mathematical model for the wellbore heat loss of single-pipe and dual-pipe well configurations [17]. However, this model neglected the effect of an oil/gas flow pattern and friction loss on the wellbore heat loss and pressure drops. Arthur et al. considered a horizontal well with a parallel dual pipe well configuration in the Mocan heavy oilfield [14]. They developed a mathematical model for a steam circulation process in this parallel dual-pipe horizontal well. It can be applied to simulate the steam quality, enthalpy, and pressure effectively along a horizontal wellbore. Castrup discussed the wellbore heat loss of a multipipe well configuration [18]. Considering the problems of control and allocation of a steam injection rate in different pipes, for a multilayered heavy oil reservoir, a conventional single-pipe well configuration with a small radius is more attractive than a multipipe well configuration with a large radius. Gu et al. proposed a mathematical model for tubing heat loss and annulus heat loss in a concentric dual-pipe well configuration [19]. This model can be used to evaluate fluid thermodynamic properties in tubing and annulus during a concentric steam injection process. Xiong et al. developed a stand-alone fully implicit thermal wellbore simulator [10]. In this simulator, a series of correlations proposed from computational fluid dynamics runs were embedded. These correlations can be used for an accurate heat-loss estimation from a wellbore to its surrounding formations in thermal wells with a single pipe or multiple pipes. Furthermore, in a version of CMG's commercial simulator, the addition of the FlexWell model can also effectively address heat loss calculations in steam injection wells with a multiple-pipe well configuration.

3.2 Configuration of vertical steam injection wells

For a steam injection operation in heavy oil reservoirs, the whole steam injection system includes a surface steam injector, surface pipelines, and a steam injection well. High pressure-high temperature saturated steam produced from a surface steam generator (boiler) can enter a formation by a surface pipeline and a steam injection wellbore. Because of the effect of heat transfer behavior in these two stages, steam properties in the formation are significantly different from those in the boiler outlet. In addition, the configuration of a steam injection wellbore is usually the most important factor for wellbore heat loss.

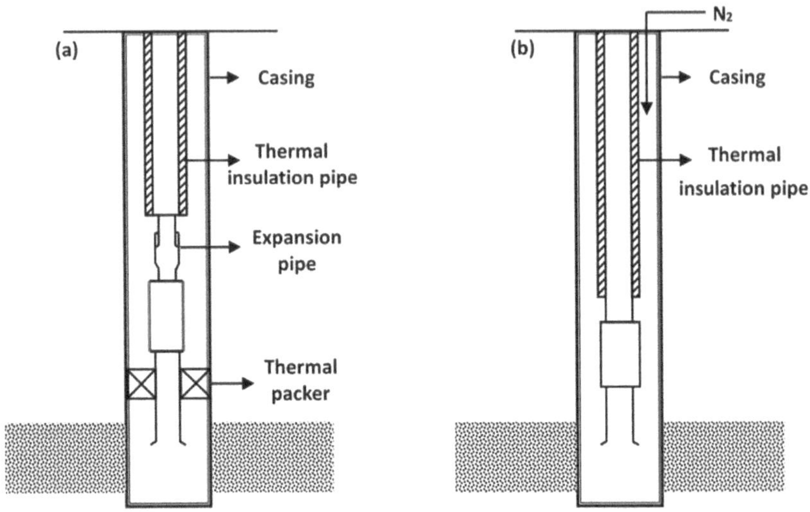

FIGURE 3.1 Configurations of vertical steam injection wells: (A) well configuration by thermal recovery packer, and (B) thermal insulation pipe by N_2 injection in annulus.

Commonly used configurations of a vertical steam injection well include two forms, as shown in Fig. 3.1: The first is a string configuration of a thermal insulation pipe, an expansion pipe, and a thermal recovery packer; the second one is a thermal insulation pipe by nitrogen (N_2) injection in annulus.

3.2.1 Thermal insulation pipes

A thermal insulation pipe is actually a concentric dual pipe. The annular enclosed space in this concentric dual pipe is usually filled with thermal insulation materials. During steam injection, the inner surface temperature in this pipe is higher than that in the outer surface. Therefore, because of a temperature difference between the inner and outer surfaces, the phenomenon of thermal stress accumulation can be observed. To address this problem, a prestressed thermal insulation pipe and a corrugated thermal insulation pipe are proposed [20].

3.2.2 Thermal recovery packers

A thermal recovery packer can be used to prevent fluid breakthrough between different wellbore segments. For a steam injection process in a horizontal well, the application of a thermal recovery packer can also protect casing and reduce heat loss. Commonly used thermal recovery packers include slip-type thermal packers and thermosensitive packers. The former packers work by the hydraulic pressure and the latter work by the expansion and contraction of a thermosensitive element.

3.2.3 N_2 thermal insulation process in annulus space

An N_2 thermal insulation process is another important well configuration in steam injection wells. N_2 is a thermostable inert gas. It has low thermal conductivity. During a steam injection process, N_2 can be filled in the annulus space between the tubing and casing. This configuration can effectively reduce the wellbore heat loss rate between the wellbore and formation.

3.3 Configuration of horizontal steam injection wells

Horizontal wells have become the primary option for developing most heavy oil reservoirs in the world. In this section, different configurations of onshore and offshore horizontal wells are provided.

3.3.1 Onshore horizontal wellbore configuration

For the development of heavy oil reservoirs, compared with vertical wells, horizontal wells can improve the steam conformance along their wellbores and also improve heat transfer behavior between a wellbore and a reservoir. However, for long horizontal wells, because of the effect of reservoir heterogeneity and variable mass flow behavior along their wellbores, a low reservoir–production ratio and a nonuniform steam injection profile usually hinders the performance of the thermal horizontal wells with a conventional single-pipe configuration. Thus, concentric and parallel dual pipes horizontal wells are proposed [21]. Fig. 3.2 shows the configurations of three commonly used onshore horizontal steam injection wells. For a dual-pipe well configuration, there are two pipes in a wellbore; both can be used in steam injection. Their relationship can be concentric or parallel, as shown in Fig. 3.2. In some field operations, one pipe is used for steam injection and the other is used for oil production. These two steam injection pipes are respectively placed at the heel-end and toe-end of a horizontal wellbore. During operations, according to the actual distribution of a steam injection profile along a horizontal wellbore, the steam injection parameters of each pipe may be appropriately changed through a steam regulator on its surface.

3.3.2 Offshore horizontal wellbore configuration

As onshore heavy oil reserves continue to decline, offshore heavy oil resources have become an important alternative for the world energy supply. Compared with the development of onshore heavy oil reservoirs, a complicated offshore environment is one of the most important factors affecting the normal

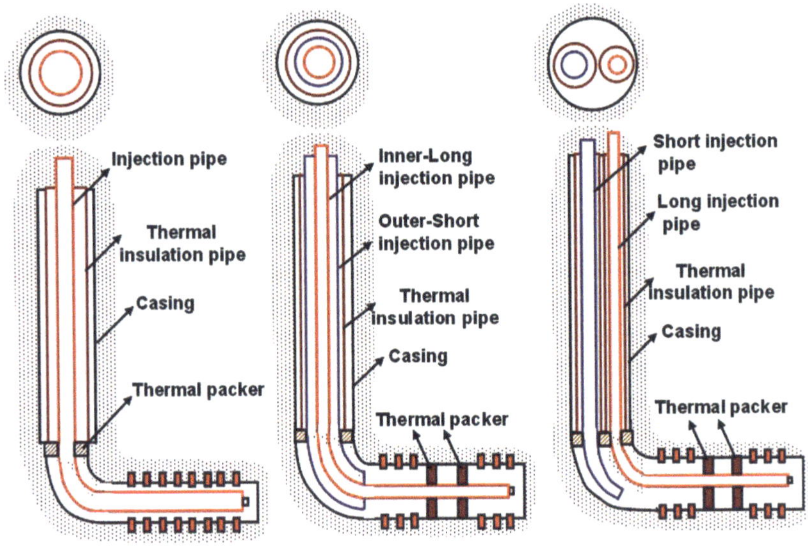

FIGURE 3.2 Schematic diagram for the configurations of three commonly used onshore horizontal steam injection wells: (A) single pipe; (B) concentric dual-pipe; (C) parallel dual-pipe.

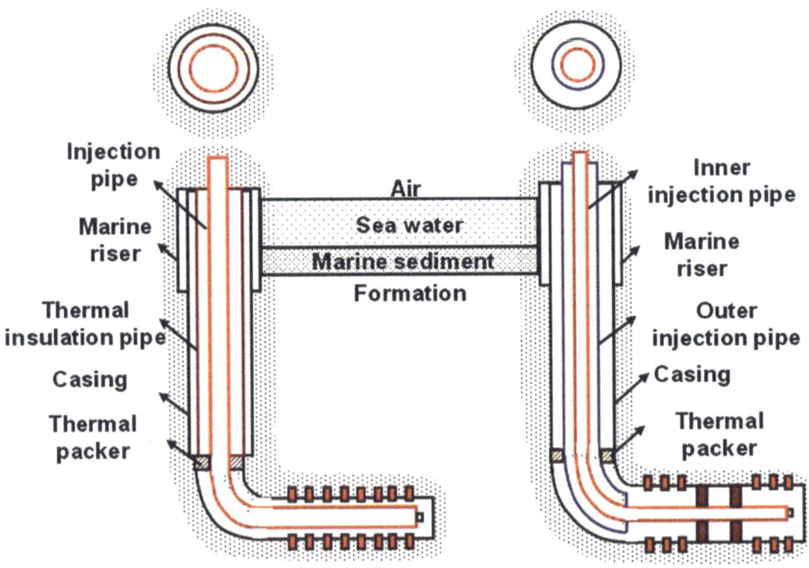

FIGURE 3.3 Schematic diagram of steam injection well configurations in offshore heavy oil reservoirs: (A) single pipe and (B) concentric dual pipe.

exploitation of offshore heavy oil reservoirs. As a wellbore configuration, compared with an onshore horizontal well, a surface marine riser will be one of the most distinct characteristics. Fig. 3.3 shows a typical single-pipe and a dual pipe wellbore configuration in offshore heavy oilfields. Furthermore, the effects of flowing seawater and marine sediment at the sea bottom on wellbore heat loss are nonnegligible.

3.4 Types of heat transfer

For thermal recovery processes in heavy oil reservoirs, heat transfer behavior is a key issue affecting wellbore heat loss. There are three major mechanisms by which heat is transferred between a wellbore and a reservoir: heat conduction through tubing and rock materials, heat convection by fluid movement, and heat radiation in wellbore annulus [20,22]. For the three types of heat transfer mechanisms, heat conduction and convection mechanisms have a more critical role in wellbore heat loss.

3.4.1 Heat conduction

Heat conduction refers to the transfer behavior of heat energy from a high-temperature section to a low-temperature section. For radial heat transfer in a wellbore, heat conduction is the basic form of heat transmission. For heat conduction in a cylinder, Fourier's equation is applied. A heat flux rate caused by heat conduction is related to the cylinder geometry and thermal conductivity:

$$Q_{HCond} = -\lambda A \frac{dT}{dr} = -2\pi r \lambda l \frac{dT}{dr} \tag{3.1}$$

$$\frac{Q_{HCond}}{2\pi l} \int_{r_1}^{r_2} \frac{dr}{r} = -\lambda \int_{T_1}^{T_2} dT \tag{3.2}$$

$$Q_{HCond} = \frac{\Delta T}{\frac{1}{2\pi\lambda l} ln \frac{r_2}{r_1}} \tag{3.3}$$

$$q_{Hcond} = \frac{Q_{HCond}}{l} = \frac{\Delta T}{\frac{1}{2\pi\lambda} ln \frac{r_2}{r_1}} \tag{3.4}$$

$$R_{HCond} = \frac{1}{2\pi\lambda} ln \frac{r_2}{r_1} \tag{3.5}$$

where q_{Hcond} refers to the heat flux in a unit cylinder length within unit time, kcal/(h·m); l is the cylinder length, m; ΔT is the temperature difference between the outer surface and inner surface of a cylinder ($T_1 - T_2$), °C; r_1 and r_2 are the inner and outer radii of the cylinder, respectively, m; R_{HC} is the heat resistance for heat conduction; and λ is the thermal conductivity, kcal/(h·m·°C).

For transient heat transfer behavior along a wellbore, heat conduction is always the primary heat transfer mechanism between the wellbore and its surrounding formation. It usually includes heat conduction processes in metal pipes and cement rings. Simultaneously, in porous media, heat conduction is also the main mechanism of heat transfer; this behavior will be introduced in the next chapter.

3.4.2 Heat convection

Heat convection is the heat transfer behavior caused by moving fluids. During fluid movement, it can contact with other types of fluids or solid surfaces (e.g., tubing, casing, and rock) with different temperatures. Thus, heat convection behavior between moving fluids and solid surfaces is activated. For a heat convection process in a cylinder, a heat flux rate caused by heat convection is related to the cylinder radius and thermal convection coefficient:

$$Q_{HConv} = h_c A \Delta T = 2\pi r l h_c \Delta T \tag{3.6}$$

$$q_{HConv} = \frac{Q_{HConv}}{l} = 2\pi r h_c \Delta T = \frac{\Delta T}{\frac{1}{2\pi r h_c}} \tag{3.7}$$

$$R_{HConv} = \frac{1}{2\pi r h_c} \tag{3.8}$$

where q_{Hconv} refers to the heat flux caused by heat convection in a unit cylinder length within unit time, kcal/(h·m); l is the cylinder length, m; r is the inner radius of the cylinder, m; and h_c is the thermal convection coefficient, kcal/(h·m^2 °C).

For wellbore heat loss calculations, the heat convection mechanism mainly works within steam injection pipes.

3.4.3 Heat radiation

Heat radiation refers to the heat transfer behavior that is moved from one place to another by electromagnetic radiation waves or rays [23]. For wellbore heat loss, heat radiation is observed in a wellbore annulus space. For a heat radiation process in a cylinder space, the heat flux is:

$$Q_{HRad} = h_r A \Delta T = 2\pi r l h_r \Delta T \tag{3.9}$$

$$h_r = \varepsilon \sigma_0 \left[\left(T_1^2 + T_2^2 \right) \left(T_1 + T_2 \right) \right] \tag{3.10}$$

$$q_{HRad} = \frac{Q_{HRad}}{l} = 2\pi r h_r \Delta T = \frac{\Delta T}{\frac{1}{2\pi r h_r}} \tag{3.11}$$

where q_{HRad} refers to the heat flux caused by heat radiation in a unit cylinder length within unit time, kcal/(h·m); l is the cylinder length, m; r is the inner radius of the cylinder, m; and h_r is the thermal radiation coefficient, kcal/(h·m^2·°C).

3.5 Wellbore heat loss models in pure steam injection processes

Before the development of a wellbore heat loss model, a wellbore is first discretized. The length of each control element is Δz, the pressure drop is Δp, and the steam quality drop is Δx. Then, based on each microcontrol element along a wellbore, as shown in Fig. 3.4, three different wellbore heat loss models for an onshore horizontal wellbore configuration (Fig. 3.2) can be developed [21].

3.5.1 Assumptions

The mathematical models are based on the following assumptions:

1) Only the radial heat transfer process is considered;
2) Heat transfer from a steam injection pipe to a cement ring is a steady-state process, and heat transfer from a cement ring to a formation is an unsteady-state process;

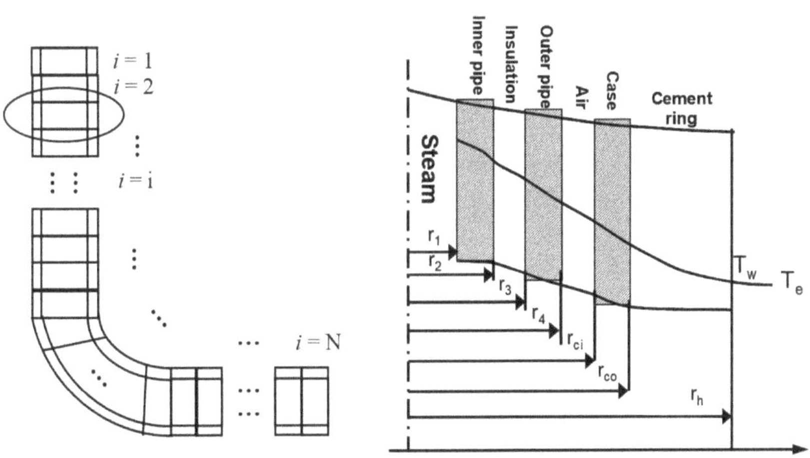

FIGURE 3.4 Schematic diagram of wellbore microcontrol elements.

3) For a dual-pipe well configuration, the radial location of the two pipes keeps unchanged along the wellbore;
4) A thermal packer is applied to prevent intersegment fluid breakthrough, and the annulus space is filled with N_2.

3.5.2 Pressure drop model

1) Mass conservation equation:

$$\rho_1 v_1 A = \rho_2 v_2 A = i_s \tag{3.12}$$

2) Momentum conservation equation:

$$Adp = \rho_m g A dz \cos\theta + \rho_1 v_1^2 A - \rho_2 v_2^2 A - \tau_f \tag{3.13}$$

$$\tau_f = \frac{f \pi D \rho_m}{8} v^2 dz \tag{3.14}$$

where ρ_i is the fluid density, kg/m^3, $i = 1, 2$, m; v_i is the fluid velocity, m/s, $i = 1, 2, m$; A is the inner cross-sectional area of the injection pipe, m^2; θ is the dip angle, ; τ_f is the friction loss of steam through a microcontrol element; f is the frictional resistance coefficient in a gas–liquid two-phase fluid process; and D is the inner diameter of the pipe, m.

These equations are independent of a wellbore configuration. Therefore, for the different wellbore configurations in Fig. 3.2, all of these equations are valid.

3.5.3 Heat transfer models of single and dual-pipe well configurations

Based on the method of equivalent flow resistance, a circuit diagram can be applied in the heat transfer process of the three different wellbore configurations in Fig. 3.2. It indicates that the transient fluid flow and heat transfer processes between a wellbore and a formation can be simulated by the electric circuit rule. The heat transfer resistance in single-pipe and concentric dual-pipe wellbore configurations is in a serial connection. For a parallel dual-pipe wellbore configuration, it is in a parallel-serial connection. Therefore, heat transfer models for the three different wellbore configurations can be developed.

3.5.3.1 Single-pipe wellbore configuration

Based on a steady-state heat transfer assumption, the heat flux rate from a steam injection pipe to a cement ring is:

$$dQ = \frac{(T_s - T_h)}{R_{C1}} dz \tag{3.15}$$

$$R_{C1} = \frac{1}{2\pi r_2}\left[\frac{r_2}{\lambda_{tub}}\ln\frac{r_2}{r_1} + \frac{r_2}{\lambda_{ins}}\ln\frac{r_3}{r_2} + \frac{r_2}{\lambda_{tub}}\ln\frac{r_4}{r_3} + \frac{r_2}{r_4(h_c + h_r)} + \frac{r_2}{\lambda_{cas}}\ln\frac{r_{co}}{r_{ci}} + \frac{r_2}{\lambda_{cem}}\ln\frac{r_h}{r_{co}}\right] \tag{3.16}$$

where R_{CI} is the thermal resistance in a single wellbore configuration; T_s is the steam temperature, K; T_h is the temperature in the cement ring, K; r_i is the pipe radius, m, $i = 1, 2, 3, 4$, ci, co, h; λ_{tub} is the heat conductivity coefficient of an injection pipe, W/(m·K); λ_{ins} is the heat conductivity coefficient of the insulation pipe, W/(m·K); λ_{cas} is the heat conductivity coefficient of the casing pipe, W/(m·K); λ_{cem} is the heat conductivity coefficient of the cement ring, W/(m·K); h_r is the radiation heat transfer coefficient, W/(m^2·K); and h_c is the convective heat transfer coefficient, W/(m^2·K).

For the time-dependent unsteady heat transfer process from a cement ring to a formation, the heat flux rate can be expressed as:

$$dQ = \frac{2\pi\lambda_e(T_h - T_e)}{f(t)} dz \tag{3.17}$$

where T_e is the reservoir temperature, K; λ_e is the heat conductivity coefficient of the formation, W/(m·K); and $f(t)$ is the heat transfer time function. Hasan and Kabir proposed the following form of this function [5]:

$$f(t) = \begin{cases} 1.1281\sqrt{\tau_D}(1 - 0.3\sqrt{\tau_D}) & \text{while} \quad \tau_D \le 1.5 \\ (0.4063 + 0.5\ \ln\ \tau_D)(1 + 0.6/\tau_D) & \text{while} \quad \tau_D > 1.5 \end{cases} \tag{3.18}$$

3.5.3.2 Concentric dual-pipe wellbore configuration

For a concentric configuration, two different steam injection pipes can be considered as a whole system. It indicates that heat transfer between a wellbore and a formation is dominated only by the outer pipe. Moreover, a transient heat transfer equilibrium can be observed between the inner and outer pipes. Fig. 3.5A shows the specific heat transfer procedure in a concentric dual-pipe wellbore configuration. Therefore, the heat flux rate from a steam injection pipe to a cement ring in a microcontrol element can be derived as:

$$dQ = \frac{(T_o - T_h)}{R_{C2}} dz \tag{3.19}$$

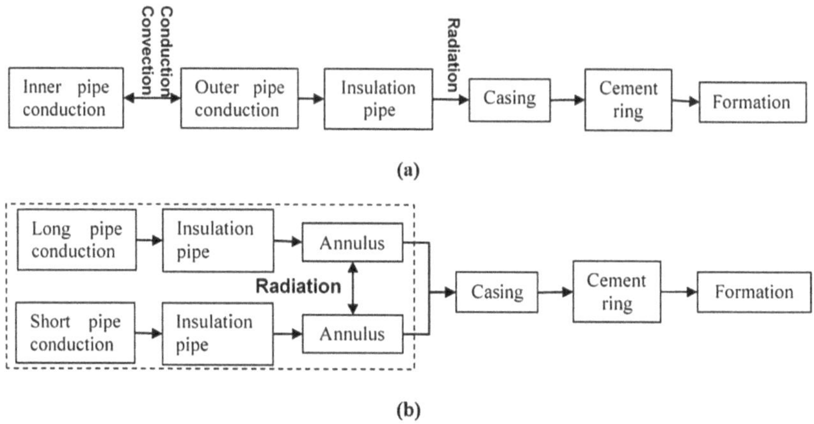

FIGURE 3.5 Heat transfer process in dual-pipe wellbore configurations. (A) The heat transfer process of concentric dual-pipe configuration. (B) The heat transfer process of parallel dual-pipe configuration.

$$R_{C2} = \frac{1}{2\pi r_4}\left[\frac{r_4}{\lambda_{tub}}\ln\frac{r_4}{r_3} + \frac{r_4}{\lambda_{ins}}\ln\frac{r_{cii}}{r_4} + \frac{r_4}{\lambda_{tub}}\ln\frac{r_{cio}}{r_{cii}} + \frac{1}{h_c + h_r} + \frac{r_4}{\lambda_{cas}}\ln\frac{r_{co}}{r_{ci}} + \frac{r_4}{\lambda_{cem}}\ln\frac{r_h}{r_{co}}\right] \tag{3.20}$$

where R_{C2} is the thermal resistance in a concentric dual-pipe wellbore configuration.

Eqs. (3.17) and (3.18) for the heat transfer behavior from a cement ring to a formation are still valid for this wellbore configuration [21].

3.5.3.3 Parallel dual-pipe wellbore configuration

For a parallel configuration specifically, heat transfer from a steam injection pipe to a formation includes different parts:

1) heat conduction in both the steam injection pipe and the thermal insulation pipe;

2) heat radiation in the annulus space between the thermal insulation pipe and casing;

3) heat conduction in the casing and cement ring;

4) unsteady-state heat transfer between the cement ring and formation.

Fig. 3.5B shows the specific heat transfer procedure in a parallel dual-pipe wellbore configuration. It is in a parallel-serial connection.

In this wellbore configuration, self-heat conduction and convection processes in each pipe are first performed. Then, the heat energy is lost to the annulus space between the insulation pipe and casing through heat radiation. After that, a radial heat front can reach the formation through heat conduction between the casing and cement ring [24]. Fig. 3.6 shows the structure of this wellbore configuration.

Compared with the other two configurations, it is an eccentric heat transfer process. To simplify heat transfer in this configuration, a concept of a virtual equivalent pipe is introduced.

Heat transfer in this virtual pipe is equivalent to the comprehensive effect of parallel dual pipes. Also, the size of this virtual pipe depends on the size of the dual pipes:

$$\begin{cases} r_{ei} = \dfrac{r_{li} + r_{si}}{2} \\ r_{eo} \ln\dfrac{r_{eo}}{r_{ei}} = r_{lo} \ln\dfrac{r_{lo}}{r_{li}} + r_{so} \ln\dfrac{r_{so}}{r_{si}} \end{cases} \tag{3.21}$$

where r_{ei} and r_{eo} are the inner and outer radii of the virtual equivalent pipe, respectively.

Therefore, the heat transfer resistance in the virtual equivalent pipe (to the inner surface of the casing pipe) can be expressed as,

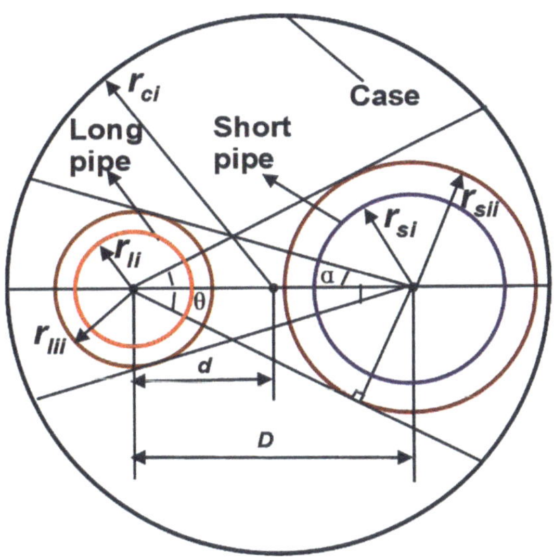

FIGURE 3.6 Parallel dual-pipe wellbore configuration.

$$\frac{1}{R_e}=\frac{1}{R_l}+\frac{1}{R_s} \tag{3.22}$$

$$R_l=\frac{1}{2\pi r_{lo}}\left[\frac{r_{lo}}{h_{ml}r_{li}}+\frac{r_{lo}}{\lambda_{tub}}\ln\frac{r_{lo}}{r_{li}}+\frac{r_{lo}}{\lambda_{ins}}\ln\frac{r_{lii}}{r_{lo}}+\frac{r_{lo}}{\lambda_{tub}}\ln\frac{r_{lio}}{r_{lii}}+\frac{1}{\sigma_{lc}h_{rlc}+h_{clc}}+\frac{1}{\sigma_{ls}h_{rls}+h_{cls}}\right] \tag{3.23}$$

$$R_s=\frac{1}{2\pi r_{so}}\left[\frac{r_{so}}{h_{ms}r_{si}}+\frac{r_{so}}{\lambda_{tub}}\ln\frac{r_{so}}{r_{si}}+\frac{r_{so}}{\lambda_{ins}}\ln\frac{r_{sii}}{r_{so}}+\frac{r_{so}}{\lambda_{tub}}\ln\frac{r_{sio}}{r_{sii}}+\frac{1}{\sigma_{sc}h_{rsc}+h_{csc}}+\frac{1}{\sigma_{sl}h_{rsl}+h_{csl}}\right] \tag{3.24}$$

$$\sigma_{ls}=\theta/\pi;\ \sigma_{lc}=1-\sigma_{ls};\ \sigma_{sl}=\alpha/\pi;\ \sigma_{sc}=1-\sigma_{sl} \tag{3.25}$$

where σ is the radiation shape factor; and α, θ is the dip angle, (Fig. 3.6).

Based on Eq. 3.22, the total heat transfer resistance and heat flux rate in a parallel configuration can be derived as:

$$R_{C3}=\frac{1}{2\pi r_{eo}}\left[R_e+\frac{r_{eo}}{\lambda_{cas}}\ln\frac{r_{co}}{r_{ci}}+\frac{r_{eo}}{\lambda_{cem}}\ln\frac{r_h}{r_{co}}\right] \tag{3.26}$$

$$dQ=\frac{(T_s-T_w)}{R_{C3}}dz \tag{3.27}$$

where R_{C3} is the thermal resistance in a parallel dual-pipe wellbore configuration.

Simultaneously, Eqs. (3.17) and (3.18) for heat transfer from a cement ring to a formation are still valid. Thus, the wellbore heat loss rate in a parallel configuration can be derived.

3.5.4 Steam quality model

To evaluate steam quality along a wellbore, an energy conservation equation in each microcontrol element can be developed:

$$\frac{dQ}{dz}+\frac{dW}{dz}=-i_s\frac{dH_m}{dz}-i_s\frac{d}{dz}\left(\frac{v_m^2}{2}\right)+i_sg \tag{3.28}$$

where

$$\frac{dH_m}{dz}=L_v\frac{dx}{dz}+\frac{dH_w}{dp}\frac{dp}{dz}+\frac{dL_v}{dp}\frac{dp}{dz}x$$

$$\frac{d}{dz}\left(\frac{v_m^2}{2}\right)=\frac{i_s}{A^2}\frac{1}{\rho_m}\frac{d}{dz}\left(\frac{1}{\rho_m}\right)$$

$$\frac{dW}{dz}=\frac{\tau_f v_m}{dz}=\frac{\tau_f(v_1+v_2)}{2dz}$$

3.5.5 Intermediate parameters treatment

3.5.5.1 Thermophysical properties of a formation

The thermophysical properties of a formation have an important effect on wellbore heat loss. In an actual heavy oil reservoir, they usually depend on the mineral composition of the formation rock, porosity, fluid saturation, temperature, and pressure. Microfractures in the rock also affect the thermophysical properties. The first important parameter is a thermal conductivity coefficient of a formation. A power-law based correlation can be applied [25,26]:

$$\lambda_e=a\left[\varphi_0+C_f(P-P_0)\right]^b \tag{3.29}$$

where λ_e is the thermal conductivity coefficient of the formation, W/(m K); a and b are the constants, $a=6.893$, $b=-0.3069$; φ_0 is the original porosity of the formation; P is the current formation pressure, MPa; P_0 is the original formation pressure, MPa; and C_f is the formation compressibility, MPa^{-1}.

Then, based on a relationship between thermal conductivity and thermal diffusivity, we have:

$$\alpha_e=\frac{\lambda_e}{\rho C_P}=\frac{a\left[\varphi_0+C_f(p-p_0)\right]^b}{\rho_a e^{C_a(P-P_0)}\cdot(c+dT)} \tag{3.30}$$

where α_e is the thermal diffusivity of a formation, m^2/s; ρ_a is the rock density, kg/m^3; C_a is the total compressibility, MPa^{-1}; T is temperature, K; and c and d are the constants, $c=0.813$, $d=9.797\times10^{-4}$.

3.5.5.2 Frictional resistance coefficient in gas–liquid two-phase flow

For a frictional resistance coefficient in gas–liquid two-phase flow in a wellbore, the methods of correlations, experimental tests, and Moody's plates can be applied [27–29]. Beggs-Brill model is a commonly used correlation method. It is based on an estimation of the flow pattern of gas–liquid two-phase flow in a wellbore. The gas–liquid two-phase flow can be divided into four different flow patterns: separated flow, transition flow, intermittent flow and dispersed flow [30]. During a simulation process, the pattern of gas–liquid two-phase flow within a microcontrol element is usually identified first. The liquid holdup ratio, E_l, can be determined according to the specific criteria of different flow patterns. Then, the Reynolds number in a gas–liquid two-phase flow process can be derived:

$$N_{\mathrm{Re}} = \frac{Dv\left[\rho_l E_l + \rho_g (1 - E_l)\right]}{\mu_l E_l + \mu_g (1 - E_l)} \tag{3.31}$$

Thus, we see that:

$$f = \begin{cases} \dfrac{64}{N_{\mathrm{Re}}} & N_{\mathrm{Re}} \leq 2000 \\ \left[1.14 - 2\lg\left(\Delta + 21.25 N_{\mathrm{Re}}^{-0.9}\right)\Delta\right]^{-2} & N_{\mathrm{Re}} > 2000 \end{cases} \tag{3.32}$$

$$\Delta = \varepsilon / D \tag{3.33}$$

where N_{Re} is the gas–liquid two-phase Reynolds number; D is the inner diameter of a steam injection pipe, m; ρ_l is the fluid density, kg/m^3; Δ is the relative roughness of a pipe surface; and ε is the roughness of the pipe surface.

3.5.5.3 Simplification of annulus flow

For simplicity, the concepts of an equivalent diameter and equivalent roughness are introduced for fluid flow in an annulus space between a steam injection pipe and casing:

$$D_r = D_o - D_i \tag{3.34}$$

$$\varepsilon = \varepsilon_i \left(\frac{D_i}{D_o + D_i}\right) + \varepsilon_o \left(\frac{D_o}{D_o + D_i}\right) \tag{3.35}$$

where D_r is the equivalent diameter, m; D_i is the outer diameter of the inner pipe, m; D_o is the inner diameter of the outer pipe, m; ε is the equivalent roughness; ε_i is the roughness of the outer surface of the inner pipe; and ε_o is the roughness of the inner surface of the outer pipe.

3.5.5.4 Correlation for saturated steam

Eq. (3.36) can be applied for a relationship between the saturated temperature and saturated pressure of steam [31]:

$$T_s = 210.2376 p_s^{0.21} - 30 \tag{3.36}$$

3.5.6 Case study

Based on these wellbore heat loss models of three wellbore configurations, the heat transfer and pressure drop characteristics of steam along different horizontal wells can be simulated. The used wellbore configurations and thermophysical properties data are shown in Table 3.1.

TABLE 3.1 Wellbore configuration and thermophysical properties data.

Parameter	Value	Parameter	Value
Well depth/m	800	Steam quality/decimal	0.8
Length of horizontal wellbore/m	300	Steam injection rate/t/d	120
Surface temperature/°C	22	Cement radius/m	0.120
Temperature gradient/°C/m	0.032	TC of pipe/W·(m·K)$^{-1}$	45.7
Injection time/d	5	TC of insulation/W·(m·K)$^{-1}$	0.034
Injection pressure/MPa	5.0	TC of cement/W·(m·K)$^{-1}$	0.350
Conventional single-pipe configuration			
OD of injection pipe/m	0.0603	ID of insulation pipe/m	0.0883
ID of injection pipe/m	0.0507	OD of casing/m	0.1685
OD of insulation pipe/m	0.1016	ID of casing/m	0.1524
Concentric dual-pipe configuration			
OD of inner pipe/m	0.0603	OD of insulation pipe/m	0.1429
ID of inner pipe/m	0.0507	ID of insulation pipe/m	0.1296
OD of outer pipe/m	0.1016	OD of casing/m	0.1685
ID of outer pipe/m	0.0883	ID of casing/m	0.1524
Parallel dual-pipe configuration			
OD of short pipe/m	0.0603	OD of long insulation pipe/m	0.1143
ID of short pipe/m	0.0507	ID of long insulation pipe/m	0.1010
OD of short insulation pipe/m	0.1016	OD of casing 1/m	0.2445
ID of short insulation pipe/m	0.0883	ID of casing 1/m	0.2244
OD of long pipe/m	0.0730	OD of casing 2/m	0.1685
ID of long pipe/m	0.0620	ID of casing 2/m	0.1524

ID, inner diameter; *OD*, outer diameter; *TC*, thermal conductivity.

3.5.6.1 Differences among three configurations

Based on the data in Table 3.1, the heat transfer and pressure drop characteristics for the three different wellbore configurations in Fig. 3.2 are simulated. For the concentric and parallel wellbore configurations, the steam

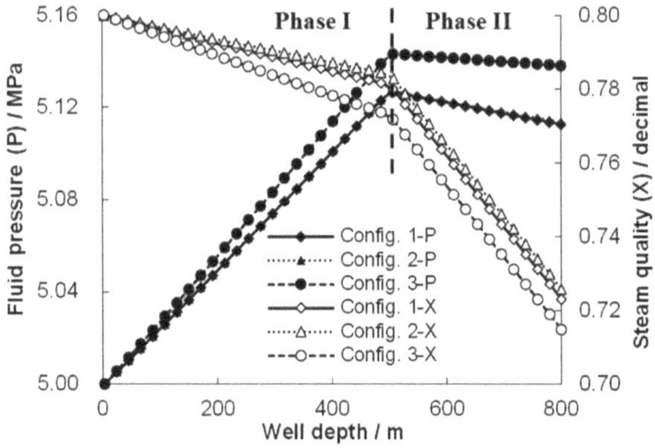

FIGURE 3.7 Simulation results of three steam injection pipe configurations. Configuration (Config.) 1 represents a conventional single pipe wellbore configuration; Config. 2 represents a concentric dual-pipe wellbore configuration. Config. 3 represents a parallel dual-pipe wellbore configuration.

injection rate in each pipe is set to 120 t/d. The simulation results are shown in Fig. 3.7. As shown, as the well depth increases, heat transfer in the three well configurations presents a two-phase flow, which is greatly different from that in a vertical well. Phase I represents the fluid flow and transfer in a vertical segment, and phase II represents the fluid flow and heat transfer in a horizontal segment. For the three wellbore configurations, the pipe diameter of a parallel configuration is the largest. It indicates that the heat transfer area is the largest. Therefore, under the same conditions, the heat flux rate in a parallel configuration is the highest, and the steam quality along an injection pipe is also the lowest. A lower steam quality can also result in an increase in steam density. Thus, the gravity energy loss is increased and the wellbore pressure is increased. As show in Fig. 3.7, for a parallel wellbore configuration, steam quality along the long pipe is the lowest and steam pressure is the highest. On the other hand, steam quality in a concentric wellbore configuration is the highest, and with regard to the pressure drop characteristics there is little difference between single-pipe and concentric dual-pipe configurations. That is because of the extra steam supplement from the short pipe in a concentric wellbore configuration, compared with a single-pipe configuration.

3.5.6.2 Results of concentric configuration

Fig. 3.8 shows the simulation results for a concentric wellbore configuration under different steam injection rates. Under the same injection rate, fluid pressure in the inner pipe is much higher than that in the outer pipe because of the size difference of the two steam injection pipes. Simultaneously, steam injection in the outer pipe is further offset by the heat loss rate in the inner pipe. Because of the effect of friction loss, as the steam injection rate

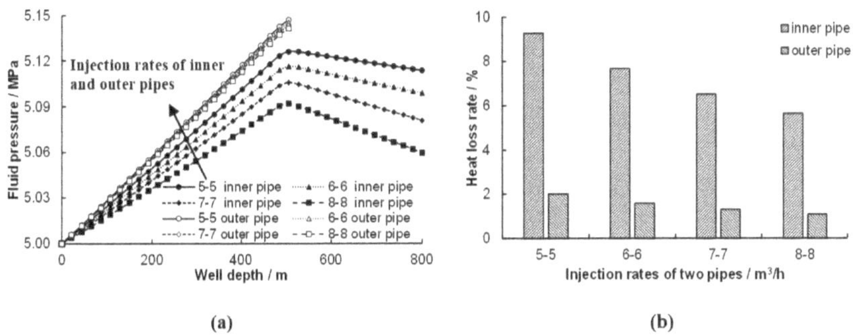

FIGURE 3.8 Simulation results of concentric wellbore configuration: (A) fluid pressure along the wellbore; (B) heat loss rate.

increases, the fluid pressure in both the inner pipe and outer pipe is reduced. Fig. 3.8B shows that the heat loss rate in the horizontal wellbore segment is higher than in the vertical segment. It mainly results from the existence of the heat insulation pipe in the vertical section. The total heat flux in the vertical section is only about one-third that of the horizontal section. Also, as the injection rate increases, the heat loss rates in both the vertical and horizontal sections are reduced.

3.5.6.3 Results in a parallel configuration

Fig. 3.9 shows the simulation results in a parallel wellbore configuration under different steam injection rates. For a parallel wellbore configuration, the long pipe corresponds to the inner pipe in a concentric configuration and the short pipe corresponds to the outer pipe. Compared with the concentric one, the steam pressure in the short pipe in the parallel one is slightly lower than that in the concentric one, which results from the size difference between the parallel

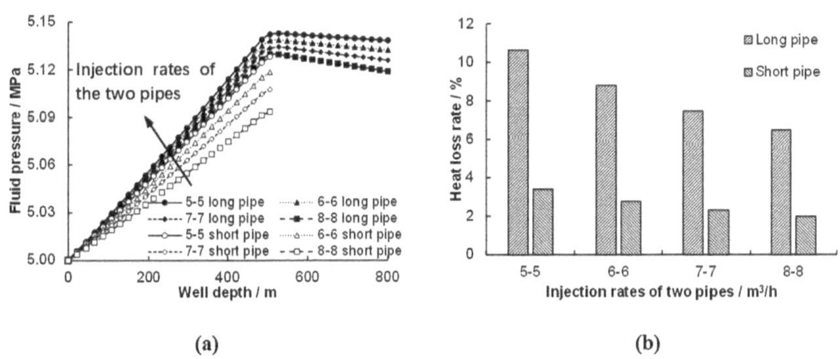

FIGURE 3.9 Simulation results of parallel wellbore configuration. (A) Pressure along the wellbore; (B) Heat loss rate.

and concentric configurations. Fig. 3.9B shows that the heat loss rate in the parallel wellbore configuration is higher than that in the concentric configuration. For the simulation plan of injection rates at 5−5 m^3/h, the heat loss rate in the long pipe in a parallel wellbore configuration is about 10.6%. Compared with the heat loss rate in the inner pipe in the concentric wellbore configuration, it is increased by about 1.4%. In the three different wellbore configurations, the heat loss rate in the parallel one is the highest, and that in the concentric one is the lowest. For field applications, to reduce a heat loss rate effectively, the concentric pipe configuration is highly recommended.

3.5.7 Optimization of operation parameters

For a steam injection process in heavy oil reservoirs, keeping a high steam quality condition under well bottom hole conditions is a prerequisite for a successful operation. High steam quality indicates high heat energy. It can highly benefit an oil viscosity reduction. In this section, based on the wellbore heat loss models of the three horizontal wellbore configurations, the operation parameters in a pure steam injection process are optimized to reduce a heat loss rate effectively and obtain higher steam quality under well bottom hole conditions. The basic data in Table 3.1 are applied. The simulation results are shown in Fig. 3.10.

First, for a steam injection rate, according to Fig. 3.10A, as the injection rate increases, the wellbore heat loss rate is gradually reduced. Among the three different pipe configurations, the wellbore heat loss rate in a parallel dual-pipe well configuration is the highest, and that in the concentric dual pipe well configuration is the lowest. The difference between single-pipe and concentric dual-pipe configurations is relatively small. For a field operation, to improve the steam quality under well bottom hole conditions, a high steam injection rate is recommended.

Based on a saturated vapor pressure curve, there is a one-to-one correspondence between the pressure and temperature of saturated steam. Therefore, a different steam injection pressure indicates a different steam temperature. The simulation results of different steam injection pressures are shown in Fig. 3.10B. As the steam injection pressure (temperature) increases, the temperature difference between steam and the formation increases and the heat flux rate also gradually increases. Thus, the wellbore heat loss rate is increased. For the three different wellbore configurations, the heat loss rate in the parallel dual pipe well configuration is the highest, and that in the concentric configuration is the lowest. For a field operation, steam injection pressure (temperature) is usually related to the formation and fluid properties, including formation pressure and oil viscosity.

The simulation results at different steam injection times are shown in Fig. 3.10C. As the steam injection process continues, the formation temperature around a wellbore increases. Thus, the heat flux rate is reduced and the

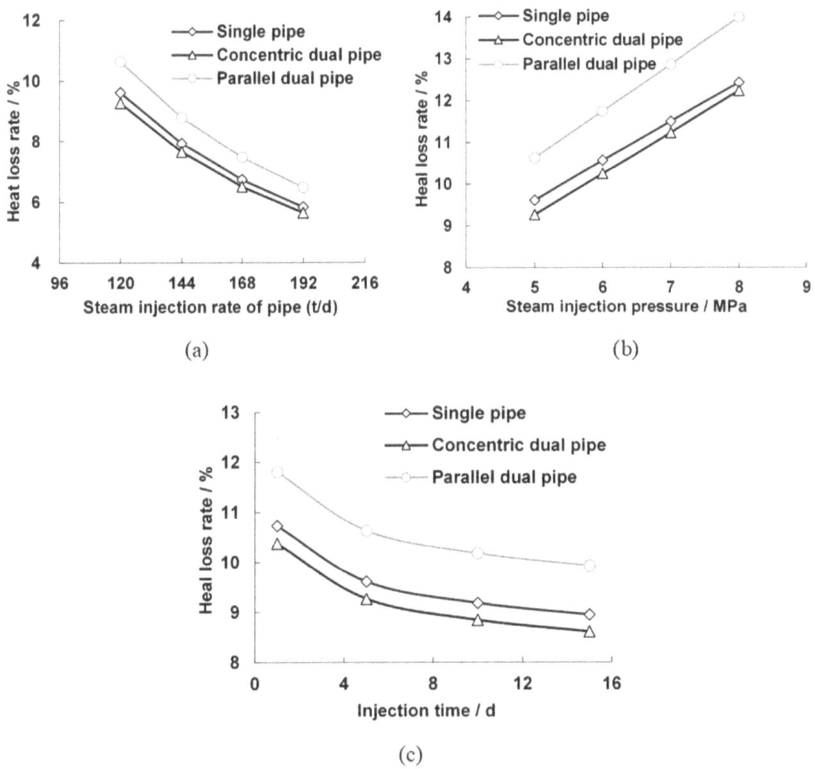

FIGURE 3.10 Heat loss rates under different operation parameters: (A) effect of steam injection rate; (B) effect of steam injection pressure; (C) effect of injection time.

wellbore heat loss rate is gradually reduced. In Fig. 3.10C, because the injection time is greater than 5 days, the relationship between the heat loss rate and steam injection time is reduced. On the other hand, a decrease in the wellbore heat loss rate also indicates an increase in steam quality. Therefore, gravity loss in the gas–liquid two-phase flow process along a wellbore is reduced and the steam pressure at the well bottom hole conditions is usually reduced.

A dual-pipe wellbore configuration is one of the most commonly used configurations for a profile control process in heavy oil reservoirs. Compared with other steam injection parameters at the wellhead conditions, changes in an injection pipe size are more acceptable. Under a given steam injection condition, as the size of a steam injection pipe increases, a heat transfer area increases and thus the heat flux rate is increased, especially for a horizontal segment. To obtain higher steam quality under well bottom hole conditions, a steam injection pipe with a smaller size is recommended during steam injection processes.

3.6 Wellbore heat loss models for steam-NCG coinjection process

As discussed in Chapter 1, although all the three hybrid enhanced oil recovery (EOR) processes can be applied to improve the recovery performance of heavy oil reservoirs, only non-condensable gas (NCG) additives can sometimes be coinjected with steam. Moreover, considering the effect of a high-temperature condition, solvent and chemical additives are usually injected into a formation before or after steam injection. For steam-NCG coinjection, conventional wellbore heat loss models are no longer effective. In this section, a wellbore heat loss model for steam-NCG co-injection is developed [32].

For a steam-NCG mixture, according to the phase state of steam, the fluid flow process of steam-NCG along a wellbore can be classified into two categories: single gas-phase flow process (i.e., superheated steam in which steam quality equals 100%) and gas–liquid two-phase flow process. In a single gas-phase flow process, a wellbore heat loss model mainly involves calculations of fluid pressure, temperature, and a heat loss rate. In a gas–liquid two-phase flow process, a model involves calculations of fluid pressure, temperature, steam quality, and a heat loss rate. Based on a discretized microcontrol element along a wellbore, as shown in Fig. 3.4, wellbore heat loss models in the steam-NCG coinjection processes can be developed.

3.6.1 Assumptions

It is assumed that the steam in well head conditions is superheated. As the well depth increases, the presence of wellbore heat transfer can induce the condensation of steam, and the enthalpy of steam is gradually reduced, which indicates that a transition from superheated steam to saturated steam is observed. In steam-NCG coinjection, a mixture of N_2 and CO_2 is simulated. Besides the assumptions regarding a pure steam injection made in Section 3.5.1, the following assumptions are required for developing a wellbore heat loss model in steam-NCG coinjection:

(1) The mixture of steam and NCG is a compressible fluid, and the effects of temperature and pressure on fluid properties are considered.
(2) The fluid injection process is a constant-mass flow process, and the composition of this mixture keeps unchanged along a wellbore.

3.6.2 Models for single gas-phase flow process

3.6.2.1 Pressure drop model

The pressure drop model in Section 3.5.2 is still valid in this section. When the kinetic energy loss is neglected, a pressure drop model can be expressed as:

$$\frac{dp}{dz} = \rho_m g \cos\theta - \frac{f_g \rho_m v^2}{2D} \tag{3.37}$$

where f_g is the frictional resistance coefficient in a single gas-phase flow process.

3.6.2.2 Heat transfer model

The flow process of a steam-NCG mixture along a wellbore is considered to be a steady-state flow process. For each wellbore microcontrol element, according to the energy conservation equation, the heat flux rate can be expressed as:

$$dQ = m_t g dz \cos\theta - m_t v dv + dH_m \tag{3.38}$$

where m_t is the total mass flux rate, kg/s; and H_m is the fluid enthalpy:

$$dH_m = C_p m_t dT = C_p m_t (T_1 - T_2) \tag{3.39}$$

Thus, combining Eqs. (3.38) and (3.40), we have

$$T_2 = T_1 + [m_t g dz \cos\theta - m_t v dv - dQ]/C_p m_t \tag{3.40}$$

3.6.3 Models for gas-liquid two-phase flow process

3.6.3.1 Pressure drop model

As discussed in Section 3.5.5, a model in a gas–liquid two-phase flow process is different from a single gas-phase flow process. Considering the effect of friction loss, a pressure drop model is usually based on the estimation of flow pattern:

$$\frac{dp}{dz} = \rho_m g \cos\theta + \rho_m v dv - \frac{f \rho_m v^2}{2D} \tag{3.41}$$

3.6.3.2 Steam quality model

For heat transfer behavior along a wellbore, steam quality is an important parameter in a gas–liquid two-phase flow process. According to the theory of fluid phase equilibrium, the molar fraction of steam can be expressed as:

$$C'_{H_2O} = \frac{p_s(T)}{p} \tag{3.42}$$

On the other hand:

$$C'_{H_2O} = \frac{\frac{G_{H_2O}}{M_{H_2O}}}{\frac{G_{H_2O}}{M_{H_2O}} + \frac{G_{CO_2}}{M_{CO_2}} + \frac{G_{N_2}}{M_{N_2}}} = \frac{\frac{G_{H_2O}}{M_{H_2O}}}{\frac{G_{H_2O}}{M_{H_2O}} + C_1} \tag{3.43}$$

$$C_1 = \frac{G_{CO_2}}{M_{CO_2}} + \frac{G_{N_2}}{M_{N_2}} \tag{3.44}$$

where G_i is the mass flux rate of component i, kg/s, and $i = N_2, CO_2, H_2O$; M_i ($i = N_2, CO_2, H_2O$) is the molecular weight of component i, kg/mol.

Therefore, based on Eqs. (3.42) and (3.43),

$$C'_{H_2O} = \frac{p_s(T)}{p} = \frac{\frac{xG_{H_2O}}{M_{H_2O}}}{\frac{xG_{H_2O}}{M_{H_2O}} + C_1} \tag{3.45}$$

In Eq. 3.45, taking a derivative with respect to the well depth, z, an equation for the steam quality in a gas–liquid two-phase flow process can be obtained:

$$\frac{dx}{dz} = \frac{C_1 M_{H_2O}}{G_{H_2O}} \frac{1}{(p-p_s)^2}\left(p\frac{dp_s}{dT}\frac{dT}{dz} - p_s\frac{dp}{dz}\right) \tag{3.46}$$

3.6.3.3 Heat transfer model

Based on the energy conservation theory, we have:

$$m_t g \cos\theta - dQ - \frac{dH_m}{dz} - m_t v\frac{dv}{dz} = 0 \tag{3.47}$$

where H_m is the enthalpy of the fluid mixture, J/Kg:

$$H_m = G_{CO_2}H_{CO_2}(T) + G_{N_2}H_{N_2}(T) + xG_{H_2O}H_S(T) + (1-x)G_{H_2O}H_W(T) \tag{3.48}$$

Taking a derivative with respect to the well depth (z) for this equation:

$$\begin{aligned}\frac{dH_m}{dz} &= G_{CO_2}\frac{dH_{CO_2}}{dz} + G_{N_2}\frac{dH_{N_2}}{dz} + \\ &\left[xG_{H_2O}\frac{dH_S}{dT} + (1-x)G_{H_2O}\frac{dH_W}{dT}\right]\frac{dT}{dz} + L_v G_{H_2O}\frac{dx}{dz}\end{aligned} \tag{3.49}$$

Thus, we obtain:

$$C_2\frac{dT}{dz} - L_v C_1\frac{M_{H_2O}}{(p-p_s)^2}p_s\frac{dp}{dz} - m_t g\sin\theta + q_l = 0 \tag{3.50}$$

where:

$$\begin{aligned}C_2 &= G_{CO_2}\frac{dH_{CO_2}}{dT} + G_{N_2}\frac{dH_{N_2}}{dT} + \\ &xG_{H_2O}\frac{dH_S}{dT} + (1-x)G_{H_2O}\frac{dH_W}{dT} + L_v C_1 G_{H_2O}\frac{M_{H_2O}}{(p-p_s)^2}p\frac{dp_s}{dT}\end{aligned} \tag{3.51}$$

For heat flux rates and heat transfer coefficients of a steam-NCG mixture in the three different wellbore configurations in Fig. 3.2, the models in Section 3.5.3 are still valid. Therefore, integrating Eqs. (3.41), (3.46), and (3.50), heat transfer and pressure drop characteristics in a gas–liquid two-phase flow process for steam-NCG coinjection can be obtained.

3.6.4 Intermediate parameters treatment

Compared with pure steam injection, calculations of the physical properties of a fluid mixture along a wellbore is a key issue in a steam-NCG coinjection process. For the intermediate parameters' treatment during wellbore heat loss calculations, besides the formulas in Section 3.5.5, treatment methods on the density and viscosity of the fluid mixture are introduced in this section.

3.6.4.1 Density of a fluid mixture

For the density of a fluid mixture, equation of state (EOS) modeling is the most commonly used method [33,34]. In this section, one of the cubic equation of states, the Redlich-Kwong (R–K) equation, is selected to describe the pressure-volume-temperature (PVT) properties of a steam-NCG mixture.

In the R–K EOS model, the compressibility factor of a gas mixture can be expressed as:

$$Z_m = \frac{V_m}{V_m - b_m} - \frac{\Omega_a}{\Omega_b} \frac{b_m}{V_m - b_m} F_m \tag{3.52}$$

where:

$$b_m = \sum_i y_i b_i = \sum_i y_i \cdot \Omega_b \frac{RT_{ci}}{P_{ci}} \tag{3.53}$$

Wilson's model can be applied for function F_m in Eq. (3.52), as shown below:

$$F_m = \sum_i y_i F_i \tag{3.54}$$

where:

$$F_i = 1 + (1.57 + 1.62\omega_i)\left(T_{ri}^{-1} - 1\right) \tag{3.55}$$

$$T_{ri} = \frac{T}{T_{ci}} \tag{3.56}$$

Then, based on R–K EOS modeling, the PVT properties and density of a steam-NCG mixture in each wellbore microcontrol element can be obtained.

3.6.4.2 Viscosity of a fluid mixture

For the viscosity of a fluid mixture, the viscosity of each component should be obtained. Then, based on a mixing rule, the viscosity of the fluid mixture can be calculated. Considering the effect of pressure on fluid viscosity, a correction measure at a high-pressure condition is required.

1) Viscosity of single component fluids at a low-pressure condition

Theoretical calculations and reduced state evaluations are the two methods for the viscosity of a single component fluid under a low-pressure condition. The method of reduced state evaluations includes Golubev's, Thodos', Reichenberg's, and Lucas' models [34,35]. For the viscosity of CO_2 or N_2 under a low-pressure condition, Golubev's model can be applied:

$$\mu_i = \begin{cases} \mu_{ci} T_{ri}^{0.0965} & \text{while,} \quad T_{ri} \leq 1 \\ \mu_{ci} T_{ri}^{0.71+\frac{0.29}{T_{ri}}} & \text{while,} \quad T_{ri} > 1 \end{cases} \tag{3.57}$$

where μ_{ci} is the critical viscosity of CO_2 or N_2 under a low pressure condition:

$$\mu_{ci} = \frac{3.5 M_i^{0.5} P_{ci}^{2/3}}{T_{ci}^{1/6}} \tag{3.58}$$

A correlation can be applied for the viscosity of steam. It is a function of temperature and a specific volume [20].

2) Viscosity mixing rule

A mixing rule of a root-squaring model can be applied:

$$\mu_m = \frac{\sum_i y_i \mu_i M_i^{0.5}}{\sum_i y_i M_i^{0.5}} \tag{3.59}$$

3) Gas-viscosity correction method under a high-pressure condition

Under a high-pressure condition ($T_r > 1.0$), gas viscosity is significantly different from that under a low-pressure condition. A residual-viscosity rule is applied to calculate gas viscosity under a high-pressure condition:

$$\left(\mu_m - \mu_m^o\right)\xi_m = 1.08\left[\exp(1.439\rho_{rm}) - \exp\left(-1.111\rho_{rm}^{1.858}\right)\right] \tag{3.60}$$

where:

$$\xi_m = T_{cm}^{1/6} / M_m^{1/2} p_{cm}^{2/3} \tag{3.61}$$

$$\rho_{rm} = \rho/\rho_{cm} = \rho/\left(p_{cm} / Z_{cm} R T_{cm}\right) \tag{3.62}$$

3.6.5 Case study

Based on these above wellbore heat loss models for a steam-NCG mixture, the heat transfer and pressure drop characteristics in different horizontal well configurations can be simulated. The wellbore configurations and thermophysical properties data are shown in Table 3.1. The fluid injection temperature is 250°C, which shows that the steam component in this fluid mixture is superheated steam. The molar fractions of CO_2, N_2, and steam H_2O are 0.49:0.09:0.42.

Fig. 3.11 shows simulation results in the three different wellbore configurations. As the measured depth increases, an obvious three-stage behavior can be observed. For the results of fluid pressure, a changing tendency of declining-rising-declining can be found. It is significantly different from the results in a pure steam injection process in Fig. 3.7. For a steam-NCG mixture, the first declining trend is caused by the single gas phase flow behavior in a vertical wellbore. After injection of the steam-NCG mixture, the density is relatively lower than pure steam. Therefore, friction loss is higher than gravity loss, and fluid pressure is reducing. The second stage of rising is caused by the heat transfer behavior. It results in the transformation of a fluid phase state from the single gas phase flow to the gas–liquid two-phase flow. This changing behavior can increase gravity loss. The last stage of declining is caused by the fluid flow process in a horizontal wellbore. The friction loss will dominate flow behavior. On the other hand, by comparing the differences in the three different well configurations, it can be observed that the steam quality in a parallel dual-pipe configuration is the lowest, and that in the single-pipe configuration is the highest. It is caused by a difference in the wellbore size. A larger wellbore size can induce a higher heat loss rate.

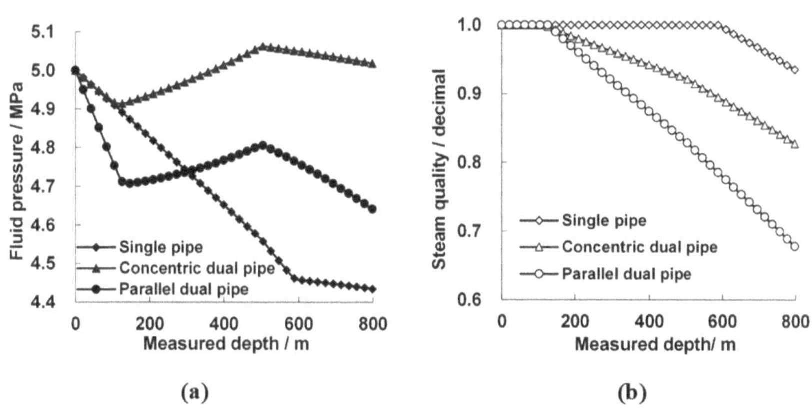

FIGURE 3.11 Simulation results in three different wellbore configurations: (A) fluid pressure; (B) steam quality.

3.6.6 Optimization of operation parameters

For actual steam-NCG coinjection, optimal operation parameters are required to improve the recovery process of heavy oil reservoirs effectively. Therefore, in this section, based on the developed mathematical models, a series of simulation runs are performed to optimize operation parameters in a steam-NCG coinjection process. The basic data in Table 3.1 are applied and the simulation results are shown in Fig. 3.12.

For the fluid injection rate, the results are shown in Fig. 3.12A. In the three different well configurations, as the fluid injection rate increases, the steam quality is gradually increased. On the other hand, from the simulation results, it can be also observed that the fluid pressure is reduced as the fluid injection rate increases. It is caused by an increase in friction loss for a higher fluid injection rate. Comparatively, the steam quality in the parallel dual-pipe well configuration is the lowest, and that in the single-pipe well configuration is the highest. It indicates the highest heat loss rate from the parallel dual-pipe one. The simulation results show that the fluid injection rate should not be less than 120 m^3/d.

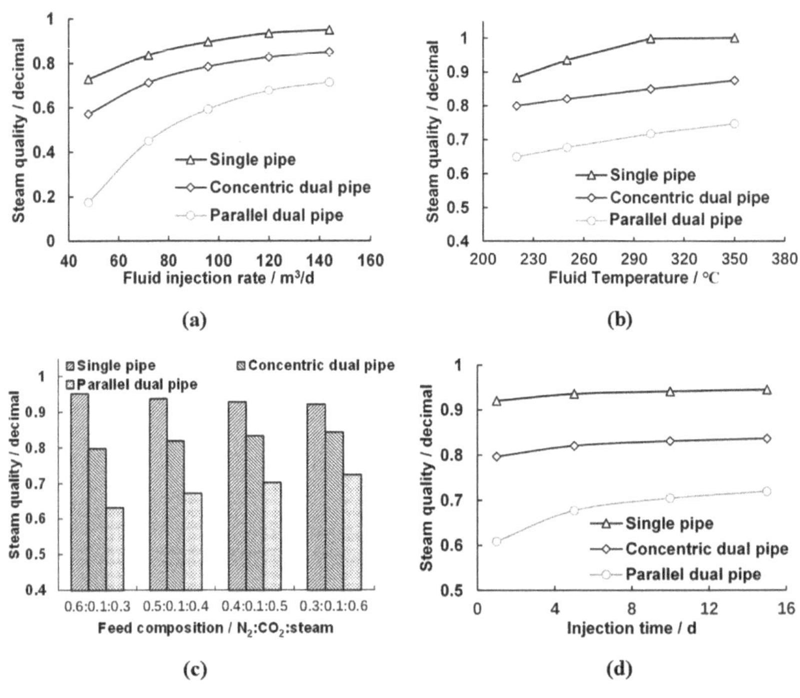

FIGURE 3.12 Steam qualities under different operation parameters: (A) effect of fluid injection rate; (B) effect of fluid temperature; (C) effect of feed composition; (D) effect of injection time.

Second, the fluid temperature affects the enthalpy of a steam-NCG mixture. As the fluid temperature increases, the enthalpy increases. The simulation results from different fluid temperatures are shown in Fig. 3.12B. Similar to the results in Fig. 3.12A, as the fluid temperature increases, the steam quality in the three different well configurations is also increased, whereas that in the parallel dual-pipe well configuration is the lowest.

Third, the feed composition alters the fluid properties and heat energy carried by a fluid mixture. As the fraction of the steam component increases, the total heat energy carried by steam is increased. The simulation results at different feed compositions are shown in Fig. 3.12C. As shown, as the fraction of the steam component increases, the steam quality is gradually increased. The steam quality in the parallel dual-pipe well configuration is the lowest.

Finally, fluid injection time is simulated; the results are shown in Fig. 3.12D. As the fluid injection time increases, the reservoir temperature in the vicinity of a wellbore is gradually increased. Therefore, the heat loss rate is reduced and the steam quality is increased. Similarly, the steam quality in the parallel dual-pipe well configuration is the lowest. From the simulation results, in field tests it is recommended that the fluid injection time should not be fewer than 5 days.

3.7 Wellbore heat loss models for offshore wellbore configurations

As discussed in Section 3.3.2, because of the effects of flowing seawater and marine sediments, wellbore configurations in an offshore heavy oil reservoir are different from those in an onshore heavy oil reservoir. In this section, wellbore heat loss models for two different offshore wellbore configurations are discussed. The wellbore configurations are shown in Fig. 3.3.

3.7.1 Model development

The most remarkable difference between offshore and onshore wellbore configurations is the presence of a marine riser. A marine riser is in the environment of flowing seawater. It indicates that the main difference in wellbore heat loss calculations between offshore and onshore thermal wells is the treatment of a marine riser. In this section, a heat transfer model between a marine riser and its surrounding environment is developed. According to the wellbore configurations in Fig. 3.3, it includes the heat transfer process of flowing seawater and marine sediments. The basic assumptions of this model are:

(1) The annulus space between a marine riser and casing is filled with nitrogen.
(2) The temperature of seawater and marine sediments remains constant.
(3) The heat transfer process between a steam injection pipe and seawater is a steady-state process.

Based on the discussion in Section 3.5.3, heat transfer resistance is a serial connection relationship from a steam injection pipe to flowing seawater and marine sediments. Therefore, the heat flux rate from steam injection to seawater can be obtained:

$$dQ_{sw} = \frac{(T_s - T_{sw})}{R_{sw}} dz \tag{3.63}$$

$$\begin{aligned} R_{sw} = \frac{1}{2\pi r_2} \left[\frac{r_2}{\lambda_{tub}} ln\frac{r_2}{r_1} + \frac{r_2}{\lambda_{ins}} ln\frac{r_3}{r_2} + \frac{r_2}{\lambda_{tub}} ln\frac{r_4}{r_3} + \frac{r_2}{r_4(h_c + h_r)} + \frac{r_2}{\lambda_{cas}} ln\frac{r_{co}}{r_{ci}} \right. \\ \left. + \frac{r_2}{r_{co}(h_c + h_r)} + \frac{r_2}{\lambda_{iw}} ln\frac{r_{wo}}{r_{wi}} + \frac{1}{h_{sw}} \right] \end{aligned} \tag{3.64}$$

where T_{sw} is the temperature of seawater, °C; R_{sw} is the heat transfer resistance from a steam injection pipe to seawater; λ_{iw} is the heat conductivity coefficient of a marine riser, W/(m ·°C); r_{wo} and r_{wi} are the outer and inner radii of the marine riser, m; and h_{sw} is the convective heat transfer coefficient of seawater, W/(m^2·°C).

This equation for heat transfer resistance is also valid for heat transfer from a steam injection pipe to marine sediments. Also, the heat flux rate from the steam injection pipe to marine sediments is:

$$dQ_{ms} = \frac{(T_s - T_{ms})}{R_{ms}} dz \tag{3.65}$$

where T_{ms} is the temperature of seawater, °C; and R_{ms} is the heat transfer resistance from the steam injection pipe to marine sediments.

Because the temperature of marine sediments is constant, the heat transfer resistance can be expressed as:

$$\begin{aligned} R_{ms} = \frac{1}{2\pi r_2} \left[\frac{r_2}{\lambda_{tub}} ln\frac{r_2}{r_1} + \frac{r_2}{\lambda_{ins}} ln\frac{r_3}{r_2} + \frac{r_2}{\lambda_{tub}} ln\frac{r_4}{r_3} + \frac{r_2}{r_4(h_c + h_r)} + \frac{r_2}{\lambda_{cas}} ln\frac{r_{co}}{r_{ci}} \right. \\ \left. + \frac{r_2}{r_{co}(h_c + h_r)} + \frac{r_2}{\lambda_{iw}} ln\frac{r_{wo}}{r_{wi}} \right] \end{aligned} \tag{3.66}$$

For models of pressure drop and steam quality, the equations in Section 3.5 still hold. Therefore, based on the mathematical models, the transient heat loss and pressure drop behavior in a steam injection process in different offshore wellbore configurations can be obtained.

3.7.2 Case study

Based on the wellbore heat loss model for offshore wellbore configurations, the heat transfer and pressure drop characteristics of steam and a steam-NCG mixture along the different wellbore configurations are simulated. The wellbore configurations and thermophysical properties data are shown in Table 3.2.

TABLE 3.2 Physical parameters of different offshore wellbore configurations.

Parameter	Value	Parameter	Value
Measured depth/m	800	Wellbore length of marine sediment section/m	10
Length of horizontal wellbore/m	300	Steam injection pressure/MPa	5.0
Air temperature/°C	22	Steam injection rate/t/d	120
Average temperature of sea water/°C	20	Cement radius/m	0.120
Temperature gradient/°C m^{-1}	0.032	TC of casing and tubing/W $(m\ K)^{-1}$	45.7
Steam injection time/d	5	TC of insulation material/W $(m\ K)^{-1}$	0.034
Length of expansion coefficient/m	50	TC of cement ring/W $(m\ K)^{-1}$	0.350
Wellbore length of seawater section/m	30	TC of formation/W $(m\ K)^{-1}$	1.73
Conventional single-pipe wellbore configuration			
OD of steam injection pipe/m	0.0889	OD of casing/m	0.1397
ID of steam injection pipe/m	0.076	ID of casing/m	0.12426
OD of thermal insulation pipe/m	0.1143	OD of marine riser/m	0.1778
ID of thermal insulation pipe/m	0.1016	ID of marine riser/m	0.1562
Concentric dual-pipe wellbore configuration			
OD of inner steam injection pipe/m	0.0889	OD of casing/m	0.1397
ID of inner steam injection pipe/m	0.076	ID of casing/m	0.12426
OD of outer steam injection pipe/m	0.1143	OD of marine riser/m	0.1778
ID of outer steam injection pipe/m	0.1016	ID of marine riser/m	0.1562

ID, inner diameter; *OD*, outer diameter; *TC*, thermal conductivity.

3.7.2.1 Pure (saturated) steam injection process

Fig. 3.13 shows the simulation results in a pure steam injection process in different offshore wellbore configurations. Based on the wellbore configurations, the whole wellbore segment can be classified into three stages. Heat transfer behavior in stage I reflects wellbore heat loss in a marine riser. Stage II is the vertical wellbore part, and stage III is the horizontal wellbore part. In Stage II, because of gravity loss, the fluid pressure increases with the well depth. Also, in stage III, because of friction loss, fluid pressure in the horizontal wellbore is reduced with the wellbore length. For a single-pipe wellbore configuration, a heat insulation pipe is usually applied to reduce wellbore heat

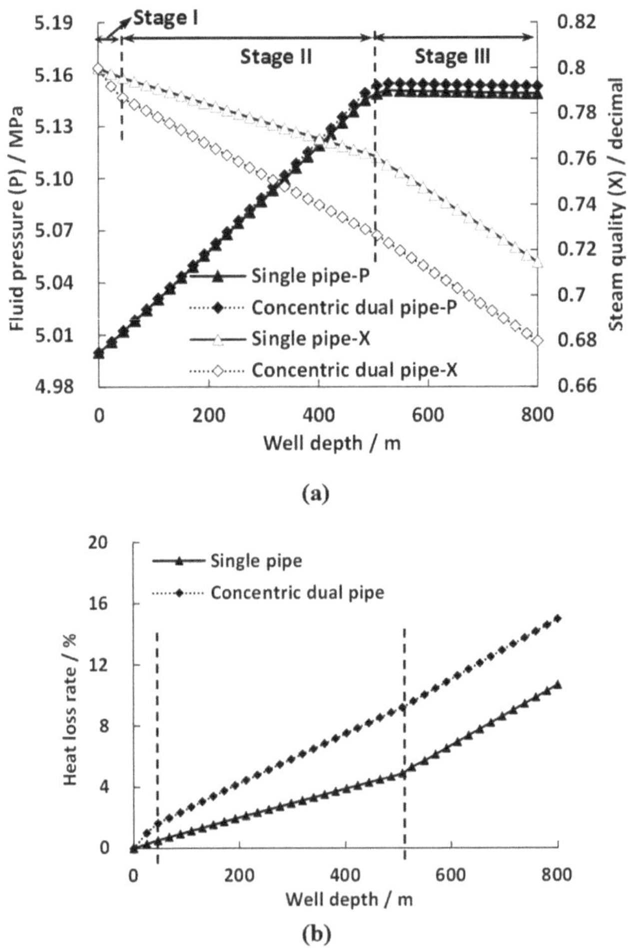

FIGURE 3.13 Simulation results of steam injection process in offshore wellbore configurations. (A) Fluid pressure and steam quality; (B) heat loss rate.

loss. For a concentric dual-pipe wellbore configuration, considering the effects of a wellbore size and steam injection technology, a heat insulation pipe is rarely applied. Fig. 3.13B shows that because of the performance of the heat insulation pipe, the heat loss rate in the single-pipe wellbore configuration is lower and the steam quality is higher. Under the same conditions, the steam quality in the concentric dual-pipe wellbore configuration is lower. It indicates that gravity loss is higher, and thus fluid pressure is higher.

3.7.2.2 Steam-NCG coinjection process

Using this mathematical model, wellbore heat loss in a steam-NCG coinjection in the two offshore wellbore configurations is discussed. The basic data are shown in Table 3.2. The fluid injection temperature is 250°C, which indicates that the steam component in this fluid mixture is superheated steam. The molar fractions of CO_2, N_2, and steam H_2O are 0.49:0.09:0.42.

Fig. 3.14 shows the simulation results in a steam-NCG coinjection in different offshore wellbore configurations. Under the same operation conditions, compared with saturated steam, the fluid density of a steam and NCG mixture is smaller, and thus the gravity-loss gradient along the wellbore in the steam-CNG coinjection is significantly lower than that in the saturated steam injection. It indicates that the fluid pressure in the steam-CNG coinjection is lower than that in the pure steam injection. Similar to the results of pure steam injection, for a steam-NCG injection, under the same operation conditions, the heat loss rate in a concentric dual-pipe wellbore configuration is also higher than that in a single-pipe wellbore configuration. It is mainly caused by the heat insulation pipe in the single-pipe wellbore configuration. Moreover, the results of fluid pressure and steam quality show an obvious three-stage behavior. The first stage is the pressure reduction in the vertical wellbore caused by single gas-phase flow. In this stage, friction loss dominates fluid flow. Thus, as the well depth increases, the fluid pressure is reduced. The second stage is the pressure increase stage in the vertical wellbore. In this stage, a transition from single gas-phase flow to gas—liquid two-phase flow can be observed. Also, gravity loss significantly increases. Thus, as the well depth increases, fluid pressure increases. The last stage is fluid flow in the horizontal wellbore. In this stage, the friction loss dominates again, and thus the fluid pressure is reduced.

3.8 Discussion on wellbore heat loss

Based on this discussion, wellbore heat loss is a nonnegligible issue for developing heavy oil and oil sand reservoirs. Especially for conventional saturated steam injection, wellbore heat loss can significantly affect the effectiveness of thermal recovery. High wellbore heat loss can induce poor recovery performance. For an actual thermal well, both formation/fluid

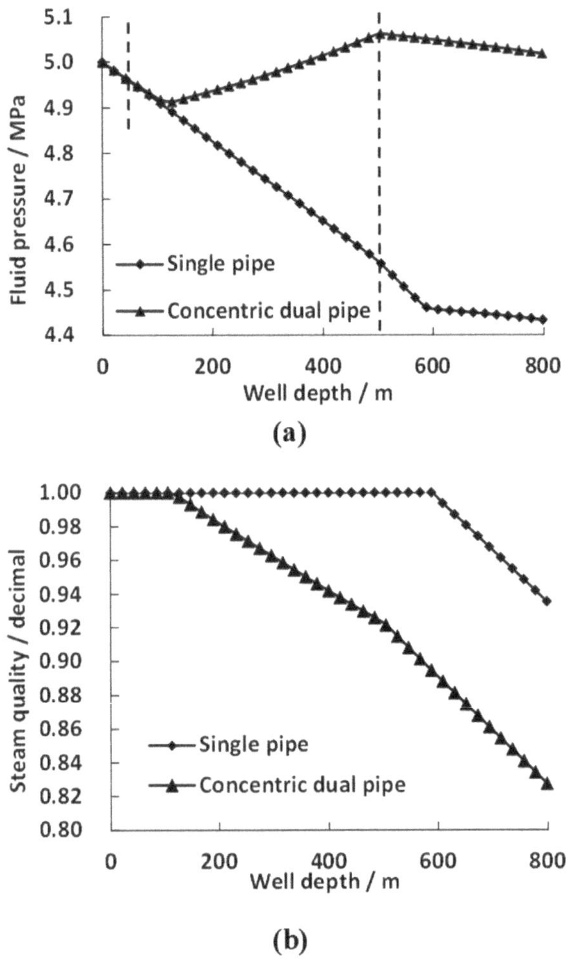

FIGURE 3.14 Simulation results in a steam-NCG coinjection process in offshore wellbore configurations. (A) Fluid pressure; (B) steam quality.

properties and wellbore configurations can affect wellbore heat loss. For a deep heavy oil reservoir (>1000 m), wellbore heat loss is even a dominating factor. Therefore, considering the effect of wellbore heat loss on the performance of thermal recovery, a hot fluid with higher heat energy and a wellbore configuration with a lower overall heat transfer coefficient have been applied. In some field tests, both can be applied to the same well.

For these types of hot fluids, commonly used fluids include superheated steam, supercritical steam, and hybrid steam additives. Superheated steam refers to steam with a temperature higher than its saturated temperature under a predefined pressure. The difference between superheated steam temperature and saturated temperature is the superheated degree. The quality of superheated steam is 1.0. For an actual injection process with superheated steam, its temperature is gradually reduced and the steam quality always remains 1.0. Once the temperature is reduced to the saturated temperature, the superheated steam will switch to saturated steam. For an injection process of saturated steam, the temperature keeps constant and steam quality is reduced. For superheated steam injection, higher steam quality can be obtained under bottom hole conditions. It can significantly improve recovery processes of heavy oil and oil sand reservoirs, for which conventional saturated steam injection is no longer economic. Second, supercritical steam is steam in which temperature and pressure are above the critical temperature (374.3°C) and critical pressure (22.12 MPa) of water. Compared with saturated steam and superheated steam, supercritical steam has a higher enthalpy and a higher diffusivity. For supercritical steam injection, the mechanism of near-miscible flooding is more significant. Moreover, an interaction between supercritical steam and heavy oil/rock can also benefit recovery processes. However, considering strict requirements regarding operation conditions, supercritical steam injection processes currently occur only in laboratory-scale tests. Third, for a hybrid steam additive coinjected process, it mainly refers to a steam-NCG process discussed. As discussed in Section 3.6, the wellbore heat loss rate in steam-NCG coinjection is much lower than that in saturated steam. Compared with the two types of hot fluids, steam-NCG coinjection is easily operated based on conventional steam-based facilities. Moreover, the wellbore heat loss rate in steam-NCG coinjection can be reduced compared with a pure steam injection process.

On the other hand, the wellbore configurations usually include thermal insulation pipes, concentric/parallel dual pipes, and FCDs. All of these measures can improve the heat transfer behavior between a wellbore and a formation and also steam conformance along the wellbore. For thermal insulation pipes, the wellbore heat loss can be reduced by a low wellbore heat transfer coefficient. For concentric/parallel dual-pipe wellbore configurations, wellbore heat loss can be reduced by increasing the steam injection rate. Another important purpose of a dual-pipe configuration is to improve steam conformance along a wellbore. For FCDs, by optimization of an FCD number and deployment, the steam properties and steam conformance under bottom hole conditions can be significantly improved.

This chapter addressed the problems of wellbore heat loss under different conditions. Based on this discussion, compared with a pure steam injection operation, hybrid steam-NCG processes can significantly improve fluid properties under bottom hole conditions and higher steam quality can be obtained, which benefits EOR processes for heavy oil reservoirs. In field operations, although fluid properties perform better, they usually do not indicate a good recovery performance. A steam injection profile along a (horizontal or vertical) wellbore is another important issue that affects hybrid EOR processes. Once hot fluids enter a formation by perforations or a slotted liner, fluids interact with a reservoir. Especially for commonly used horizontal wells in hybrid EOR processes, fluid flow along horizontal wellbores is a variable mass flow process. It is a coupled flow behavior between the reservoir and wellbores. This issue will be covered in the next chapter.

References

[1] Ramey HJ. Wellbore heat transmission. J Petrol Technol 1962;14(4):427–35.
[2] Satter A. Heat losses during flow of steam down a wellbore. J Petrol Technol 1965;17(7):845–51.
[3] Horne RN, Shinohara K. Wellbore heat loss in production and injection wells. J Petrol Technol 1978;31(1):116–8.
[4] Meyer BR. Heat transfer in hydraulic fracturing. SPE Prod Eng 1989;4(4):423–9.
[5] Hasan AR, Kabir CS. Heat transfer during two-phase flow in wellbores: part I—formation temperature. In: SPE 22866 presented at the SPE annual technical conference and exhibition, Dallas, Texas, USA; October 6–9, 1991.
[6] Wu YS, Pruess K. An analysis solution for wellbore heat transmission in layered formations. SPE Reservoir Eng 1990;5(4):531–8.
[7] Shiu KS, Beggs HD. Predicting temperatures in flowing oil wells. J Energy Resour Technol 1980;102(1):2–11.
[8] Sagar RK, Doty DR, Schmidt Z. Predicting temperature profiles in a flowing well. SPE Prod Eng 1991;6(4):441–8.
[9] Hagoort J. Ramey's wellbore heat transmission revisited. SPE J 2004;9(4):465–74.
[10] Xiong W, Bahonar M, Chen Z. Development of a thermal wellbore simulator with focus on improving heat-loss calculations for steam-assisted-gravity-drainage steam injection. SPE Reservoir Eval Eng 2016;19(2):305–15.
[11] Shoushtari MA, Al-Kayiem HH, Irawan S, et al. Developing novel wellbore heat transfer strategies for HPHT wells. In: Paper SPE 154768 presented at SPE Middle East health, safety, security, and enviroment conference and exhibition. Abu Dhabi: UAE; April 2–4, 2012.
[12] Gao G, Jalali Y. Prediction of temperature propagation along a horizontal well during injection period. SPE Reservoir Eval Eng 2008;11(1):131–40.
[13] Nuong AN. Thermal transient analysis applied to horizontal wells. In: Paper SPE 117435 presented at the 2008 SPE international thermal operations and heavy oil symposium, Calgary, Alberta, Canada; October 20–23, 2008.
[14] Arthur JE, Best DA, Lesage RP. A model describing steam circulation in horizontal wellbores. SPE Prod Facil 1993;8(4):263–8.
[15] Dong X, Liu H, Chen Z, et al. Enhanced oil recovery techniques for heavy oil and oilsands reservoirs after steam injection. Appl Energy 2019;239:1190–211.
[16] Griston S, Willhite GP. Numerical model for evaluating concentric steam injection wells. In: Paper SPE 16337 presented at the SPE California reginal meeting, Ventura, California, USA; April 8–10, 1987.
[17] Barua S. Computation of heat transfer in wellbores with single and dual completions. In: Paper SPE 22686 presented at the SPE annual technical conference and exhibition, Dallas Texas, USA; October 6–9, 1991.

[18] Castrup SG. Recommended practices for slim-hole steam injectors. SPE Prod Facil 2001;16(3):166–75.

[19] Gu H. Mass and heat transfer model and application of wellbore/formation coupling during steam injection in SAGD process. China University of Petroleum (Beijing); 2013 [Ph.D. dissertation].

[20] Liu HQ. Principle and design of thermal oil recovery processes. Beijing: Petroleum Industry Press; 2013.

[21] Dong X, Liu H, Chen Z. Mathematical Modeling of heat transfer and pressure drops in single- and dual-pipe horizontal well. J Therm Sci Eng Appl 2017;9(1). 011016-10.

[22] Chen Y. Thermal recovery processes of steam injection. Dongying: Petroleum University Press; 2002.

[23] https://simple.wikipedia.org/wiki/Thermal_radiation.

[24] Dong X, Liu H, Pang Z, Zhang C, Yi Y. Model for steam properties in parallel-tubing horizontal wells of heavy oil reservoir. J Cent S Univ 2014;45(3):939–45.

[25] Özkahraman HT, Selver R, Cislk E. Determination of the thermal conductivity of rock from P-wave velocity. Int J Rock Mech Min Sci 2004;41(4):703–8.

[26] Singh TN, Sinha S, Singh VK. Prediction of thermal conductivity of rock through physico mechanical properties. Build Environ 2007;42(1):146–55.

[27] Yuan EX. Engineering fluid mechanics. Beijing: Petroleum Industry Press; 2006.

[28] Izgec B. Transient fluid and heat flow modeling in coupled wellbore/reservoir systems. College Station, Texas, USA: Texas A&M University; 2008 [Ph.D. thesis].

[29] Moody LF. An approximate formula for pipe friction factors. Trans ASME 1947;69(12):1005–11.

[30] Beggs HD, Brill JP. A study of two-phase flow in inclined pipes. J Petrol Technol 1973;25(5):607–17.

[31] Liu HQ, Fan YP, Zhao DW, Zhang Y. Theory and method of thermal recovery. Dongying: University of Petroleum Press; 2000.

[32] Dong X, Liu H, Wang C. Flow and heat transfer characteristics of multi-thermal fluid in a dual-string horizontal well. Numer Heat Tran A 2014;66(2):185–204.

[33] Somerton WH. Thermal properties and temperature-related behavior of rock/fluid systems. Elsevier Science Publishers H.V.; 1992.

[34] Tong JS. Thermal properties of fluids. Beijing: China Petrochemical Press; 1996.

[35] Lu ZD. Design handbook of chemical process. 2nd ed. Beijing: Chemical Industry Press; 1996.

Chapter 4

Heat and mass transfer behavior between wellbores and reservoirs

4.1 Flow behavior of heavy oil in porous media

4.1.1 Introduction to heavy oil properties in porous media

As discussed in Chapter 1, a thermal recovery process is the primary recovery process for heavy oil reservoirs. Because of the unique properties of heavy crude oil, a large difference in rheology behavior can be observed between heavy oil and conventional oil. Under reservoir conditions, the flow process of heavy crude oil in porous media is nonlinear [1,2]. Therefore, the successful development of a heavy oil reservoir must address two issues. First, effective oil flow from a reservoir to a borehole needs to be established. Second, the heavy crude oil needs to be effectively lifted from a bottom hole to the surface [1–3]. In these two issues, the key is to improve the flowability and rheologic behavior of heavy oil in porous media. The unique temperature-sensitive flow of heavy oil in a reservoir affects the productivity of thermal wells and has an important role in the distribution of a steam injection profile along a wellbore.

Numerous studies on the flow and rheologic behavior of heavy crude oil have been performed. Most of them are concentrated on three aspects. First, the saturate, aromatic, resin, and asphaltene fractions of heavy oil can tremendously affect the flow behavior of heavy crude oil in porous media. They are the leading cause of the high viscosity of heavy oil [4–7]. Based on an experimental study, a critical asphaltene concentration can be observed [2,5]. Once the critical concentration is achieved, asphaltene particles start to overlap. Such molecule level-changing behavior can dramatically increase oil viscosity and alter the elastic characteristics of heavy oil. Then, by removing asphaltene from heavy oil, a reduction of two to three orders of magnitude in viscosity can be observed [5,6]. In hybrid enhanced oil recovery (EOR),

Hybrid Enhanced Oil Recovery Processes for Heavy Oil Reservoirs.
https://doi.org/10.1016/B978-0-12-823954-4.00010-2

through the performance of additives, heavy oil viscosity can be significantly reduced by improving the aggregation behavior of asphaltene [6]. Second, heavy oil is considered to be a temperature-sensitive Bingham fluid, and the behavior of heavy oil in porous media is closely related to temperature [8]. For a Bingham fluid, heavy oil cannot flow when stress is below the yield stress. It starts to move only when the stress is above the yield. Heavy oil starts to present the behavior of a Newtonian fluid when temperature increases [9,10]. Third, heavy oil can also show the behavior of a thixotropic fluid with a shear-shinning property [11–13]. The behavior of a threshold pressure gradient (TPG) and the thixotropic property of heavy oil were also observed in the laboratory [9]. The shear stress of heavy oil decreased with the shear time. Also, this unique behavior of heavy oil is related to its composition and viscosity.

For the TPG of heavy oil in porous media, a temperature-dependent behavior can be observed [14,15]. It is significantly different from the TPG behavior in a low permeability reservoir. The former is caused by the unique characteristics of fluid (heavy oil), and the latter is caused by the low permeability characteristics of porous media (formations). In the flow process of heavy oil in porous media, raising the temperature can effectively address the problem of TPG and shift the flow state from a non-Newtonian viscoelastic fluid to a Newtonian fluid. A critical temperature can be observed, and defined as the transition temperature of a flow state for heavy oil. Specifically, in steam assisted gravity drainage (SAGD), this critical temperature is an important sign to convert from a preheating phase to a normal SAGD phase. Simultaneously, in field operations, the temperature in a reservoir and wellbore should be kept above the critical temperature to achieve the effective exploitation of heavy oil reservoirs. In this section, an experimental method is introduced to address the critical temperature of heavy crude oil in porous media.

4.1.2 Experimental tests on heavy oil flow behavior in porous media

Commonly used methods to study oil flow behavior include flow tests in capillary tubes and porous media. The former is based on capillary tube bundles in porous media. In this method, the reservoir rock is supposed to be a porous material by a series of capillary tubes; it is far from real fluid flow behavior in a reservoir. The latter is based on a real porous medium environment or reservoir rock. In this section, a sandpack model will be applied to simulate a real porous medium environment. Using this method, a series of fluid flow experiments will be performed to study the non-Newtonian behavior of heavy oil in porous media [16].

4.1.2.1 Experimental method

The experimental setup is similar to the schematic in Fig. 2.8 in Chapter 2. The sandpack model used in this section has a 2.5-cm inside diameter and 35-cm length. During experiments, a pressure drop between the inlet end and outlet end of the sandpack model will be recorded. The heavy oil sample is the wellhead dead oil from three typical wells in the Karamay oil field, China National Petroleum Corp. (CNPC), denoted as 1#, 2#, and 3#. Fig. 4.1 shows the viscosity−temperature curves of the three oil samples. Heavy oil 1# has the highest oil viscosity and 3# has the lowest. Quartz sands with a uniform grain size and good abrasion will be applied to simulate the porous medium. To simulate the porous medium with different permeabilities, quartz sands with different meshes (60−80, 100−120, and 140−160 mesh) are applied.

For the experimental method, a steady-state fluid flow method is applied. The detailed experimental procedure is thus:

- ➢ Sandpack model preparation to simulate a predefined permeability condition;
- ➢ Connecting experimental apparatuses;
- ➢ Gas tightness testing to check for leaks;
- ➢ Distilled water injection to test pore volume and porosity;
- ➢ Oil sample injection to establish an irreducible water saturation condition;
- ➢ As the injected heavy oil reaches 2−3 pore volumes, a single oil-phase flow state is supposed;

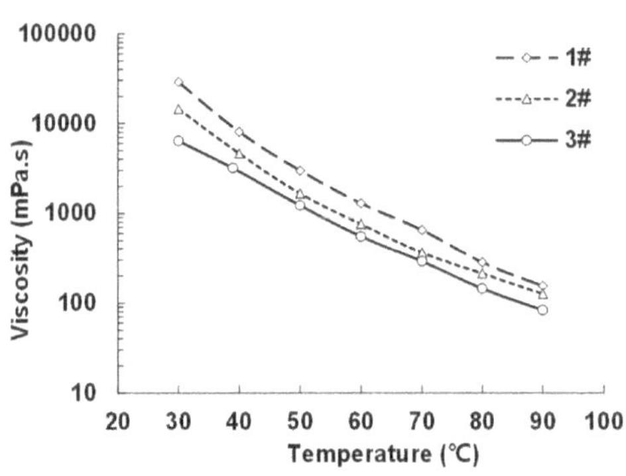

FIGURE 4.1 Viscosity−temperature curves of three heavy oil samples.

➢ Activating an oil flow test and adjusting its oil flow rate from a small value to the desired high value; and

➢ For each test, recording the pressure difference across the sandpack model once a steady flow state is achieved.

4.1.2.2 Experimental results

Based on this procedure, the flow behavior of heavy oil in a porous medium can be discussed. A Newtonian fluid can flow in a porous medium once the pressure gradient is above zero. However, for heavy oil, it is a typical temperature-related Bingham fluid, and heavy oil starts to flow only when the pressure gradient is above the yield pressure gradient or TPG. As temperature increases, the TPG is gradually decreased. During experiments, the curves for the relationship between an oil flow rate and the pressure gradient across the sandpack model can be obtained. Based on the intercept of each curve, the relationship between TPG and temperature (oil viscosity) can be achieved. Fig. 4.2 presents the experimental results of the three oil samples under the same permeability condition (3660×10^{-3} μm^2). As shown, for a given permeability, as the oil flow rate increases, the pressure gradient increases. Simultaneously, as temperature increases, the oil viscosity is reduced and the pressure gradient is also reduced. Fig. 4.2A shows that the TPG of heavy oil 1# at 30°C is about 0.202 MPa/m. When the temperature increases to 70°C, the TPG of this oil is reduced to 0. By the method of linear interpolation, we can see that the critical transition temperature of this heavy oil is about 58°C under this permeability condition. Similarly, the critical transition temperatures of heavy oil 2# and 3# under this permeability condition can be also achieved. Fig. 4.2B and C illustrate that compared with heavy oil 1#, the pressure gradients of heavy oil 2# and 3# are lower because of their lower oil viscosities.

On the other hand, to study the effect of reservoir permeability on the TPG of heavy oil in a porous medium, two other permeability conditions (9180×10^{-3} μm^2 and 1320×10^{-3} μm^2) are simulated by the quartz sands with different meshes. Then, based on the experimental results, as the permeability is reduced, the flow resistance is increased, and thus the TPG of heavy oil is also increased. Simultaneously, the critical transition temperature of heavy oil is also increased as the reservoir permeability is reduced. When temperature is higher than this critical temperature, the overlapping phenomena of asphaltene will be destroyed [5]. This temperature is related to the oil composition and also the reservoir properties.

Fig. 4.3 gives the critical transition temperature of the three different oil samples under different permeability conditions. As shown, the critical temperature of oil sample 1# is the highest, and the critical temperature of oil sample 3# is the lowest. This is mainly caused by the difference in the oil

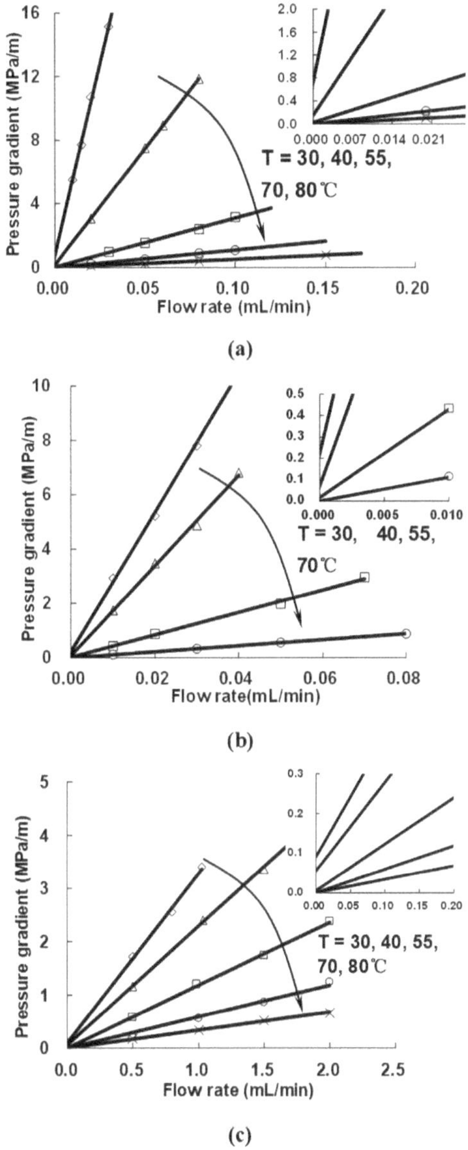

FIGURE 4.2 Experimental results for heavy oil flow behavior in a porous medium at different temperatures. (A) Oil sample 1#. (B) Oil sample 2#. (C) Oil sample 3#.

viscosity. For an extraviscous crude oil, a higher temperature is required to reduce the oil viscosity greatly. However, for the heavy oil with a low viscosity, a relatively lower temperature is also acceptable to maintain the flow of heavy oil in a porous medium.

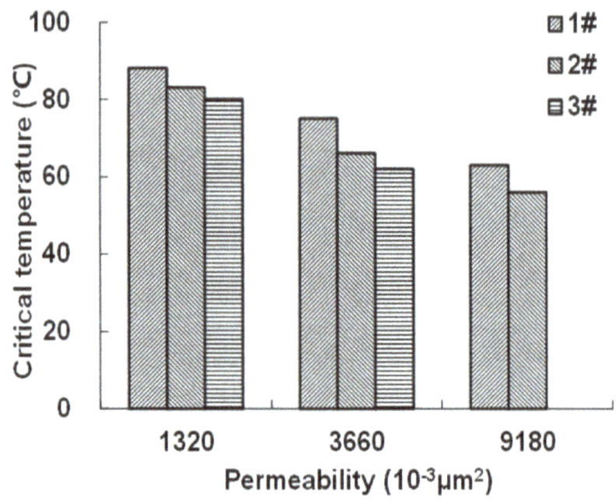

FIGURE 4.3 Critical transition temperature.

Moreover, from these experimental results, the flow behavior of heavy oil in porous media is related to the oil and rock properties (oil viscosity and permeability) and also the operation condition (temperature). Fig. 4.4 shows the relationship between the TPG and oil mobility (K/μ_o). As oil mobility increases, the TPG of heavy oil in porous media is reduced. Moreover, a power relationship can be observed between the TPG and oil mobility. Under a very low oil mobility condition, the TPG can be tremendously increased.

On the other hand, based on the tested TPG results under different conditions, the dependence of critical temperature on the oil mobility can be obtained, as shown in Fig. 4.5. With increases in oil mobility, the critical temperature is logarithmically reduced. This indicates that the permeability and oil viscosity will respectively present a negative correlation and positive correlation with critical temperature.

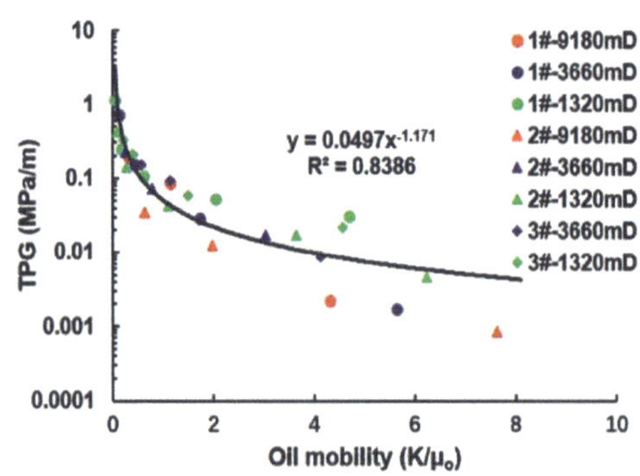

FIGURE 4.4 Relationship between threshold pressure gradient (TPG) and oil mobility for three different oil samples under different permeability conditions.

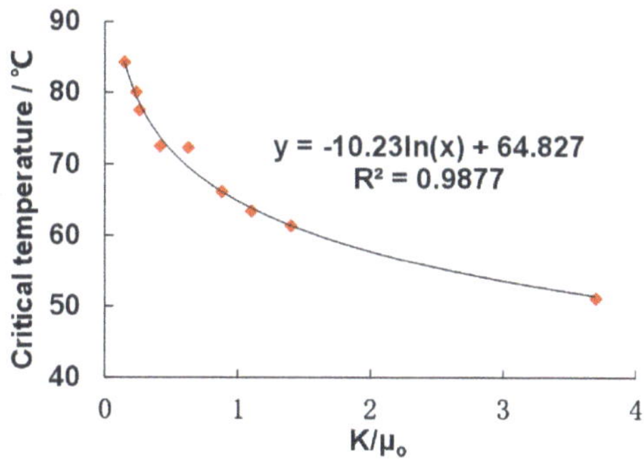

FIGURE 4.5 Relationship between critical temperature and oil mobility.

4.2 New productivity models for thermal wells

Based on these experimental results, the phenomena of TPG and critical temperature in porous media can be observed for heavy oil. This unique behavior of heavy oil is significantly related to the oil composition and reservoir properties. For the actual development of heavy oil reservoirs, it can also affect the normal production of thermal wells. In this section, based on the TPG and critical temperature, models for the productivity of thermal wells in heavy oil reservoirs are developed. Both vertical and horizontal wells are considered.

4.2.1 Productivity model for vertical wells

To develop vertical wells in heavy oil reservoirs, an inverted nine-spot pattern is a relatively common well pattern in steam injection, and the controlling area of a nine-spot well pattern is higher than other patterns (e.g., five-spot and seven-spot patterns) [17]. Fig. 4.6 shows a schematic diagram of an inverted

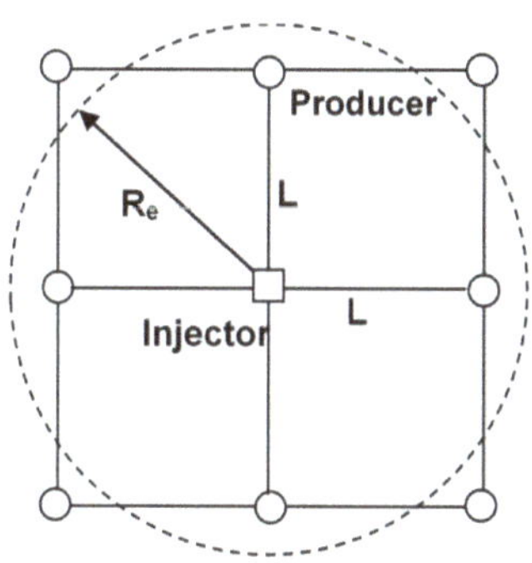

FIGURE 4.6 Schematic diagram of inverted nine-spot well pattern.

nine-spot well pattern. Below L is the distance between an injection well and an edge well, S is the controlling area of the injector, and R_e is the equivalent radius of the circular control area (Fig. 4.6).

Fig. 4.6 shows that the controlling area of the injector in the inverted nine-spot well pattern is:

$$S = (2L)^2 = \pi R_e^2 \tag{4.1}$$

Therefore,

$$R_e = 2L\sqrt{\frac{1}{\pi}} = 1.1284\text{L} \tag{4.2}$$

After a consideration of the TPG of heavy oil, the oil flow rate in a porous medium can be obtained from Eq. (4.3):

$$\begin{cases} \overrightarrow{v} = 0 \quad \text{while} \dfrac{dp}{dr} < \lambda, \quad \text{and} \\ \overrightarrow{v} = -\dfrac{K}{\mu(T)}\left(\dfrac{dp}{dr} - \lambda\right) \quad \text{while} \quad \dfrac{dp}{dr} > \lambda \end{cases} \tag{4.3}$$

where K is the permeability of the porous medium; λ is the TPG of heavy oil; and $\mu(T)$ is the oil viscosity. Therefore, for vertical wells, when the TPG is considered, well productivity can be expressed as:

$$dQ = vA = 2\pi rh\frac{K}{\mu(T)}\left(\frac{dp}{dr} - \lambda(T)\right) \tag{4.4}$$

Integrating this formula,

$$Q_{vw} = \frac{2\pi Kh[p_i - p_w - \lambda(T)(R_e - R_w)]}{\mu(T)\ln(R_e/R_w)} = \frac{2\pi Kh[\Delta p - \lambda(T)(R_e - R_w)]}{\mu(T)\ln(R_e/R_w)} \tag{4.5}$$

Thus, Eqs. (4.4) and (4.5) can be used for to evaluate the productivity of thermal vertical wells. This model fully considers the non-Newtonian flow behavior of heavy oil in porous media.

4.2.2 Productivity model for horizontal wells

A productivity model for horizontal wells for heavy oil reservoirs also needs to be improved by considering TPG. Joshi proposed a productivity model for a horizontal well in homogeneous reservoirs by assuming an elliptical drainage volume [18]:

$$Q_{hw} = \frac{2\pi K_o h\Delta p}{\mu\left\{\ln\left[\frac{a+\sqrt{a^2-(L/2)^2}}{L/2}\right] + \frac{h}{L}\ln\left(\frac{h}{2R_w}\right)\right\}} \tag{4.6}$$

where

$$a = (L/2)\left[1/2 + \sqrt{1/4 + \frac{1}{(0.5L/R_e)^4}}\right]^{0.5} \tag{4.7}$$

where $L/2$ is the half-length of a horizontal segment:

$$L/2 = \sqrt{a^2 - b^2} \tag{4.8}$$

Similarly, the equivalent oil drainage radius of a horizontal well can be expressed by:

$$R_e = \sqrt{A/\pi} = \sqrt{ab} \tag{4.9}$$

Based on a modification of Joshi's model, a productivity model for a horizontal well in heavy oil reservoirs can be developed. Joshi's model can be effectively applied in single-phase fluid flow in isotropic oil reservoirs. For heavy oil reservoirs, when the TPG and thermal effects are considered, this model can be modified to:

$$Q_{hw} = \frac{2\pi K_h h[\Delta p - \lambda(T)(R_e - R_w)]}{\mu(T)\left\{\ln\left[\frac{a+\sqrt{a^2-(L/2)^2}}{L/2}\right] + \frac{h}{L}\ln\left(\frac{h}{2R_w}\right)\right\}} \tag{4.10}$$

Eq. (4.10) can be used to evaluate the productivity of thermal horizontal wells in heavy oil reservoirs.

A relationship between vertical well productivity and horizontal well productivity is given by:

$$\delta = \frac{Q_{hw}}{Q_{vw}} = \frac{\ln(R_e/R_w)}{\ln\left[\frac{a+\sqrt{a^2-(L/2)^2}}{L/2}\right] + \frac{h}{L}\ln\left(\frac{h}{2R_w}\right)} \tag{4.11}$$

4.2.3 Evaluation on productivity of thermal wells

After considering the TPG behavior of heavy oil in porous media, the productivity of thermal wells is discussed in this section. For wells in heavy oil reservoirs, because of the effect of TPG behavior, their productivity may be low. Specifically, when a formation pressure gradient is below the TPG of heavy oil or the formation temperature is below the critical temperature, the productivity of a well may be 0. Therefore, a new concept of threshold temperature of thermal wells can be defined. When the reservoir temperature is higher than the threshold temperature of a thermal well, this well can start normal production.

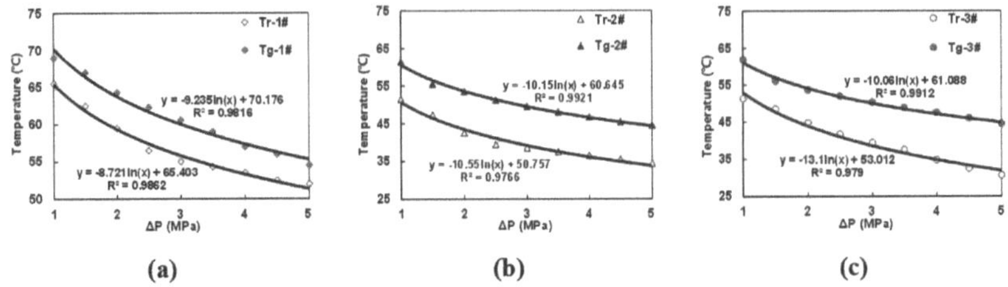

FIGURE 4.7 Simulation results of vertical well for three different oil samples. T_r refers to threshold temperature ($Q_o = 0$ m^3/d), and T_g refers to the temperature of a given productivity ($Q_o = 1$ m^3/d). (A) 1#; (B) 2#; and (C) 3#.

First, for vertical wells, the properties of a typical heavy oil reservoir from the Karamay oil field, China, are applied. The reservoir thickness is 6 m, the horizontal permeability is 3200×10^{-3} μm^2, the well spacing is 80 m, and the wellbore diameter is 0.06 m. Therefore, from the basic reservoir data, the productivity of a vertical well in this heavy oil reservoir can be calculated. Based on Eqs. (4.4) and (4.5), both the threshold temperature (corresponding to $Q_o = 0$ m^3/d) and the temperature of a given productivity ($Q_o = 1$ m^3/d) for this vertical well are calculated. Fig. 4.7 shows the simulation results of the vertical well under different pressure drops (ΔP) between the injection and production wells.

Second, for horizontal wells, another heavy oil reservoir from the Karamay oil field is applied. The reservoir thickness is 10 m, the horizontal permeability is 3200×10^{-3} μm^2, the drainage area of this horizontal well is about 4.2×10^4 m^2, and the wellbore diameter is 0.032 m. Thus, the productivity of a horizontal well in this heavy oil reservoir can be calculated. Based on Eq. (4.10), both the threshold temperature (corresponding to $Q_o = 0$ m^3/d) and the temperature of a given productivity ($Q_o = 10$ m^3/d) for this horizontal well are calculated. Fig. 4.8 shows the simulation results of this well under different pressure drops (ΔP).

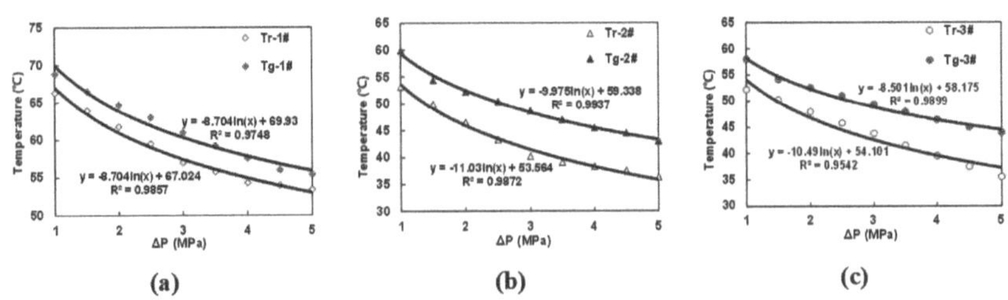

FIGURE 4.8 Simulation results of horizontal well for three different oil samples. T_r refers to threshold temperature ($Q_o = 0$ m^3/d), and T_g refers to the temperature of a given productivity ($Q_o = 10$ m^3/d). (A) 1#; (B) 2#; and (C) 3#.

As shown, for both vertical and horizontal wells, as the displacement pressure drop increases, both the threshold temperature and the given productivity temperature are reduced. A logarithmic relationship between the pressure drop and temperature can be observed. Comparatively, the temperature required for oil sample 1# is the highest, and that for oil sample 3# is the lowest. This is caused by the viscosity difference among the three oil samples. When the oil sample viscosity increases, the well productivity is reduced when the other parameters remain unchanged. Moreover, based on Eq. (4.11), the productivity of a horizontal well is about 3.09 times than that of a vertical well. Therefore, with the same well productivity, the temperature required for a horizontal well will be lower than that for a vertical well.

4.3 Experimental tests for steam conformance along wellbores

Although the productivities of vertical and horizontal wells in heavy oil reservoirs can be calculated from Eqs. (4.5) and (4.10), these models are valid only in an ideal single oil phase flow process. For an actual well in the field, an accurate estimate of the productivity of thermal wells is challenging. Especially for a horizontal well in heavy oil reservoirs, the behavior of heat and mass transfer of fluids along the horizontal wellbore can have an important effect on the recovery of the well. These behaviors are affected by well completion and also heat transmission characteristics between a formation and a horizontal well. In steam injection, transient flow behavior of steam along a horizontal wellbore is a gas–liquid two-phase flow process, and heat transfer between a wellbore and a reservoir also needs to be considered [19,20]. In this section, steam conformance along a horizontal wellbore is experimentally addressed [21–22].

4.3.1 Experimental method

The cylindrical wellbore model used in Section 2.2 is applied to the experimental discussion of steam conformance along a horizontal wellbore with different configurations. In this section, both a single-pipe well configuration and concentric pipe well configuration are simulated. The schematic of the experimental setup is shown in Fig. 4.9. Using this model, a full cyclic steam stimulation (CSS) process using single-pipe and concentric dual-pipe well configurations will be experimentally simulated. During CSS experiments, the pressure, temperature, and oil production rate along the wellbore are recorded to discuss steam conformance in different well configurations.

The oil sample used is a wellhead dead oil (2200 mPa s under the standard conditions) from the Henan oil field, Sinopec, China. Water is distilled water,

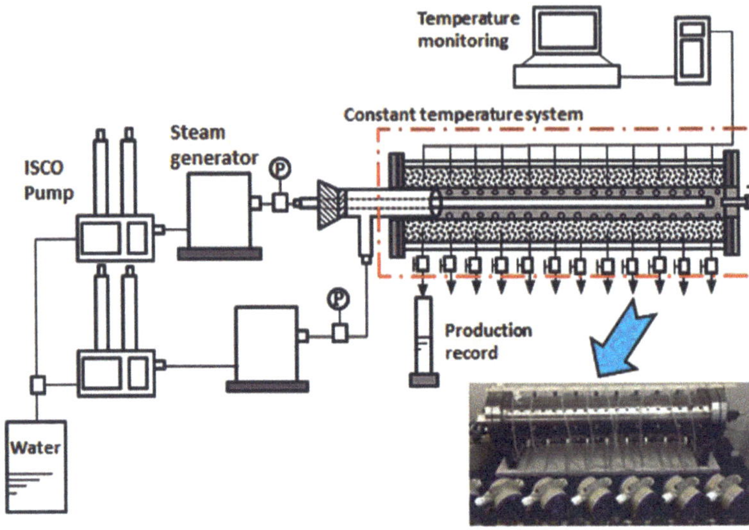

FIGURE 4.9 Schematic of experimental setup. *ISCO*, Teledyne ISCO pump.

and quartz sands with 80–100 mesh are used to simulate the porous medium. The experimental procedures are:

(1) sand filling to simulate permeability conditions;
(2) connecting the experimental apparatuses to check for leaks;
(3) water injection to test the pore volume and porosity;
(4) oil injection to establish the original oil saturation conditions;
(5) steam injection to simulate CSS.
(6) data monitoring for the distribution of pressure, temperature, and liquid production along the wellbore.

4.3.2 Experimental results

4.3.2.1 General behavior of hot fluids flow along a wellbore

During experiments, pure saturated steam is injected; the injection rate is 60 mL/min. A fluid injection tubing is placed around the heel-end section of the wellbore model. Well completion is a perforated method. The simulated porous medium environment is under a homogeneous condition with 80-mesh quartz sands, and its permeability is about 2.84×10^{-3} μm^2. Fig. 4.10 shows the obtained temperature and pressure distributions inside the wellbore. As the hot fluids flow along the wellbore, both the temperature and pressure inside the

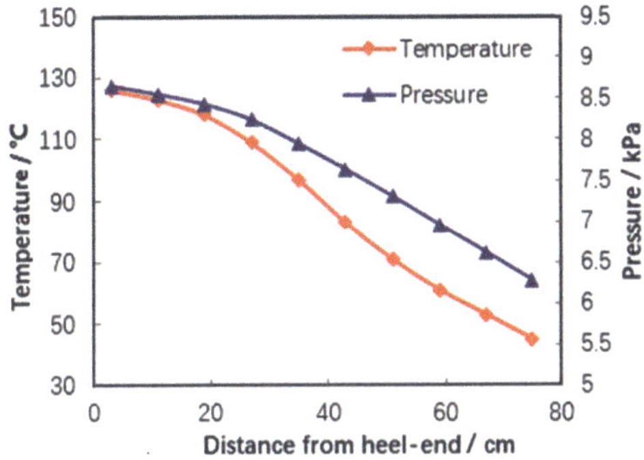

FIGURE 4.10 Temperature and pressure distributions along the horizontal wellbore.

wellbore are reduced. Mass and heat transfer behaviors between the reservoir and horizontal wellbore are important reasons for the decrease in fluid temperature. Fig. 4.11 gives the formation temperature distribution in the axial and radial directions. As shown, for the wellbore segment closed to the heel end, because the heel-end injection method is applied, both the formation temperature and heating radius around the heel-end section are much higher. This is caused by the fluid outflow position used in this test. For heel-end injection, the heating effect around the heel end is significantly better than that for the toe end. In some situations, most fluid in a wellbore enters into a formation before it reaches the toe-end section. In addition, a seriously nonuniform steam injection profile is observed. Thus, the reservoir section close to the toe end does not contribute to total oil production.

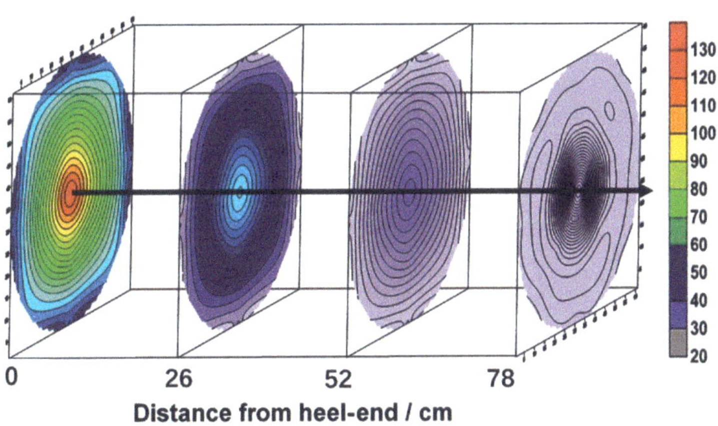

FIGURE 4.11 Steam heating areas along the horizontal wellbore.

4.3.2.2 Effect of well configuration

In this section, steam conformance in three different well configurations is simulated, including single-well heel-end steam injection, single-well toe-end steam injection, and concentric dual-pipe steam injection. During experiments, the total steam injection volume for the three cases is the same (10 mL/min × 30 min). For the concentric dual-pipe experiment, the steam injection rate in the long and short pipes is one-half the steam injection rate in the single-pipe well configuration. The experimental results are shown in Fig. 4.12. For all three experiments, the temperature around the steam outflow points is always higher. Steam injected preferentially enters the reservoir at a wellbore interval close to the steam outflow points. In the dual-pipe well configuration, both the heel-end and toe-end sections have a higher temperature. Compared with the single-pipe well configuration, bimodal distributions of temperature and oil production can be observed in the dual-pipe one. The results of pressure distribution have a similar observation. The dual-pipe well configuration can enhance the heating volume (oil drainage volume) by about 35% compared with the single-pipe well configuration. Although total oil production does not change significantly, the effective wellbore interval is much different. The concentric dual-pipe well configuration can significantly improve steam conformance along the horizontal wellbore.

4.3.2.3 Effect of hot fluid type

In hybrid EOR processes, multicomponent and multiphase fluids are applied to improve recovery performance for heavy oil reservoirs. In particular, hybrid thermal noncondensable gases (NCGs) can enhance heavy oil recovery by improving the steam injection profile along a horizontal wellbore. In this section, the steam conformance of pure steam injection and steam-NCG co-injection is comparatively studied. During experiments, the flow rates of both pure steam and a steam-NCG mixture (steam plus NCG, with 15% CO_2 + 85% N_2) are set to 60 mL/min. The results are shown in Fig. 4.13. The fluid pressure and temperature in the steam-NCG co-injection process is higher because of the improvement of NCG on the fluid physical properties.

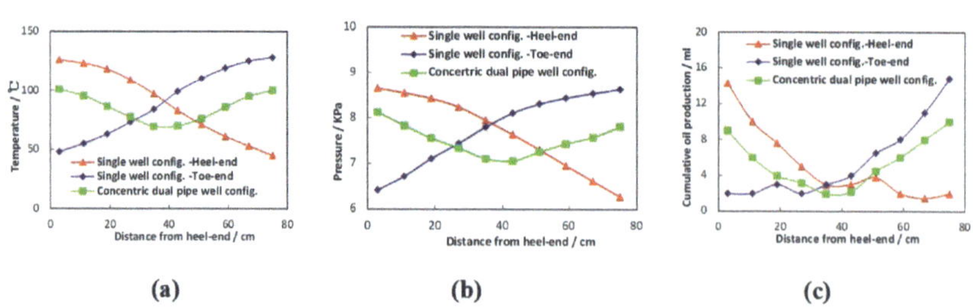

FIGURE 4.12 Experimental results of three different well configurations (*config.*) within one cyclic steam stimulation cycle: (A) temperature distribution after soaking; (B) pressure distribution after soaking; (C) cumulative oil production distribution.

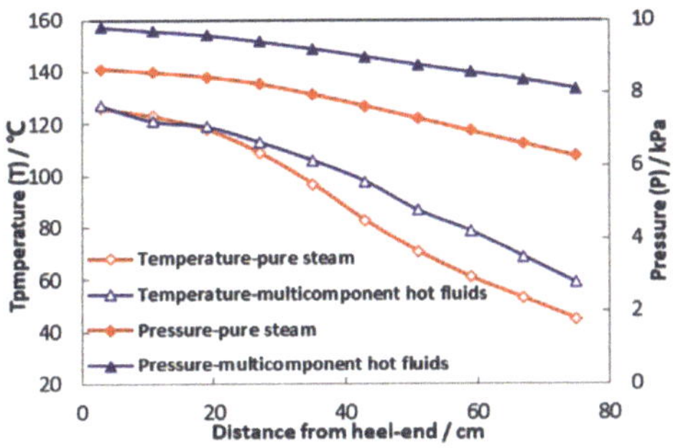

FIGURE 4.13 Effect of hot fluid type.

Compared with pure saturated steam, the fluid density and viscosity of multicomponent hot fluids are much lower. Therefore, the heat loss rate and pressure drop along the horizontal wellbore are reduced. Simultaneously, from the results of temperature, distribution along the wellbore is more uniform for steam-NCG co-injection. This indicates an improvement in steam conformance.

4.4 Mathematical models for pure steam injection processes

Based on this experimental observation, in this section, heat and mass transfer behavior between a wellbore and a reservoir in pure steam injection will be studied through a semianalytical model. In steam injection, fluid flow behavior along a horizontal well is a variable mass/temperature fluid flow process. Main causes include reservoir heterogeneity, perforation flow, and a pressure drop along the wellbore. Commonly used wellbore configurations of a horizontal well in heavy oil reservoirs include a single-pipe well configuration and dual-pipe well configuration. For a single pipe, a heel-end steam injection mode and a toe-end steam injection mode can be applied, as shown in Fig. 4.14A. For a dual pipe, a concentric steam injection mode and a parallel steam injection mode can be applied, as shown in Fig. 4.14B. In this section, the mathematical models for them will be developed [21,23].

4.4.1 Assumptions

The assumptions for the mathematical models are:

(1) During a steam injection process, steam injection parameters in a heel-end section remain unchanged;

(2) Changes in reservoir properties (e.g., porosity and permeability) caused by pressure and temperature are not considered;

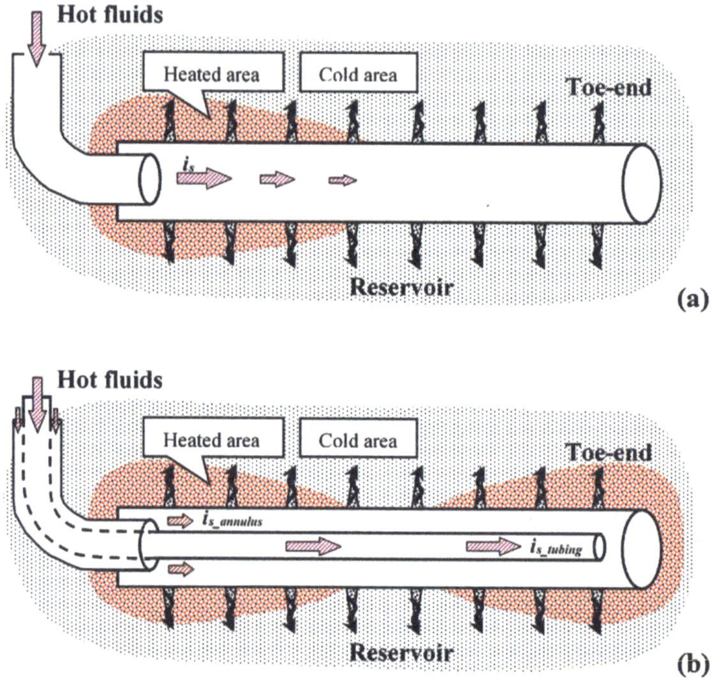

FIGURE 4.14 Commonly used completion modes of a thermal horizontal well: (A) single-pipe steam injection mode; (B) concentric dual-pipe steam injection mode.

(3) Only a radial heat transfer process is considered;
(4) Heat transfer from a horizontal wellbore to a cement ring is a steady state, and heat transfer from a cement ring to a formation is an unsteady state.

4.4.2 Model development

As in Chapter 3, before the model development, a whole horizontal wellbore is first discretized. A schematic of a wellbore microcontrol element is shown in Fig. 4.15.

4.4.2.1 Mass conservation equation

$$\rho_1 v_1 A - \rho_2 v_2 A - q_{mi} = 0 \tag{4.12}$$

where q_{mi} is the steam injection rate within a wellbore microcontrol element:

$$q_{mi} = J_i I_i \Delta p = J_i I_i (p_i - p_e) \tag{4.13}$$

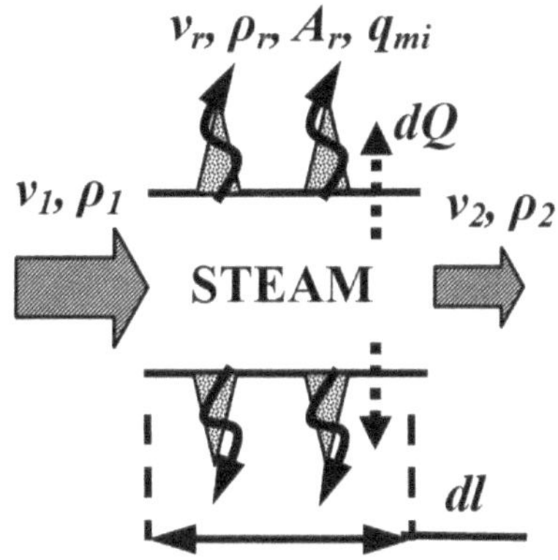

FIGURE 4.15 Schematic of a wellbore microcontrol element.

where p_i is the steam injection pressure; p_e is the reservoir pressure; J_i is a liquid production index [24]; and I_i a steam adsorption index:

$$J_i = \frac{2\alpha\pi\sqrt{\frac{K_h}{K_v}}K_v\left(\frac{K_{ro}}{\mu_o B_o} + \frac{K_{rw}}{\mu_w B_w}\right)dl}{\ln\frac{0.571A_i^{0.5}}{r_w} - \frac{3}{4} + S} \tag{4.14}$$

$$I_i = \frac{2\ \ln\left(\frac{A_i}{r_w^2}\right) - 3.86}{\ln\left(\frac{A_i}{r_w^2}\right) - 2.71 - \ln E_h} \tag{4.15}$$

where A_i is the oil drainage area; r_w is the horizontal wellbore radius; and E_h is the thermal effect constant [25]:

$$E_h = \begin{cases} \frac{1}{t_D}\left[e^{t_D} erfc\sqrt{t_D} + 2\sqrt{t_D/\pi} - 1\right] & t_D \le t_{CD} \\ \frac{1}{t_D}\left[\begin{array}{l}\left(e^{t_D} erfc\sqrt{t_D} + 2\sqrt{t_D/\pi} - 1\right) - \\ \sqrt{(t_D - t_{CD})/\pi}\left(\begin{array}{l}\frac{1}{1+h_D} + \\ \frac{t_D - t_{CD} - 3}{3}e^{t_D} erfc\sqrt{t_D} - \frac{t_D - t_{CD}}{3\sqrt{\pi t_D}}\end{array}\right)\end{array}\right] & t_D > t_{CD} \end{cases} \tag{4.16}$$

$$t_D = \frac{4\lambda t}{M_R L^2}, \quad h_D = \frac{L_v x}{C_w \Delta T} \tag{4.17}$$

M_R refers to the specific heat capacity of a formation:

$$M_R = (1-\varphi)\rho_R C_R + \varphi\left(S_o \rho_o C_o + S_w \rho_w C_w + S_g \rho_g C_g\right) \tag{4.18}$$

4.4.2.2 Momentum conservation equation:

$$Adp = \rho_1 v_1^2 A - \rho_2 v_2^2 A - \rho_r v_r^2 A_r - \tau_f \tag{4.19}$$

where v_r is the steam mass rate through perforations during steam injection; and τ_f is the friction force of steam in a wellbore microcontrol element:

$$v_r = \frac{q_{mi}}{\rho_r A_r}, \quad \tau_f = \frac{f_{tp} \pi D \rho_m}{8} v_m^2 dl \tag{4.20}$$

For fluid flow along a horizontal wellbore, friction loss dominates a wellbore pressure drop. For a horizontal well with the well completion method of perforation, friction losses caused by both an unperforated wellbore segment and a perforated wellbore segment need to be considered [23]. To simulate friction losses, an important step is to address calculations of a frictional resistance coefficient, f_{tp}:

$$f_{tp} = f_{main} + f_{perf} \tag{4.21}$$

where f_{main} is the frictional resistance coefficient in an unperforated wellbore segment and f_{perf} is the frictional resistance coefficient in perforations.

Considering the different phase states of injected steam, the calculation methods for f_{main} under different conditions are different. For superheated steam, steam flow along a wellbore is a single gas flow process. For saturated steam, steam flow is a gas–liquid two-phase flow process. For single gas flow, the evaluation method of the Reynolds number can be applied to calculate the frictional resistance coefficient:

$$R_{eg} = 1.2732 \frac{m}{2r\mu_g} \tag{4.22}$$

$$f_{main} = \begin{cases} \dfrac{64}{R_{eg}} & while, \quad R_{eg} \le 2000 \\[2ex] \left[1.14 - 2\lg\left(\dfrac{\varepsilon}{D} + \dfrac{21.25}{R_{eg}^{0.9}}\right)\right]^{-2} & \text{while,} \quad R_{eg} > 2000 \end{cases} \tag{4.23}$$

For gas–liquid two-phase flow, the three methods of experimental tests, correlation, and Moody's plate can be applied to calculate the frictional resistance coefficient [26,27]. In this section, the commonly used Beggs-Brill method is applied [28]. In Beggs-Brill's method, the gas−liquid two-phase flow process along a horizontal wellbore is classified into three different flow regimes of separated, intermittent, and dispersed flow. Based on the specific flow properties of steam in each wellbore microcontrol element, the frictional resistance coefficient can be derived. More details can be found in Section 3.5.5.

On the other hand, for the frictional resistance coefficient in perforations, the method proposed by Su and Gudmundsson is applied [29,30]:

$$B=\sqrt{\frac{8}{f_{perf}}}-2.5\ln\left(\frac{\mathrm{Re}}{2}\sqrt{\frac{f_{perf}}{8}}\right)+3.75 \tag{4.24}$$

$$\frac{\Delta u}{u^{*}}=7.0\left(\frac{d}{D}\right)\left(\frac{n}{12}\right) \tag{4.25}$$

$$\sqrt{\frac{8}{f_{perf}}}=2.5\ln\left(\frac{\mathrm{Re}}{2}\sqrt{\frac{f_{perf}}{8}}\right)+B-\frac{\Delta u}{u^{*}}-3.75 \tag{4.26}$$

4.4.2.3 Energy conservation equation

The types of energy loss in a horizontal wellbore include a steam mass change along the wellbore, heat loss caused by the radial flow, radial heat transfer, and friction loss. Therefore, the energy conservation equation can be developed:

$$\frac{dQ}{dl}+\frac{dW}{dl}=-\frac{d}{dl}\left[i_s\left(H_m+\frac{v_m^2}{2}\right)\right]-q_{mi}\left(H_m+\frac{v_r^2}{2}\right) \tag{4.27}$$

where:

$$\frac{dH_m}{dl}=L_v\frac{dx}{dl}+\frac{dH_w}{dp}\frac{dp}{dl}+\frac{dL_v}{dp}\frac{dp}{dl}x \tag{4.28}$$

$$\frac{d}{dl}\left(\frac{v_m^2}{2}\right)=\frac{i_s}{A^2}\frac{1}{\rho_m}\frac{d}{dl}\left(\frac{1}{\rho_m}\right) \tag{4.29}$$

$$\frac{dW}{dl}=\frac{\tau_f v_m}{dl}=\frac{\tau_f(v_1+v_2)}{2dl} \tag{4.30}$$

4.4.2.4 Treatment of intermediate parameters

4.4.2.4.1 Radial heat transfer behavior

For steady-state heat transfer from casing to a cement ring, the heat transfer rate is expressed by:

$$dQ=2\pi r_{ci}U(T_s-T_w)dz \tag{4.31}$$

where the heat transfer coefficient is:

$$U = \left[\frac{r_{co}}{r_{ci}h} + \frac{r_{eo}}{\lambda_{cas}}\ln\frac{r_{co}}{r_{ci}} + \frac{r_{eo}}{\lambda_{cem}}\ln\frac{r_h}{r_{co}}\right]^{-1} \tag{4.32}$$

Unsteady heat transfer from a cement ring to a formation is:

$$dQ' = 2\pi\lambda_e \frac{T_w - T_e}{f(t)} dl \tag{4.33}$$

where f is a dimensionless time function of heat transfer, which can be obtained by a modified expression of Ramey's model [30,31]:

$$f(t) = \begin{cases} 1.1281\sqrt{\tau_D}(1 - 0.3\sqrt{\tau_D}) & \tau_D \le 1.5 \\ (0.4063 + 0.5\ \ln\tau_D)(1 + 0.6/\tau_D) & \tau_D > 1.5 \end{cases} \tag{4-34}$$

where τ_D is dimensionless time:

$$\tau_D = \alpha\tau / r_h^2 \tag{4.35}$$

Furthermore, during steam injection, when steam is continuously injected into a reservoir, the heating radius along a wellbore is gradually increases. Therefore, the expression of r_h in Eq. (4.35) should also consider the effect of the heating radius. Based on the Marx-Langenheim model, the heating radius can be expressed by [25]:

$$R_i = \sqrt{\frac{Q_{mvi}h\lambda}{4\pi K_{ob}\Delta T}\left[e^{t_D/\lambda^2} erfc(\sqrt{t_D}/\lambda) + \frac{2}{\sqrt{\pi}}(\sqrt{t_D}/\lambda) - 1\right]} \tag{4.36}$$

For a thick heavy oil reservoir, the overburden heat loss can be neglected, and thus:

$$R_i = \sqrt{\frac{1000Q_{mvi}(x_s L_v + H_w)t}{\pi h M(T_s - T_e)}} \tag{4.37}$$

4.4.2.4.2 Equation of steam flow in reservoirs

During steam injection, the reservoir pressure is correspondingly changed, and an unsteady-state flow equation can be applied to characterize steam flow behavior in a reservoir [32,33]:

$$p_i^2 = p_{ws}^2 + \frac{q_{mv}\mu_s(T_s + 460)}{0.703K_s\Delta l}\left(\ln\frac{R_i^2}{2r_w t_i} - 1.238\right) \tag{4.38}$$

where q_{mv} is the steam injection rate within each wellbore microcontrol element; and K_s is the effective permeability of steam in the reservoir, which is one-quarter of the reservoir permeability.

4.4.2.4.3 Constraints for steam mass flow along wellbores

The total steam mass flow rate in a horizontal wellbore remains unchanged:

$$I_{s0} = \sum_{i=1}^{N} q_{mi} \tag{4.39}$$

For the ith wellbore segment, the steam mass flow rate can be expressed by:

$$I_{si} = I_{s0} - \sum_{j=1}^{i-1} q_{mj} = I_{si-1} + q_{mi} \tag{4.40}$$

4.4.3 Simulation procedure

Based on this model, using the microelement analysis method and iterative method, heat transfer behavior and pressure drops of steam along a perforated horizontal well can be simulated. The length of each microcontrol element is Δl. It is assumed that a pressure drop in a microcontrol element is Δp, and the quality-drop of steam is Δx. For a heel-end well completion mode, the simulation procedure is started from the heel end of a horizontal wellbore. A simulation flowchart is shown in Fig. 4.16 [21].

4.4.4 Case study

4.4.4.1 Laboratory-scale simulation

From this mathematical model and simulation algorithm, the flow behavior of steam along a horizontal wellbore can be simulated. A laboratory-scale simulation is first performed. The basic data are shown in Table 4.1. The simulation results can be compared against experimental data in Section 4.3.2 to confirm the accuracy of this model. The comparison results are shown in Fig. 4.17. Excellent agreement can be found between the simulation results and experimental data. This model can be used to evaluate the performance of different horizontal well configurations [21]effectively.

4.4.4.2 Field-scale simulation

On the other hand, based on data in Table 4.2, the solution of the model is also compared against field-scale simulation results from the commercial simulator, CMG STARS. The well configuration is a heel-end injection mode. Also, a refined grid system around the wellbore is applied in this CMG model. The results are shown in Fig. 4.18. In the field, only segments close to the heel-end section can be heated effectively. As shown in Fig. 4.18, the simulation results show that only the first part of the horizontal interval is heated. Moreover, the pressure and heating radius are in good agreement with CMG's results. Here,

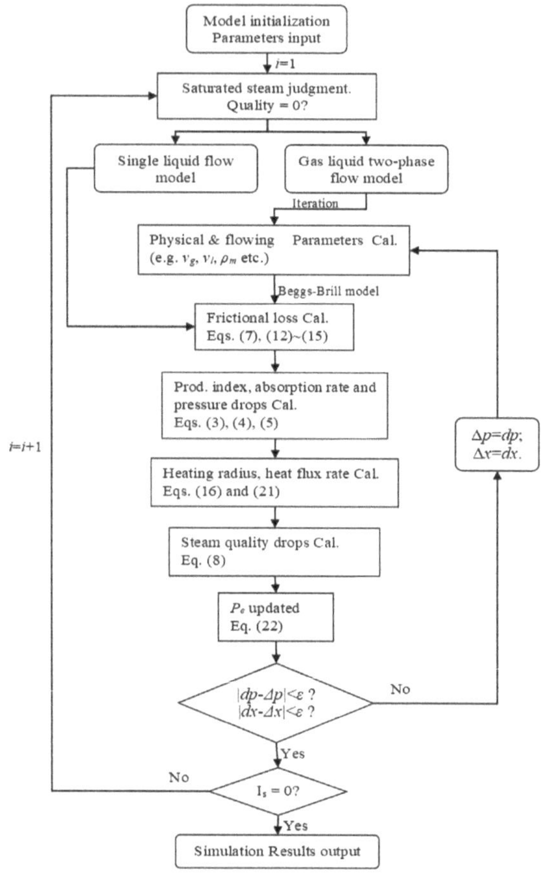

FIGURE 4.16 Calculation (Cal.) flowchart for a mathematical model.

the concept of an effective heat length of a horizontal well is introduced to evaluate the heating effect of steam injection along the horizontal wellbore, as shown in Fig. 4.18B. It refers to the length of a horizontal wellbore that steam can cover. The higher the effective heat length, the better the performance of a steam injection operation. The results also indicate that the physical parameters of steam injected along the horizontal interval are changed nonlinearly. This is because the model considers the influence of a phase change and flow-pattern characteristics of gas–liquid two-phase flow along the horizontal wellbore. The results show that the variable-mass flow of saturated steam along the horizontal interval has three flow patterns: intermittent flow (0–78 m), transition flow (78–161 m), and separation flow (161–187 m) [21].

TABLE 4.1 Laboratory-scale parameters used to validate the model.

Parameter	Value	Parameter	Value
Wellbore length/cm	80	Porosity/%	0.38
Wellbore radius/cm	6	Oil viscosity/mPa s@S.C.	2200
Radius of perforation hole/cm	3	Steam injection rate/mL/min	40
Perforation density/m^{-1}	250	Steam injection time/min	30
Surface temperature/°C	25	Thermal conductivity of wellbore model/W/(m °C)	53
Steam temperature/°C	130	Thermal conductivity of reservoir/W/(m °C)	2.745
Reservoir permeability/10^{-3} μm^2	3000	Thermal diffusion coefficient/(m^2/h)	0.7×10^{-6}

S.C., surface condition.

4.4.4.3 General behavior of different well configurations

Based on data in Table 4.2, a steam injection profile along a horizontal well with different well configurations is simulated. The results are shown in Fig. 4.19. The field scale simulation results are inconsistent with the laboratory-scale results in Fig. 4.17. For the heel-end and toe-end methods, only the reservoir around the heel-end and toe-end sections can be effectively heated and exploited. When the end of tubing ends in the middle of the horizontal well, the middle segment of the horizontal well can be effectively

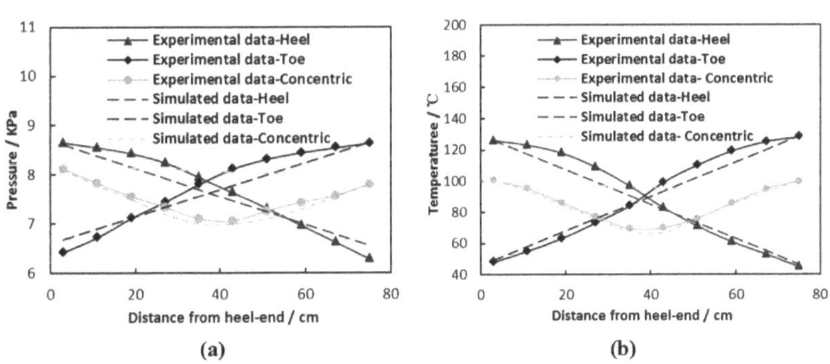

FIGURE 4.17 Comparison of experimental data and simulated results: (A) pressure; (B) temperature.

TABLE 4.2 Field-scale parameters of horizontal wellbore.

Parameter	Value	Parameter		Value
Reservoir depth/m	500	Injection pressure/MPa		5.0
Length of horizontal well/m	300	Steam quality/decimal		0.7
Inner radius of casing/m	0.0885	Injection rate/t/d		144
Outer radius of casing/m	0.1250	Cement radius/m		0.12
Inner radius of tubing/m	0.03800	Oil viscosity @ R.C./cp		5000
Outer radius of tubing/m	0.04445	Reservoir pressure/MPa		4.8
Perforation density/m^{-1}	5	Thermal diffusivity of formation/m^2/h		0.7×10^{-6}
Surface temperature/°C	22	Thermal conductivity /W $(m\ K)^{-1}$	Case and tube	45.7
Temperature gradient/°C/m^{-1}	0.032		Cement ring	0.350
Injection time/d	5		Formation	1.73

R.C., reservoir conditions.

heated and exploited. Comparatively, because the cross-sectional area of the annulus space is smaller than that of the injection tubing, the friction loss in the toe-end method is higher. Therefore, for the toe-end well configuration, under the same condition, a change in steam quality along the horizontal well is more significant, and the effective heating length is also slightly shorter. For field tests, to improve a steam injection profile effectively along a horizontal well, besides the application of a dual-pipe well configuration, we can change

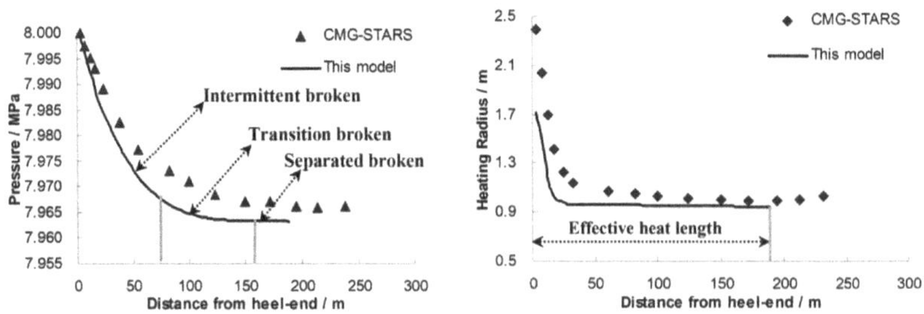

(a) (b)

FIGURE 4.18 Results of model and CMG thermal simulator: (A) pressure; (B) heating radius.

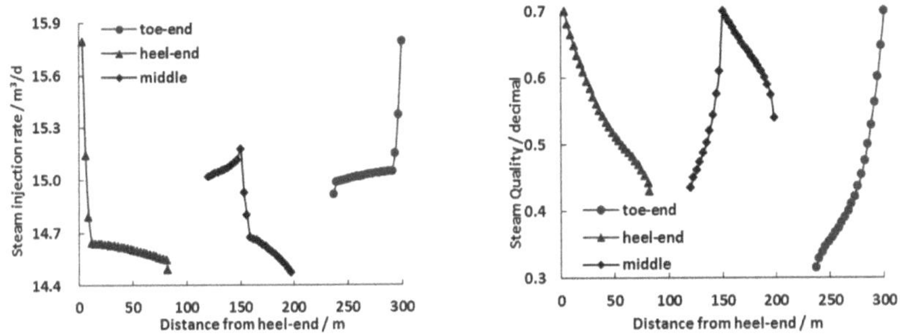

(a) (b)

FIGURE 4.19 Distribution of steam injection rate and steam quality along horizontal well with different well configurations. (A) Average steam injection rate; (B) Steam quality.

a well configuration method cycle-by-cycle. Thus, the whole reservoir along the horizontal well can be effectively heated and exploited.

4.4.5 Sensitivity analysis

Based on the basic parameters in Table 4.2, using a heel-end steam injection method, the effect of steam injection parameters on a steam injection profile along a horizontal wellbore is studied, including the steam injection rate, steam quality, and steam injection pressure (temperature).

First, for a steam injection rate, its effect on a distribution of steam quality along the horizontal wellbore is shown in Fig. 4.20A. As the steam injection rate increases, the effective heat length of the horizontal well gradually increases. It indicates that the wellbore length of the effective contribution to oil production increases. For steam quality at a given location along the wellbore, with a given steam injection rate, steam quality is slightly reduced.

Second, for the effect of the quality of injected steam, the results are shown in Fig. 4.20B. Steam quality dominates the steam enthalpy. The higher the steam quality, the higher the heat energy carried by steam. Higher heat energy always benefits a recovery process. From the simulation results, when the

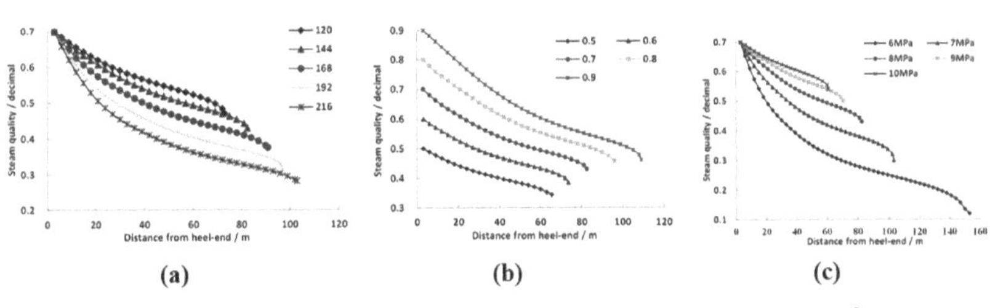

(a) (b) (c)

FIGURE 4.20 Results of sensitivity analysis: (A) effect of steam injection rate (m^3/d); (B) effect of steam quality; (C) effect of steam pressure.

steam quality increases, the effective heat length of the horizontal well increases and the recovery performance can be significantly improved.

Third, for the effect of steam injection pressure, the results are shown in Fig. 4.20C. Steam injection pressure mainly affects the steam injection rate for each wellbore segment. As the steam injection pressure increases, the pressure difference between steam pressure and reservoir pressure increases, and thus the steam injection rate for the corresponding wellbore segment is increased. The steam injection pressure is below the formation fracture pressure. Fig. 4.20C shows that for a given steam injection rate, as the steam injection pressure increases, the effective heat length of the horizontal well is reduced. Simultaneously, as the steam injection pressure increases, the steam quality is also increased.

Furthermore, the sensitivity of steam injection time is also studied. Results indicate that as the steam injection time increases, the total steam injection volume is increased, and thus both the heating radius and the effective heat length of the horizontal well are increased. However, considering the steam injection capability of a reservoir, when the steam injection time increases, a reducing tendency of the steam outflow rate for each horizontal wellbore segment can be observed.

4.5 Mathematical models for steam–noncondensable gas co-injection process

Based on these models for pure steam injection, in this section, heat and mass transfer behaviors in a steam-NCG co-injection process are studied. As discussed in Chapter 3, compared with a pure steam injection process, a hybrid steam-NCG process can significantly improve the thermodynamic properties of hot fluids along a wellbore. Similarly, because of the high compressibility and expansibility of NCG, the co-injection of steam and NCG can effectively reduce oil viscosity, increase the heating radius, recover the formation energy, and enhance heavy oil recovery [34,35]. For the fluid flow behavior of a steam-NCG mixture along a horizontal wellbore, the complicated mass transfer and phase transition behaviors can change as in a pure steam injection process. Therefore, in this section, considering the effect of NCG, the coupling behavior of heat and mas transfer between a wellbore and a reservoir in steam NCG co-injection is investigated. A semianalytical mathematical model will be developed after considering fluid outflow, wellbore pressure drops, and a fluid thermal effect [36].

4.5.1 Assumptions

Assumptions for the mathematical model are:

(1) During co-injection, the physical properties of a steam-NCG mixture at the heel-end section of a horizontal wellbore remain unchanged;

(2) Variations in reservoir properties (e.g., porosity and permeability) caused by pressure and temperature are not considered;
(3) Only a radial heat transfer process is considered;
(4) Heat transfer from a horizontal wellbore to a cement ring is a steady-state process, and heat transfer from a cement ring to a formation is an unsteady state;
(5) The molar composition of hot fluids along the wellbore keeps the same along the wellbore.

4.5.2 Model development

4.5.2.1 The model for pressure drops along a wellbore

Similar to the mathematical models in Section 4.4.2, before the model development, a horizontal wellbore is first discretized, as shown in Fig. 4.15.

For fluid flow along a horizontal wellbore, pressure drops include gravity loss, kinetic energy loss, and friction loss. Therefore, pressure drops can be expressed as:

$$\frac{dp}{dz} = \rho_m g \sin\theta + \rho_m v_m dv - \tau_f \tag{4.41}$$

$$\tau_f = \frac{f_{tp}\rho_m v_m^2}{2d} \tag{4.42}$$

where f_{tp} is the frictional resistance coefficient whose specific formula can be found in Eq. (4.21).

4.5.2.2 Model for fluid outflow profile

During steam NCG co-injection, a fluid mixture can enter a reservoir through perforations. A distribution of the fluid outflow rate along a wellbore can be expressed by:

$$q_{mi} = J_i I_i \Delta p = J_i I_i (p_i - p_e) \tag{4.43}$$

where q_{mi} is the fluid outflow rate in each wellbore microcontrol element; p_i is the injection pressure of the steam-NCG mixture; p_e is the reservoir pressure; and the expressions of a liquid production index and a fluid injection index are shown in Eqs. (4.14) and (4.15), respectively.

4.5.2.3 Model for steam quality

Compared with pure steam injection, the method for calculating steam quality in steam NCG co-injection is more complicated. According to the theory of phase equilibrium, the molar fraction of steam in a mixture can be expressed as:

$$C'_{H_2O} = \frac{p_s(T)}{p} \tag{4.44}$$

From the mass conservation equation,

$$C'_{H_2O} = \frac{\frac{G_{H_2O}}{M_{H_2O}}}{\frac{G_{H_2O}}{M_{H_2O}} + \frac{G_{CO_2}}{M_{CO_2}} + \frac{G_{N_2}}{M_{N_2}}} = \frac{\frac{G_{H_2O}}{M_{H_2O}}}{\frac{G_{H_2O}}{M_{H_2O}} + C_1} \tag{4.45}$$

$$C_1 = \frac{G_{CO_2}}{M_{CO_2}} + \frac{G_{N_2}}{M_{N_2}} \tag{4.46}$$

For saturated steam,

$$C'_{H_2O} = \frac{p_s(T)}{p} = \frac{\frac{xG_{H_2O}}{M_{H_2O}}}{\frac{xG_{H_2O}}{M_{H_2O}} + C_1} \tag{4.47}$$

Then, taking a derivative with respect to the well depth, z, a formula for the steam quality can be obtained:

$$\frac{dx}{dz} = \frac{C_1 M_{H_2O}}{G_{H_2O}} \frac{1}{(p - p_s)^2} \left(p \frac{dp_s}{dT} \frac{dT}{dz} - p_s \frac{dp}{dz} \right) \tag{4.48}$$

4.5.2.4 Energy conservation equation

During co-injection, energy losses along a horizontal wellbore include energy loss caused by fluid outflow, heat energy loss caused by heat transfer behavior, and that by the friction loss. Then, after considering the three types of energy loss along a wellbore, the energy conservation equation can be developed:

$$\frac{dQ}{dl} + \frac{dW}{dl} = -\frac{d}{dl}\left[i_s \left(H_m + \frac{v_m^2}{2} \right) \right] - q_{mi} \left(H_m + \frac{v_r^2}{2} \right) \tag{4.49}$$

where H_m is the total enthalpy of a fluid mixture, v_m is the average fluid velocity along the horizontal wellbore, and v_r is the fluid velocity entering the reservoir through perforations:

$$H_m = G_{CO_2} H_{CO_2}(T) + G_{N_2} H_{N_2}(T) + xG_{H_2O} H_S(T) + (1 - x) G_{H_2O} H_W(T) \tag{4.50}$$

$$v_m = \frac{v_1 + v_2}{2}, \; v_r = \frac{q_{mi}}{A_r \rho_m} \tag{4.51}$$

$$\frac{d}{dl}\left(\frac{v_m^2}{2}\right) = \frac{G}{A^2} \frac{1}{\rho_m} \frac{d}{dl}\left(\frac{1}{\rho_m}\right) \tag{4.52}$$

$$\frac{dW}{dl} = \frac{\tau_f v_m}{dl} = \frac{\tau_f (v_1 + v_2)}{2dl} \tag{4.53}$$

Therefore, we obtain:

$$\frac{dH_m}{dl} = G_{CO_2}\frac{dH_{CO_2}}{dl} + G_{N_2}\frac{dH_{N_2}}{dl} + \left[xG_{H_2O}\frac{dH_S}{dT} + (1-x)G_{H_2O}\frac{dH_W}{dT}\right]\frac{dT}{dl} + L_vG_{H_2O}\frac{dx}{dl} \tag{4.54}$$

Then, combined with Eq. (4.49), we have:

$$C_2\frac{dT}{dl} - L_vC_1\frac{M_{H_2O}}{(p-p_s)^2}p_s\frac{dp}{dl} - \frac{G}{A^2}\frac{1}{\rho_m}\frac{d}{dl}\left(\frac{1}{\rho_m}\right) - q_{mi}\left(H_m + \frac{v_r^2}{2}\right) - \frac{dQ}{dl} - \frac{dW}{dl} = 0 \tag{4.55}$$

where:

$$C_2 = G_{CO_2}\frac{dH_{CO_2}}{dT} + G_{N_2}\frac{dH_{N_2}}{dT} + xG_{H_2O}\frac{dH_S}{dT} + (1-x)G_{H_2O}\frac{dH_W}{dT} + L_vC_1\frac{M_{H_2O}}{(p-p_s)^2}p\frac{dp_s}{dT} \tag{4.56}$$

To calculate radial heat transfer behavior in steam-NCG co-injection, the mathematical models in Section 4.4.2 are still available. Similar to pure steam injection, as steam-NCG co-injection continues, the heating radius along a wellbore is increased. However, compared with conventional saturated steam, the enthalpy of a steam-NCG mixture is higher, and thus the heating radius is correspondingly higher. For the thermodynamic properties of a steam-NCG mixture, the models in Section 3.6.4 still hold.

4.5.3 Simulation procedure

Similar to the simulation procedure for a steam injection process in Section 4.4.3, for steam-NCG co-injection, the methods of microelement analysis and iteration can also be applied to simulate heat transfer and pressure drops of a fluid mixture along a perforated horizontal wellbore. The length of each microcontrol element is Δl. It is assumed that the pressure drop in a micro-control element is Δp and the quality drop of steam is Δx. A simulation flowchart is shown in Fig. 4.21 [36].

4.5.4 Case study

Based on this mathematical model and simulation procedure, the heat and mass transfer behaviors of a steam-NCG mixture along a horizontal wellbore can be investigated. The basic data for calculations are shown in Table 4.2. The fluid injection temperature is 250°C, which indicates that the steam component

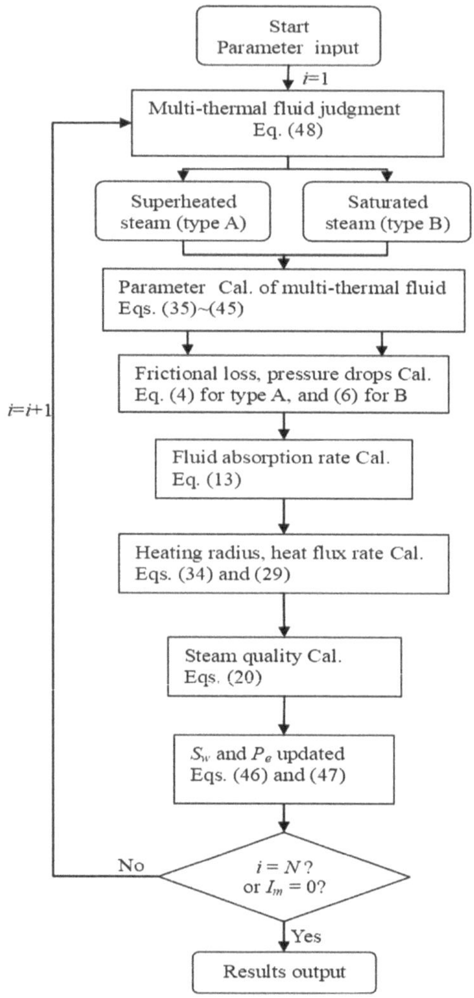

FIGURE 4.21 Simulation flowchart for a steam—noncondensable gas co-injection process. *Cal.*, calibration.

in this fluid mixture is superheated. The molar fractions of CO_2, N_2, and steam (H_2O) are 0.49:0.09:0.42.

To validate the model, its simulation results are compared against results from a commercial reservoir simulator, CMG-STARS. For the model in CMG-STARS, a hybrid gridding method is applied to improve the accuracy of calculations. The results are shown in Fig. 4.22. The simulation results of pressure and temperature from this model are in good agreement with the results of CMG. For a heel-end steam injection process, the heating

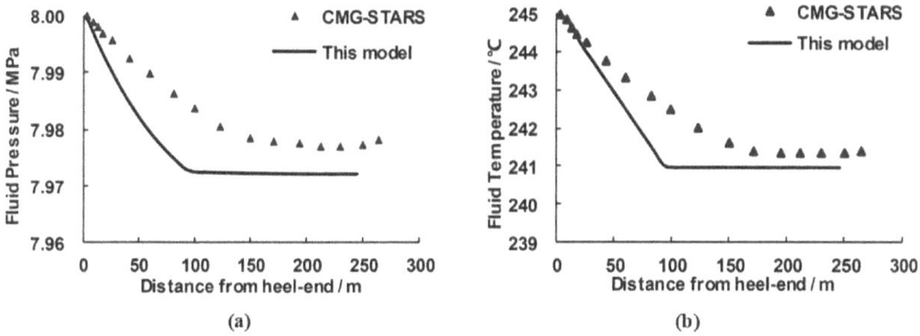

FIGURE 4.22 Simulation results versus CMG-STARS results. (A) Distribution of fluid pressure; (B) Distribution of fluid temperature.

performance in the heel-end reservoir section is better than that in the toe-end reservoir section.

4.5.5 Sensitivity analysis

In this section, from the basic parameters in Table 4.2, the effects of different sensitive factors on heat and mass transfer behaviors along a horizontal wellbore are discussed, including fluid composition, fluid pressure, fluid temperature, fluid injection rate, and injection time. The effect of the fluid injection mode is also simulated.

First, the fluid composition of a steam-NCG mixture can significantly affect the enthalpy of hot fluids. The simulation results are shown in Fig. 4.23A. As the molar fraction of the steam component increases, the

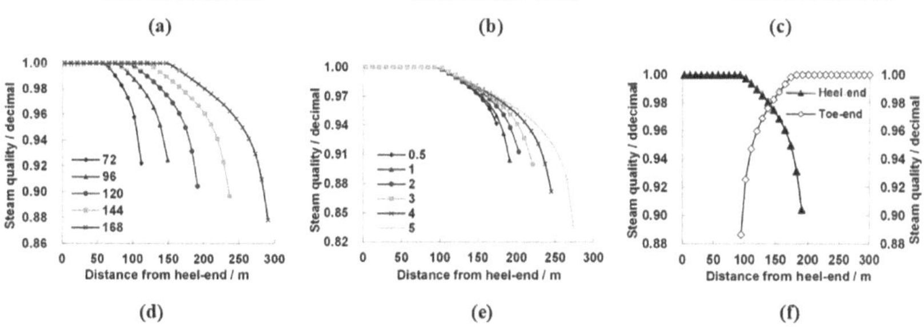

FIGURE 4.23 Results of sensitivity analysis: (A) fluid molar composition; (B) fluid injection pressure (MPa); (C) fluid temperature (°C); (D) fluid injection rate (m^3/d); (E) fluid injection time (D); (F) fluid injection mode.

effective heat length of the horizontal well is reduced because of an increase in the steam outflow rate. When the molar fraction of steam is higher than 50%, a phase change of the steam component (from superheated steam to saturated steam) can be observed. With the parameters of temperature and pressure unchanged, the molar fraction of the steam component in the fluid mixture can have an important effect on the partial pressure of steam. The higher the molar fraction of the steam component, the higher the partial pressure of steam. A higher partial pressure indicates a reduction in the superheated degree of steam. Compared with superheated steam, the enthalpy of saturated steam is lower. Therefore, the fluid outflow rate is higher. Furthermore, the effective heat length of the horizontal well is reduced.

Second, the fluid pressure can alter the actual fluid outflow rate and the phase state of the steam component. As the fluid pressure increases, the fluid outflow rate increases. Simultaneously, an increase in fluid pressure increases the partial pressure of the steam component, and thus a transformation from superheated to saturated steam can be found for the steam component in the fluid mixture. The simulation results under different fluid pressures are shown in Fig. 4.23B. As the fluid pressure increases, both the steam quality and the effective heat length are reduced. This behavior is caused by an increase in the fluid outflow rate. Once the fluid pressure is above 8.5MPa, a phase state change can be observed.

Third, similar to the effect of fluid pressure, a variation in fluid temperature affects the phase state of the steam component. Once the fluid temperature is higher than the saturated temperature of steam under the partial pressure condition, a phase state change from saturated steam to superheated steam can be observed. Then, as the temperature continuously increases, the superheated degree increases and the enthalpy also increases. The simulation results under different fluid temperatures are shown in Fig. 4.23C. As the fluid temperature increases, the effective heat length increases. Once the fluid temperature is above 230°C, a phase state change can be observed.

Fourth, the fluid injection rate is related to the capacity of a steam generator. A higher fluid injection rate indicates a higher volume of heat energy injected. It can benefit a thermal recovery process. The simulation results under different fluid injection rates are shown in Fig. 4.23D. As the fluid injection rate increases, the steam quality along the wellbore increases and the effective heat length of the horizontal well also increases. Under a high fluid injection rate, the friction loss can be increased. Thus, the fluid pressure along the wellbore is reduced and the fluid outflow rate is correspondingly reduced.

Now, with the other parameters unchanged, an increase in the fluid injection time can improve the heating performance of hot fluids injected along the horizontal wellbore. The simulation results for different fluid injection times are shown in Fig. 4.23E. As the fluid injection time increases, both the steam quality and the effective heat length increase. The effective heat length after fluid injection for 5 days is increased by about 83 m compared with that in the injection process, which lasts for 1 day.

Finally, the effect of a fluid injection mode on steam conformance along the wellbore is also simulated. The results are shown in Fig. 4.23F. The simulation results are similar to those in pure steam injection (Fig. 4.19). For the heel-end injection mode, the performance of the heel-end reservoir section is better than that of the toe-end injection mode. Comparatively, the heating effect of the toe-end injection mode can perform better than that of the heel-end injection mode. This is caused by the difference in fluid flow behavior in the heel-end and toe-end modes. For the heel-end injection mode, fluid flows along the wellbore from the heel end to the toe end, and fluid flow is performed in the tubing. However, for the toe-end injection mode, fluid flow is performed in the annulus space between tubing and casing. Because of the size difference of the tubing and annulus space, pressure drops along the horizontal well in the heel-end and toe-end injection modes are different. Therefore, the effective heat lengths of the two different injection modes are different. On the other hand, considering the effect of a steam injection mode on steam conformance along the wellbore, cyclically altering the steam injection modes can benefit recovery for heavy oil reservoirs.

4.6 Methods to improve steam conformance along wellbores

Based on these discussions, for thermal recovery using horizontal wells, steam conformance along a horizontal wellbore is an important issue for the effective development of heavy oil reservoirs. In field operations, considering heterogeneous distributions of reservoir and fluid properties along the horizontal wellbore, nonuniform distribution of a steam injection profile can be further enhanced. How to improve steam conformance effectively along a wellbore is always the top concern for EOR processes for heavy oil reservoirs. In this section, two methods are introduced: the application of a novel wellbore configuration and a hybrid fluid injection process.

4.6.1 Novel wellbore configuration

A separate injection and production scheme is a commonly used method for improving the steam conformance of a horizontal well in onshore heavy oil reservoirs. A separate injection and production scheme refers to one whose wellbore is divided into a couple of sections. During production, each wellbore section can be controlled separately. Especially for heterogeneous heavy oil reservoirs, this process can significantly improve the nonuniform distribution of a steam injection profile along a horizontal wellbore and increase the effective heat length of the horizontal well. It includes two types of operation procedures: a uniform steam injection system and a dual-pipe steam injection system. A schematic of a uniform steam injection system is shown in

Fig. 4.24A. In this system, steam injection allocation devices or flow control devices are set up along a wellbore by a given distance. During an operation, a more uniform distribution of a steam injection profile can be achieved by the performance of steam injection allocation devices [37,38]. This process has been applied in the Shengli oil field, Sinopec. For field operation tests in the Binnan oil production plant in Shengli, the application of the devices increased the oil production rate by about 12 t/d.

Fig. 4.24B gives the temperature distribution along a wellbore before and after the installation of steam injection allocation devices. Their locations have been marked in this curve. After application of the devices, an obvious improvement can be observed in the distribution of a steam injection profile along the wellbore. The temperature in the previous low temperature wellbore segment with a low oil contribution has significantly increased. For the previous high-temperature wellbore segment, after the devices were installed, the temperature was reduced.

A dual-pipe steam injection system is another type of a novel wellbore configuration in onshore heavy oil reservoirs, including concentric dual pipes and parallel dual pipes (Fig. 3.2). A concentric dual-pipe wellbore configuration can be applied in deep heavy oil reservoirs (e.g., Shengli oil field, Sinopec and Liaohe oil fields, CNPC), and a parallel dual-pipe wellbore configuration is employed in shallow heavy oil reservoirs (e.g., Xinjiang oil field, CNPC). Combining the experimental results in Fig. 4.17, a bimodal temperature distribution is the one of the most notable characteristics in a dual-pipe steam injection process. It can dramatically improve recovery for heavy oil reservoirs.

4.6.2 Hybrid fluid injection process

A hybrid fluid injection process is another important technique to improve steam conformance along a wellbore, especially for steam-NCG co-injection. Based on the discussion in Sections 4.4 and 4.5, for hybrid fluid injection, the

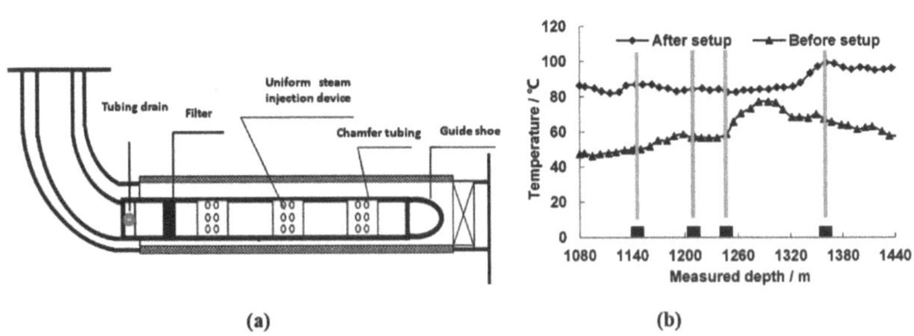

FIGURE 4.24 Uniform steam injection system in horizontal well: (A) schematic of uniform steam injection system; (B) temperature distribution before and after the setup of a uniform steam injection system.

heat loss rate is lower and the quality of the steam component is higher. This is caused by the improvement in physical properties of injected hot fluids. Compared with conventional saturated steam, the density and viscosity of a steam-NCG mixture are lower, and, therefore, friction loss in the steam-NCG co-injection process is reduced. Simultaneously, the expansion ability of a steam-NCG mixture is higher, and it can benefit recovery. On the other hand, considering the low thermal conductivity of NCG (N_2 and CO_2), co-injection also reduces the heat transfer rate between fluids and a wellbore/formation. Fig. 4.25 compares the difference between pure steam injection and steam-NCG co-injection. Under the condition of other unchanged parameters, the effective heat length in steam-NCG co-injection can be increased by about 28 m compared with pure steam injection. Moreover, the steam quality and fluid pressure along the horizontal wellbore are higher in steam-NCG co-injection.

In this chapter, the heat and mass transfer behaviors of a wellbore and a reservoir are experimentally and numerically studied. Variable mass flow along the wellbore can tremendously affect the recovery of a horizontal well in heavy oil reservoirs. The more uniform distribution of a steam injection profile can significantly benefit oil production. Accurate estimation of the distribution of a steam injection profile along a wellbore is related to the physical properties of heavy oil in porous media and also to a wellbore configuration. For conventional pure steam injection, heterogeneous steam conformance along a horizontal wellbore is an urgent issue for horizontal well-based recovery in heavy oil reservoirs. To a certain extent, the methods of a hybrid steam NCG and a novel wellbore configuration can address this issue. However, when hybrid fluids are injected into a reservoir, a series of pressure–volume–temperature (PVT) equilibrium problems arise. In the next chapter, the PVT behavior of different hybrid fluids and heavy oil will be addressed.

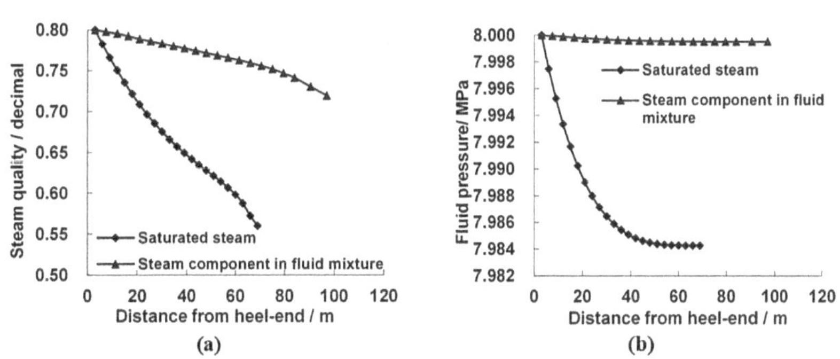

FIGURE 4.25 Difference between pure saturated steam injection and hybrid steam NCG co-injection: (A) steam quality; (B) fluid pressure.

References

[1] Chen B, Lu X. Development of heavy and high-viscosity crude oils. Beijing: Petroleum Industry Press; 1990.
[2] Henaut I, Argillier J-F, Pierre C, Moan M. In: Thermal flow properties of heavy oils, paper OTC 15278 presented at the offshore technology conference, Houston, Texas, USA. 5–8 May 2003; 2003.
[3] Steinborn R, Flock DL. The rheology of heavy crude oils and their emulations. J Can Pet Technol 1983;22:37–52.
[4] Alshmakhy A, Maini B. In: Foaminess and viscosity effects in heavy oil flow, paper CSUG/SPE 145231 presented at the Canadian unconventional resources conference, Calgary, Alberta, Canada, 15–17 November 2011; 2011.
[5] Argillier J-F, Coustet C, Henaut I. In: Heavy oil rheology as a function of asphaltene and resin content and temperature, SPE paper79496 presented at the SPE international thermal operations and heavy oil symposium and international horizontal well technology conference, Calgary, Alberta, Canada, 4–7 November 2002; 2002.
[6] Schramm LL, Kwak JCT. The rheological properties of an Athabasca bitumen and some bituminous mixtures and dispersions. J Can Pet Technol 1988;27:1–11.
[7] Ovalles C, Estrella R, Segerstrom J. In: Improvement of flow properties of heavy oils using asphaltene modifiers, paper SPE 146775 presented at the SPE annual technical conference and exhibition, Denver, Colorado, USA, 31 Octorber-2 November 2011; 2011.
[8] Wu Y-S, Pruess K, Witherspoon PA. Flow and displacement of Bingham non-Newtonian fluids in porous media. SPE Reservoir Eng 1992;7:369–76.
[9] Wang S, Huang Y, Civan F. Experimental and theoretical investigation of the Zaoyuan field heavy oil flow through porous media. J Petrol Sci Eng 2006;50:83–101.
[10] Chen M, Rossen W, Yortsos YC. The flow and displacement in porous media of fluids with yield stress. Chem Eng Sci 2005;60:4183–202.
[11] Rojas MA, Castagna J, Krishnamoorti R, Han D-h, Tutuncu A. In: Shear thinning behavior of heavy oil samples: laboratory measurements and modeling, paper 2008-1714 presented at the SEG Annual Meeting, Las Vegas, Nevada, USA, 9–14 November 2008; 2008.
[12] Christos DT. Correlation of the two-phase flow coefficients of porous media with the rheology of shear-thinning fluids. J Non-newtonian Fluid Mech 2004;117:1–23.
[13] Govier GW, Fogurasi M. The interpretation of data on the rheological behavior of thixotropic crude oils. J Can Pet Technol 1972;11:42–53.
[14] Owayed JF, Tiab D. Transient pressure behavior of Bingham non-Newtonian fluids for horizontal wells. J Petrol Sci Eng 2008;61:21–32.
[15] Yang J, Li X, Chen Z, Tian J, Huang L, Liu X. A productivity prediction model for cyclic steam stimulation in consideration of non-Newtonian characteristics of heavy oil. Acta Pet Sin 2017;38(1):84–90.
[16] Dong X, Liu H, Wang Q, Pang Z, Wang C. Non-Newtonian flow characterization of heavy crude oil in porous media. J Pet Explor Prod Technol 2013;3(1):43–53.
[17] Liu H, Jiang H, Li J. Theory and method of reservoir engineering. Qingdao: China University of Petroleum Press; 2019.
[18] Joshi SD. Augmentation of well productivity with slant and horizontal wells. J Petrol Technol 1988;40:729–39.
[19] Gao G, Jalali Y. Prediction of temperature propagation along a horizontal well during injection period. SPE Reservoir Eval Eng 2008;11:131–40.

[20] Gui P, Cunha JC, Cunha LB. In: A coupled model to simulate the fluid flow in the reservoir and horizontal wellbore. Paper 2006-124 presented at the Petroleum society's 7th Canadian international petroleum conference, Calgary, Alberta, Canada, June 13-15, 2006; 2006.

[21] Dong X, Liu H, Lu N, Wu K, Wang K, Chen Z. Steam conformance along horizontal well with different well configurations of single tubing: an experimental and numerical investigation. SPE Production & Operations; 2020. SPE-195799-PA. (in press).

[22] Dong X, Liu H, Wang C, Chen Z. Experimental investigation on the steam injection profile along horizontal wellbore. Energy Rep 2020;6:264−71.

[23] Dong X, Liu H, Chen Z. Transient fluid flow and heat transfer characteristics during co-injection of steam and non-condensable gases in horizontal wells. J China Univ Pet 2016;40(2):105−14 [in Chinese].

[24] Chen DM, Zhou JY, Li ZP, Gu BL, Dong ZG. A steam injection model for horizontal well in heavy oil reservoir with thermal recovery. J Southwest Petrol Univ 2007;29:102−6.

[25] Liu HQ. Principle and design of thermal oil recovery processes. Petroleum Industry Press; 2013.

[26] Asheim H, Kolnes J, Oudeman PA. Flow resistance correlation for completed wellbore. J Petrol Sci Eng 1992;8:97−104.

[27] Bo QL, Wu N. Experimental research on flow pattern transition of gas-liquid two-phase variable flow in horizontal wellbore. Acta Pet Sin 2004;25:83−5.

[28] Beggs HD, Brill JP. A study of two-phase flow in inclined pipes. J Petrol Technol 1973;25(5):607−17.

[29] Su Z, Gudmundsson JS. In: Pressure drop in perforated pipes: experiments and analysis. Paper SPE 28800 presented at the SPE Asia Pacific oil & gas conference, Melbourne, Australia, 7−10 November, 1994; 1994.

[30] a Su Z, Gudmundsson JS. Perforation inflow reduces frictional pressure loss in horizontal wellbores. J Petrol Sci Eng 1998;19:223−32.b Hasan AR, Kabir CS. Wellbore heat-transfer modeling and applications. J Petrol Sci Eng 2012;86−87:127−36.

[31] Hasan AR, Kabir CS. In: Heat transfer during two-phase flow in wellbores: part I-formation temperature. Paper SPE 22866 presented at the SPE annual technical conference and exhibition, Dallas, Texas, USA, 6−9 October, 1991; 1991.

[32] Jones J. Why cyclic steam predictive model get no respect. SPE Reservoir Eng 1992;7:67−74.

[33] Liu HQ, Zhang HL, Deng YZ, Liu J. Effects of injection pipe shoe setting depth on steam injection performance. J China Univ Petrol 2007;31:64−7.

[34] Dong X, Liu H, Chen Z, et al. Enhanced oil recovery techniques for heavy oil and oilsands reservoirs after steam injection. Appl Energy 2019;239:1190−211.

[35] Behzad R, Peyman P, Alireza F, Mahmood RY, Kamran H, Maryam K, Mohammad M, Ahmad D. A new approach to characterize the performance of heavy oil recovery due to various gas injections. Int J Multiphas Flow 2018;99:273−83.

[36] Dong X, Liu H, Zhang Z, Wang C. The flow and heat transfer characteristics of multi-thermal fluid in horizontal wellbore coupled with flow in heavy oil reservoirs. J Petrol Sci Eng 2014;122:56−68.

[37] Gai P, Du Y, Lu G, et al. In: Uniform steam injection technology used in thermal horizontal wells. Paper SPE 130893 presented at the CPS/SPE international oil and gas conference and exhibition, Bejing, China, 8-10 June 2010; 2010.

[38] Boone TJ, Youck DG, Sun S. In: Targeted steam injection using horizontal wells with limited entry perforations. Paper SPE 50429 presented at the SPE international conference on horizontal well technology, Calgary, Alberta, Canada, 1−4 November 1998; 1998.

Chapter 5

Fluid phase behavior of heavy oil–multicomponent and multiphase fluid mixtures

5.1 Introduction

For hybrid enhanced oil recovery (EOR) processes, when hybrid fluids are injected into reservoirs, one of the most important characteristics is the interaction between heavy oil and hybrid fluids. Through a series of physical and chemical reactions, the original phase equilibrium in heavy oil reservoirs will be damaged. For different operation processes in different heavy oil reservoirs, interactions between heavy oil and multicomponent and multiphase fluids (MMFs) are also different.

First, for a hybrid thermal noncondensable gas (NCG) process, as discussed in Chapter 1, the different operation methods usually include hybrid steam-NCG stimulation and hybrid steam-NCG flooding. Moreover, a hybrid process can be classified by different types of NCGs, including CO_2, N_2, flue gas, air, and methane. Compared with pure steam injection, improvement in heavy oil properties by steam and NCG co-injection is the dominant EOR mechanism of a hybrid process. Through interactions between heavy oil and NCGs, the density, viscosity, and swelling of fluids can be altered [1,2]. On the other hand, considering the high solubility of CO_2 in heavy oil, a significant improvement in heavy oil properties can be observed in CO_2-containing hybrid EOR processes (e.g., hybrid steam–CO_2, hybrid steam–flue gas, and hybrid steam–air) [3–5]. Furthermore, for a hybrid steam–air process, low-temperature oxidation reaction behavior between heavy oil and air can upgrade heavy oil properties to improve oil mobility and enhance recovery [5–7].

Second, for a hybrid thermal-solvent process, compared with conventional pure steam injection, besides the mechanisms of oil viscosity reduction by solvent solubility and the reactivation of residual oil, the effects of solvent extraction and asphaltene precipitation are two other important mechanisms [8,9], and the behavior of in situ oil upgrading can have an important role. For

Hybrid Enhanced Oil Recovery Processes for Heavy Oil Reservoirs.
https://doi.org/10.1016/B978-0-12-823954-4.00001-1

a hybrid thermal NCG process, the soluble behavior of NCGs in heavy oil usually dominates this process. However, for a hybrid thermal solvent process, the behavior of solvent in heavy crude oil is dominated by the phase equilibrium. It is not a simplified behavior of condensate mixing [9,10]. To describe the phase behavior of a heavy oil–steam–solvent system accurately, the liquid phase in the entire system is represented by different liquids, and the solvent concentration in such different liquids varies [11,12]. Compared with pure steam injection, when the performance of steam–solvent co-injection is considered, a condensate zone of a displacement front in a hybrid thermal solvent process can be expanded. Simultaneously, the condensate temperatures at different locations of the displacement front are different. An accurate description of the phase behavior of a heavy oil–solvent mixture can benefit an understanding of EOR mechanisms of hybrid thermal solvent processes.

Next, for a hybrid thermochemical process, its purposes are to plug a chief zone or a steam breakthrough path effectively and improve the sweep efficiency. During its operation, emulsification is one of the most significant characteristics. In situ emulsification can be widely observed in a hybrid thermochemical process [13,14]. Compared with a single fluid, the viscosity, density, and flow behavior of emulsion in porous media is more complicated. When water saturation in porous media varies, the viscosity and flow behavior of emulsion can significantly alter. The water percentage in a water–oil emulsion is usually 25–27% [15]. For the behavior of an emulsion in porous media, reservoir temperature, rock wettability, and a fluid flow velocity can have important effects on the degree of stability and emulsion of fluids in porous media [16]. Emulsion can easily form in an oil–wet porous environment than in a water–wet porous environment, and the effect of rock wettability is more significant than temperature. Finally, the recovery performance of a hybrid thermal NCG–chemical process in heavy oil reservoirs is also tested; emulsification can have an important role in improving sweep efficiency in this process [17,18].

The most important EOR mechanism of hybrid recovery in heavy oil reservoirs is an interaction between heavy oil and MMFs. It can dominate the success of the operation. In this chapter, fluid phase behavior in the three different hybrid EOR processes will be discussed.

5.2 Pressure–volume–temperature behavior of heavy oil–noncondensable gas mixture

In this section, the pressure–volume–temperature (PVT) behavior of a heavy oil-NCG mixture is experimentally discussed. Through experimental observations on the gas solubility, oil viscosity, and swelling of a heavy oil-NCG mixture, the effect of NCG on the phase behavior of heavy crude oil can be investigated [19,20].

5.2.1 Experimental method

Fig. 5.1 shows the schematic of an experimental setup. The physical properties of two oil samples are shown in Table 5.1. During experiments, the NCG is first transferred into a sample cell. Then, a certain amount of an oil sample is injected into the sample preparation container. Correspondingly, the NCG is transferred into the sample preparation container. Under a given temperature and pressure condition, this oil–gas mixture in the sample preparation container is fully stirred. After that, it is transferred into the PVT cell to test viscosity and swelling. From the cumulative degassing results, the gas solubility is derived.

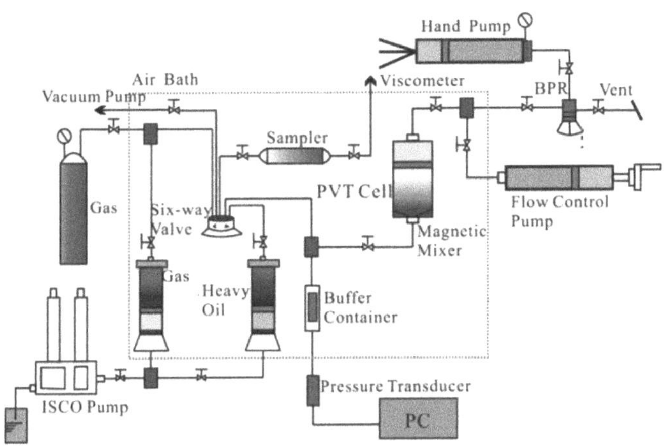

FIGURE 5.1 Pressure–volume–temperature (PVT) experiment apparatus. *BPR*, Back pressure regulator; *ISCO*, Teledyne ISCO pump; *PC*, Personal computer.

TABLE 5.1 Properties of oil samples.

Properties	Oil I	Oil II
API gravity	15.51	14.29
Reservoir temperature (°C)	45	48
Oil viscosity @ SC (cp)	1074	7843
Saturates (%)	52.56	49.68
Aromatics (%)	23.01	23.26
Resins (%)	23.07	26.21
Asphaltene (%)	1.36	0.85
Wax content (%)	—	1.1

API, American Petroleum Institute; *SC*, surface condition.

5.2.2 Experimental results

Fig. 5.2 shows the experimental results of a heavy oil-CO_2 mixture and a heavy oil-flue gas mixture. Fig. 5.2A–C is the test results of the heavy oil-CO_2 system, and Fig. 5.2D–F is the test results of the heavy oil-flue gas (15% CO_2 + 85% N_2) system. For CO_2 and flue gas, as pressure increases, both gas solubility and swelling of the oil-gas system are improved. Comparatively, solubility of flue gas in heavy oil and swelling of the heavy oil-flue gas mixture are smaller than those for the heavy oil-CO_2 mixture. The presence of CO_2 can tremendously increase oil mobility and the swelling capacity of heavy oil. On the other hand, because of the effect of temperature on gas molecular movement, high temperature is usually unhelpful for gas dissolution in heavy crude

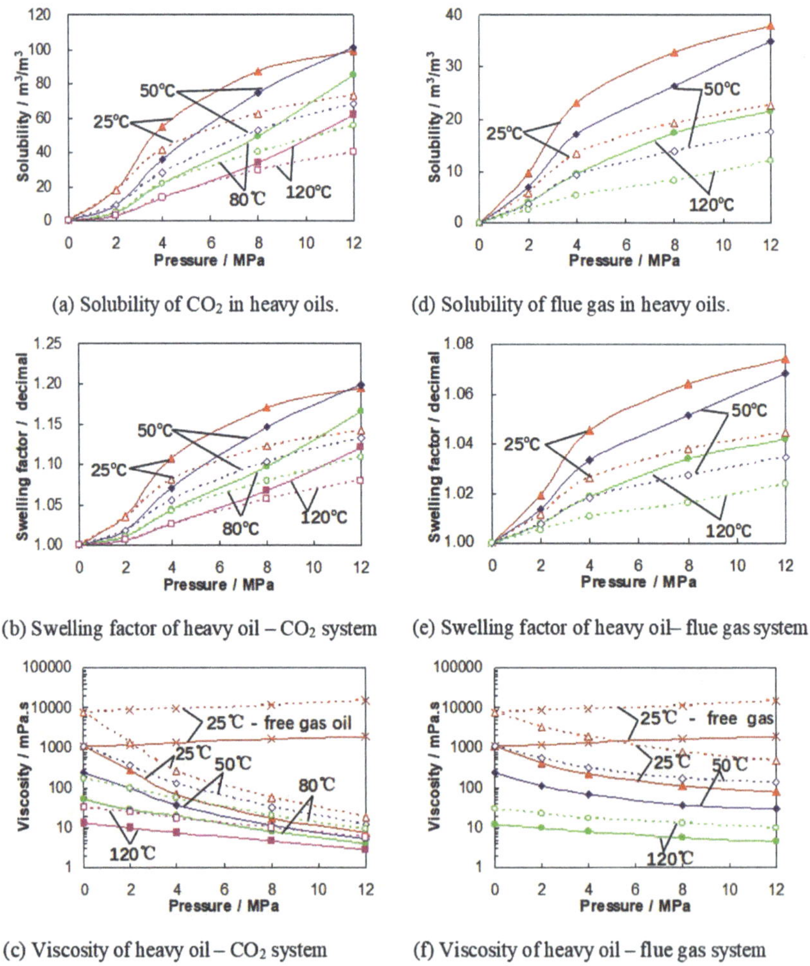

FIGURE 5.2 Pressure–volume–temperature experimental results of heavy oil-CO_2/flue gas system under different temperatures and pressures. *Solid line* refers to results of oil I and *dotted line* refers to results of oil II.

oil. Therefore, the experimental results show that as temperature increases, both gas solubility and swelling of the oil-gas system are reduced. For the results of mixture viscosity, because of the high solubility of CO_2 in heavy oil, an addition of CO_2 to crude oil can significantly reduce oil viscosity. It is an important EOR mechanism of a hybrid steam-CO_2 process. Similarly, considering the effect of pressure on gas solubility in heavy oil, as pressure increases, the viscosity-reducing ratio of heavy crude oil is gradually increased. Furthermore, comparing the results of oil I and oil II, CO_2 solubility and swelling in the oil-gas system in oil I is higher than that in oil II; this is caused by the different oil compositions of the two oil samples, in which oil sample I contains more light components than oil II (Table 5.1).

Comparing the results of CO_2 and flue gas, because of the performance of N_2, the viscosity-reducing ratio of the heavy oil-flue gas system is smaller than that of the heavy oil-CO_2 system. For a given heavy oil sample, an oil-flue gas system has higher viscosity than an oil-CO_2 system.

The dissolution of NCG in heavy oil can increase the volume of the oil phase and swelling, and reduce oil viscosity. It is useful for effective recovery in heavy oil reservoirs. CO_2 gas is highly soluble in heavy oil and tremendously affects the phase behavior of heavy oil. For hybrid thermal NCG, to improve the PVT performance of heavy crude oil effectively, based on this discussion, NCGs with a high CO_2 concentration are highly recommended.

5.2.3 Correlations for pressure–volume–temperature behavior of heavy oil–noncondensable gas mixture

For hybrid thermal NCG processes, one of the most important EOR mechanisms is to alter the PVT behavior of reservoir fluids. Specifically, it includes volume expansion, miscibility, viscosity/density improvement, and reduction of interfacial tension (IFT) . Among many different NCGs, CO_2 can significantly increase swelling and reduce oil viscosity compared with other gases (CH_4 and N_2). Thus, considering the high solubility of CO_2, some correlations have been proposed to predict the PVT behavior of a crude oil-CO_2 mixture.

5.2.3.1 Solubility of CO_2 in heavy oil

The correlation in Eq. (5.1) can be used to estimate the solubility of CO_2 in heavy crude oil; the only required parameter is the specific gravity of heavy oil [21]. It can provide a good prediction for CO_2 solubility for pressure below 20.7 MPa:

$$R_S = \frac{1}{\left[a_1 \gamma^{a_2} T^{a_7} + a_3 T^{a_4} \exp\left(-a_5 p + \frac{a_6}{p}\right)\right]} \tag{5.1}$$

where γ is the specific gravity of heavy oil; T is the temperature, °F, p is pressure, psia; $a_1 = 0.4934 \times 10^{-2}$; $a_2 = 4.0928$; $a_3 = 0.571 \times 10^{-6}$; $a_4 = 1.6428$; $a_5 = 0.6763 \times 10^{-3}$; $a_6 = 781.334$; and $a_7 = -0.2499$.

5.2.3.2 Swelling factor

The swelling factor of a heavy oil-CO_2 mixture is linearly correlated with the solubility of CO_2. Based on experimental matching, a correlation can be obtained to estimate the swelling factor of a heavy oil-CO_2 mixture [21]:

$$F_S = 1 + \frac{0.35 R_S}{1000} \tag{5.2}$$

5.2.3.3 Density

The well-known Tait equation is commonly used to reproduce the densities of heavy oil and a heavy oil-CO_2 mixture [22]:

$$\rho(T, P) = \frac{\rho_0(T, P_0)}{1 - \beta \ln\left(\frac{B + 0.001P}{B + 0.1}\right)} \tag{5.3}$$

where

$$\rho_0(T, P_0) = 784.0044 + 1.7217T - 3.3752 \times 10^{-3} T^2 \tag{5.4}$$

$$\beta = -6.1774 + 0.0213T \tag{5.5}$$

$$B = 3.7614 \times 10^4 - 1.8009 \times 10^7 T^{-1} + 2.0605 \times 10^9 T^{-2} \tag{5.6}$$

where ρ is the density of heavy oil, kg/m^3; P is pressure, kPa; and T is temperature, K.

5.2.3.4 Viscosity of heavy oil–noncondensable gas mixture

Lederer's model can be applied to predict the viscosity of a heavy oil-NCG mixture [23]:

$$\ln\mu_m = \left(1 - \frac{\alpha V_o}{\alpha V_o + V_s}\right) \ln\mu_s + \left(\frac{\alpha V_o}{\alpha V_o + V_s}\right) \ln\mu_o \tag{5.7}$$

where α is an adjustable parameter between 0 and 1, and can be obtained through regression with experimental data; μ_o and μ_s are the viscosities of heavy oil and CO_2; and V_o and V_s are the volume fractions of heavy oil and CO_2.

5.3 Oxidation reaction law of heavy oil–air system

Because of an extensive gas source and low operation costs, a hybrid steam air process in heavy oil reservoirs has attracted much attention. During air-assisted thermal recovery, according to the difference in operation temperature, three different oxidation reactions can be observed: a low-temperature oxidation (LTO) reaction (<350°C), a chemical pyrolysis reaction (350–450°C), and a high-temperature oxidation (HTO) reaction [24]. Among these three different reactions, LTO is the primary chemical reaction in a hybrid steam air process. It can fully consume oxygen gas in air and avoid the danger of mixing between hydrocarbon gases and oxygen, which may cause an explosion. Moreover, the products of the LTO reaction (e.g., CO_2) can benefit recovery. However, oil viscosity after the LTO reaction is usually higher than the initial oil viscosity, and it is unhelpful for recovery [25,26]. Especially for heavy crude oil with a higher asphaltene concentration, the liquid products of an LTO reaction can plug a pore space and reduce oil flowability. In general, a temperature range in steam-based recovery is below 300°C [26]. This indicates that after adding air, an LTO reaction can dominate a hybrid steam air process in heavy oil reservoirs, and the effect of chemical pyrolysis can be neglected. On the other hand, compared with pyrolysis, aquathermolysis of heavy crude oil is an important chemical reaction that can be observed in hybrid steam air recovery [27,28]. For aquathermolysis in heavy oil reservoirs, a series of chemical reactions will be motivated, including acid polymerization, hydrogen desulfurization, and steam-methane reforming [29]. During aquathermolysis, the heavy components of resin and asphaltene in heavy crude oil can transform into light components of saturates and aromatics, and thus oil viscosity is reduced. CO_2 gas produced from aquathermolysis can also reduce oil viscosity.

Considering the complicated interaction between heavy oil and air in hybrid EOR, a set of oxidation decomposition experiments were conducted to explore the LTO behavior of a heavy oil–air system in hybrid steam air recovery [30].

5.3.1 Experimental method

Fig. 5.3 shows an experimental setup. The experimental apparatuses include four sections: a fluid injection system, a high-pressure high-temperature (HPHT) reactor, data acquisition, and a fluid collection system. The HPHT reactor is the main oxidation cell with an inner volume of 100 .mL. The maximum temperature and pressure are 350°C and 50 .MPa, respectively. Heavy oil samples used during the experiments are dead oil from the Henan oil field, Sinopec. The physical properties are shown in Table 5.2. The air is mainly composed of 79.1% N_2 and 21.9% O_2. Distilled water is used to simulate the water component.

The detailed experimental procedure is:

- Connecting experimental apparatuses,
- Gas tightness testing to check for leaks,
- Injecting heavy oil and water into the HPHT reactor,
- Injecting air into the reactor until a predefined pressure condition is achieved,

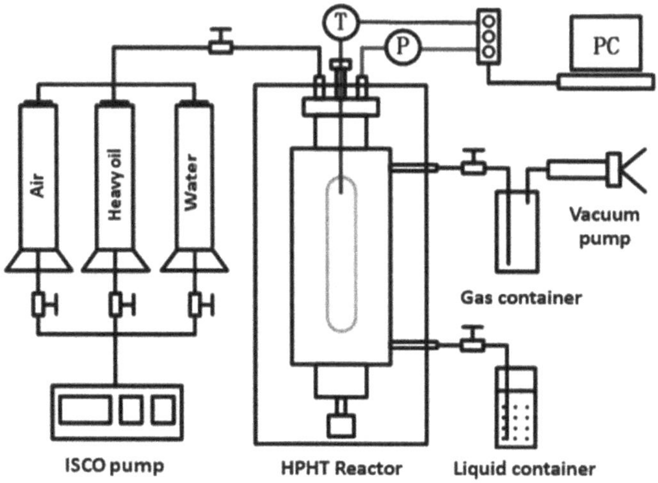

FIGURE 5.3 Experimental setup of oxidative decomposition reaction experiments. *HPHT*, high-pressure, high-temperature; *P*, pressure; *T*, temperature.

TABLE 5.2 Physical properties of heavy oil sample.

Density (25°C) (g/cm^3)	Viscosity (20°C) (mPa·s)	Saturates, aromatic, resin, and asphaltene fraction %			
		Saturates	Aromatics	Resin	Asphaltene
0.9554	14,770	47.52	27.07	19.34	5.74

- ➢ Setting the reactor temperature to the test temperature condition,
- ➢ Maintaining the reactor under the required temperature condition for about 5 days,
- ➢ Acquiring data for the pressure-changing behavior during experiments,
- ➢ Collecting gas and oil samples after the experiments, and
- ➢ Checking the gas composition, oil (saturates, aromatic, resin, and asphaltene [SARA]) fraction, and oil viscosity.

Table 5.3 shows the parameters of LTO tests in the experiments. To test the effect of a high-temperature environment on the oil properties, a pure N_2-based test (no. 12) is also performed to illustrate the aquathermolysis reaction of heavy crude oil. Through experimental observations, the effects of water saturation, temperature, and pressure on the LTO decomposition behavior of heavy oil can be obtained.

TABLE 5.3 Parameters of static oxidative decomposition reaction experiments.

No.	Oil volume, mL	Air, mL	Water saturation, %	Temperature, °C	Pressure, MPa
1	40	60	25	150	2.5
2	40	60	25	200	2.5
3	40	60	25	250	2.5
4	40	60	25	300	2.5
5	40	60	25	250	1.5
6	40	60	25	250	3.5
7	40	60	25	250	4.5
8	40	60	0	250	2.5
9	40	60	50	250	2.5
10	40	60	75	250	2.5
11	40	60	0	150	2.5
12	40	60 (pure N_2)	25	250	2.5

5.3.2 Experimental results

For each test, the system pressure will be monitored in real time for 5 days. After experiments, the gas composition, oil composition, and oil viscosity are each evaluated.

5.3.2.1 Pressure profiles

During the experiments, with high-temperature steam, the heavy crude oil can interact with O_2 in air. The consumption of O_2 will produce CO_2, CO, and other components. The thermal decomposition of heavy oil can also yield some light hydrocarbon gases (e.g., CH_4 and C_2H_6). Especially, under high temperature, an oxidation reaction will dominate phase behavior between heavy oil and air. With a certain amount of O_2 consumed, the system pressure is reduced. Fig. 5.4 presents the pressure profiles of the 12 experiments.

Fig. 5.4A shows the pressure profiles under different operation temperatures. In the initial stage, pressure is rapidly reduced. In a later stage, when the O_2 concentration is reduced, the declining rate is reduced. Comparing the

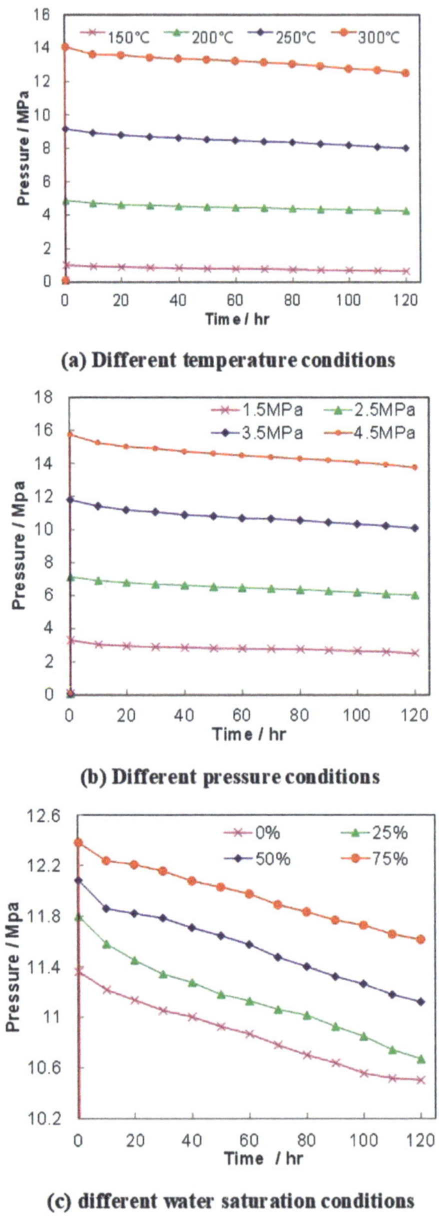

FIGURE 5.4 Pressure profiles under different operation conditions.

results at different temperature conditions, as temperature increases, the final pressure level is increased. This is caused by the expansion of heavy oil and air. Moreover, as the temperature increases, the oxidation decomposition reaction of heavy oil is promoted and the pressure drop is higher.

Fig. 5.4B shows the pressure profiles under different initial operation pressures. The results are similar to experimental results at different temperatures in Fig. 5.4A. As the operation pressure increases, the O_2 partial pressure is increased, and thus the concentrations of the reactant and activated molecules are increased. This benefits the oxidation reaction process, and the pressure drop is higher.

Fig. 5.4C shows the pressure profiles under different water saturations. Compared with the results without water (no. 8), the pressure drop in the experiment at 25% water saturation (no. 3) is higher. However, as water saturation further increases, the drop pressure is reduced. This indicates that the presence of a small amount of water can benefit the oxidation reaction between heavy oil and air.

5.3.2.2 Gas composition

After the experiments, because of the oxidation reaction behavior between heavy oil and air, gases in the reactor have changed significantly compared with the initial gas components. Table 5.4 shows the measured gas composition after the experiments. The fraction of O_2 has reduced significantly.

TABLE 5.4 Measurement results of gas compositions after reactions.

No.	Gas composition O_2, %	CO_2, %	CO, 10^{-6}	CH_4, %	Oxygen consumption ratio, %
1	16.4	1.97	401	2	22.01
2	14.9	2.66	623	3	29.19
3	9.7	3.92	604	5	54.07
4	4.8	5.29	557	7	77.51
5	15.2	1.91	527	3	27.75
6	8.1	4.02	766	6	61.72
7	6.0	4.83	838	7	71.77
8	14.4	3.42	771	2	31.58
9	12.9	2.52	563	4	38.76
10	14.8	2.21	477	4	29.67
11	13.2	3.55	702	2	37.32
12	0	3.53	256	3	–

According to the different experimental conditions, the oxygen consumption ratios are different. The experimental results show that as the temperature increases, the oxygen consumption ratio increases. The fraction of CO has a tendency to increase first and decrease later. It is caused by the conversion from CO to CO_2 at a high temperature. For the effect of pressure, as pressure increases, the strength of an oxidation reaction is enhanced. Thus, the consumption ratio of O_2 is increased and the fractions of CO_2, CO and CH_4 are correspondingly increased. Comparing the experimental results with or without water, the O_2 consumption ratio in the case of water saturation at 25% is the highest. As the water saturation further increases, the oxygen consumption ratio is reduced and the amount of gas production is also reduced. From the experimental results of no. 12, the pyrolysis of heavy crude oil can also yield the gas products of CO_2, CO, and CH_4.

5.3.2.3 Saturates, aromatic, resin, and asphaltene analysis

Table 5.5 shows the SARA fractions of heavy crude oil after oxidation. From the test results, some observations can be made.

First, results at different operation temperatures in the oxidation reaction tests show that the fraction of saturate components is significantly increased

TABLE 5.5 Results of saturates, aromatic, resin, and asphaltene (SARA) fractions of oil samples.

	SARA fraction			
No.	Saturates, %	Aromatics, %	Resin, %	Asphaltene, %
1	52.39	21.76	19.27	6.27
2	53.52	20.15	19.09	6.95
3	55.25	21.44	19.15	3.94
4	56.91	21.63	18.75	2.34
5	54.77	20.31	19.25	5.35
6	55.76	21.69	19.44	2.87
7	55.93	21.92	19.77	2.11
8	53.36	20.89	19.17	6.17
9	54.52	21.16	18.76	5.42
10	54.41	20.57	18.38	6.22
11	53.13	21.01	19.31	6.31
12	49.34	26.35	19.21	4.72

and the aromatics fraction is reduced. As the operation temperature increases, the heavy components (resin and asphaltene) show a tendency to increase first and then decrease. This is because of the breakage of carbon bonds in heavy components and the formation of long-chain saturates, because the temperature is below 200°C. A condensation reaction can further increase the production of asphaltene. On the other hand, when the temperature is higher than 250°C, the effect of an aquathermolysis reaction is enhanced. Therefore, the cracking of heavy components will further increase the fractions of saturates and aromatics.

Second, from the results under different operation pressures, after the oxidation reaction tests, the fractions of heavy components are significantly reduced. As pressure increases, the reactivity of pyrolysis for heavy components is increased. This benefits the generation of light components (saturates and aromatics).

Third, from the results at different water saturations, an increase in the fractions of heavy components can be observed for tests without water (no. 8). This indicates that the oxidation reactions of saturates and aromatics dominate the process and produce a great amount of asphaltenes. However, once the effect of water saturation is considered, a reduction in the fractions of heavy components can be observed. This is caused by the effect of high temperature on the properties of water. It benefits the cracking reaction of heavy crude oil. The results of experiment 12 further indicate that the synergistic effect of LTO and aquathermolysis reactions can reduce the fractions of heavy components.

5.3.2.4 Oil viscosity

The viscosity of heavy crude oil is highly related to its chemical composition. As discussed earlier, considering oxidation reactions, the SARA fractions of heavy oil are significantly altered and oil viscosity is changed. Fig. 5.5 displays the oil viscosities before and after oxidation experiments. From the results of the oil viscosity, the next characteristics can be obtained.

Fig. 5.5A shows the oil viscosities at different operation temperatures. When the operation temperature is lower than 200°C, oil viscosity after a test is higher than that of the initial oil sample. At this temperature range, the oxidation of light components dominates the process. Once the temperature is higher than 250°C, oil viscosity after a test is significantly reduced compared with the initial oil sample. It is caused by the domination of an aquathermolysis reaction.

Fig. 5.5B shows oil viscosities under different pressures. Under 250°C, as pressure increases, an oxidation reaction between heavy oil and air can be enhanced. Therefore, the fractions of heavy components are reduced and oil viscosity is reduced.

Fig. 5.5C shows oil viscosities at different water saturations. For an experiment without water ($S_w = 0$), oil viscosity after a test is significantly

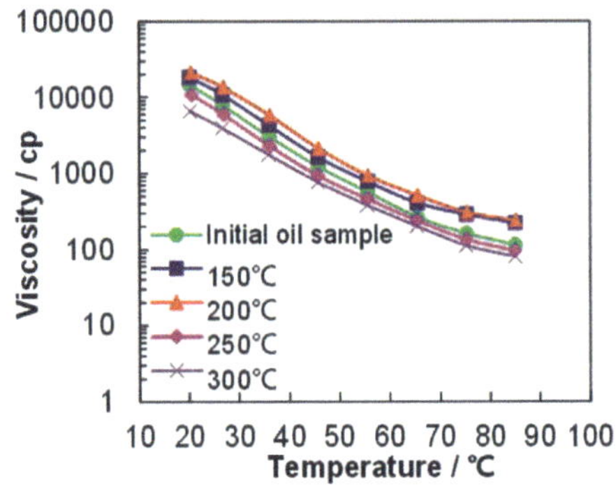

(a) Different temperature conditions

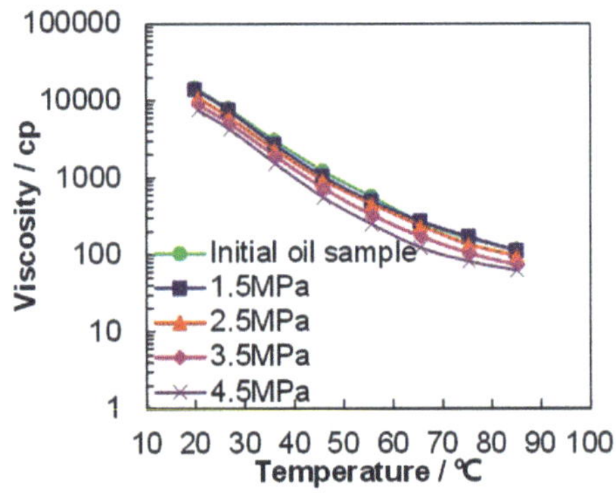

(b) Different pressure conditions

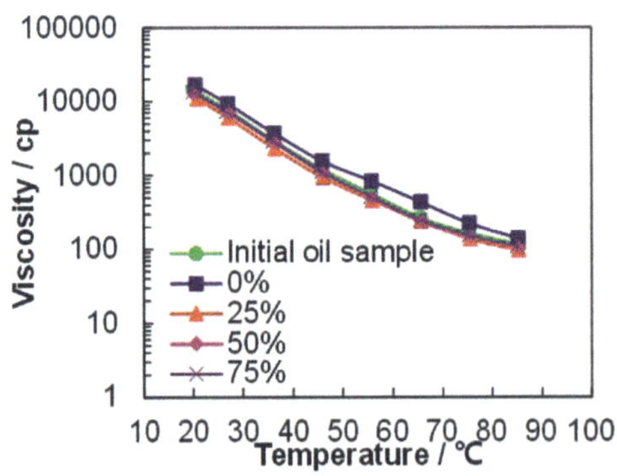

(c) Different water saturation conditions

FIGURE 5.5 Oil viscosities under different operation conditions.

increased compared with the initial oil sample. As the water saturation increases to 25%, the aquathermolysis reaction is enhanced and a sharp decrease in oil viscosity can be observed. Then, as water saturation further increases, the reduction rate of the oil viscosity is gradually reduced.

5.4 Phase behavior of heavy oil–solvent mixture

As discussed in Chapter 1, compared with other processes, solvent-based and solvent-assisted recovery processes have a lower carbon footprint. It is more attractive to improve steam-based recovery. The most important reasons for applying a solvent in heavy oil recovery are its excellent dissolubility in heavy crude oil and the mechanism of phase behavior improvement in a reservoir fluid system. In this section, the phase behavior of a heavy oil–solvent mixture will be studied.

5.4.1 Density and viscosity of heavy oil–solvent mixture

Density and viscosity are two important properties of any fluid flow behavior and reserves estimation in heavy oil reservoirs. For hybrid thermo-solvent processes, the addition of a solvent dilutes heavy crude oil and weakens intermolecular forces to reduce heavy oil viscosity based on steam injection [31]. In this section, some correlations will be introduced to estimate the density and viscosity of a heavy oil–solvent mixture.

5.4.1.1 Density

First, the density of a pure solvent or heavy oil (bitumen) usually depends on the temperature and pressure. The following correlation can be applied to estimate the density of a solvent or heavy oil:

$$\rho = [a_1 + a_2 T] exp[(a_3 T + a_4) P] \tag{5.8}$$

where T is temperature, °C; P is pressure, kPa; ρ is the density, kg/m^3; and a_1-a_4 are coefficients, which can be obtained by a history match with experimental data.

Badamchi-Zadeh et al. [32] and Kariznovi et al. [33] applied this correlation to predict the densities of a solvent or heavy crude oil; the coefficients are provided in Table 5.6. This correlation can work well to predict the densities of solvent and heavy oil under different temperature and pressure conditions. These coefficients are highly related to heavy oil samples.

Thus, for the density of a solvent–heavy oil mixture, the correlation can be applied:

$$\rho_{mix} = \frac{1}{\sum \frac{w_i}{\rho_i}} \tag{5.9}$$

TABLE 5.6 Coefficients of correlation used for densities of solvent and bitumen.

Fluid type	a_1	a_2	a_3	a_4	Reference
n-Tetradecane	776.33	−0.7025	5.792×10^{-9}	7.553×10^{-7}	[33]
Athabasca bitumen	1020.30	−0.63096	2.27016×10^{-9}	4.49743×10^{-7}	[33]
Athabasca bitumen	1020.0	−0.6317	2.521×10^{-9}	6.8072×10^{-7}	[32]

where w_i is a weight fraction of a fluid (solvent or heavy oil); and ρ_i is the density of a fluid (solvent or heavy oil).

For a single-component solvent-based mixture, such as an *n*-hexane-heavy oil mixture or an *n*-octane-heavy oil mixture, the above equation can be modified to

$$\frac{1}{\rho_{mix}} = \frac{w_s}{\rho_s} + \frac{1 - w_s}{\rho_B} \tag{5.10}$$

where ρ_s and ρ_B are the mass densities of solvent and heavy oil (or bitumen), respectively.

Considering the effect of a volume change on mixing behavior, Gao et al. proposed a formula of an excess-volume mixing rule to correct the previous equation [23]:

$$\frac{1}{\rho_{mix}} = \frac{w_s}{\rho_s} + \frac{1 - w_s}{\rho_B} - w_s(1 - w_s)\left(\frac{1}{\rho_s} + \frac{1}{\rho_B}\right)\gamma \tag{5.11}$$

where γ is an empirical parameter, which can be obtained by regression with measured densities.

5.4.1.2 Viscosity

For the viscosity of a heavy oil–solvent mixture, its commonly used models are shown in Table 5.7. From a series of tests on the viscosity of *n*-tetradecane or bitumen, Lederer's and power law models represent viscosity data better than Arrhenius', Cragoe's, and Shu's models [33]. Wu et al. applied Arrhenius' model to represent the viscosity of a heavy oil–solvent mixture in an expanding solvent–steam-assisted gravity drainage process [34]. Arrhenius' model is a commonly applied model in reservoir simulators. For a specific hybrid thermal solvent process, an optimal viscosity correlation for a heavy oil and solvent mixture is related to a solvent type and oil properties and also an operation scheme.

TABLE 5.7 Viscosity correlations for a heavy oil–solvent mixture.

Viscosity model	Correlation	Reference
Arrhenius' model	$\mu_{mix} = exp(\sum_{i=1}^{n} x_i \ln\mu_i)$	[35]
Power law model	$\mu_{mix} = [\sum_{i=1}^{n} x_i\mu_i^k]^{\frac{1}{k}},\ \sum x_i = 1$	[36]
Cragoe's model	$\frac{1}{\ln(2000\mu_{mix})} = \frac{w_s}{\ln(2000\mu_s)} + \frac{1-w_s}{\ln(2000\mu_B)}$	[37]
Lederer's model	$\ln\mu_m = (1 - \frac{\alpha V_o}{\alpha V_o + V_s})\ln\mu_s + (\frac{\alpha V_o}{\alpha V_o + V_s})\ln\mu_o$	[38]
Shu's modification of Lederer's model	$\alpha = \frac{17.04(\rho_o - \rho_S)^{0.5237} \rho_o^{3.2745} \rho_S^{1.6316}}{\ln(\frac{\mu_o}{\mu_s})}$	[39]

x_i is a molar fraction; w_s is the weight fraction of a solvent; μ_s and μ_o are the viscosities of a solvent and heavy oil; V_s and V_o are the volume fractions of a solvent and bitumen; k and α are adjustable parameters, and α is between 0 and 1.

5.4.2 Solvent extraction and asphaltene precipitation

For a hybrid thermal-solvent process, once light paraffinic solvents are co-injected into reservoirs, asphaltene-like materials can precipitate. Therefore, heavy crude oil or bitumen without asphaltenes will be produced, and this process can be regarded as a production process of in situ upgrading. Especially for the process of vapor extraction (VAPEX), solvent extraction is a key mechanism of this process for heavy oil and bitumen reservoirs, which include dry VAPEX and warm VAPEX. After a sufficient volume of solvent is injected, a solvent vapor chamber can form inside a reservoir [40]. It is similar to a steam chamber in steam-assisted gravity drainage. Because of the interaction between the solvent and heavy oil, the extraction of heavy oil and asphaltene precipitation can occur in a solvent chamber edge (or a mixing zone). The width of a moving solvent edge is usually only a few millimeters [10]. In addition, it is a steam–solvent co-injection process. Solvent extraction can benefit from in situ upgrading of heavy crude oil and increase the fractions of light components in the produced heavy oil. On the other hand, the deposited asphaltene can plug a pore space to affect normal heavy oil production as well as the transport of a fresh solvent.

For hybrid thermal-solvent processes, the most important issue is to screen an optimal solvent type. Both a pure solvent and a solvent mixture have been applied in hybrid EOR processes [41]. Commonly used solvents are light alkanes such as propane, butane, pentane, and hexane. Using a two-dimensional micromodel, Li et al. discussed the performance of a VAPEX process using different solvents, including a fully miscible solvent (toluene) and field-relevant solvents (pentane and propane) [10]. Asphaltene precipitation is observed in tests of pentane and propane, and no asphaltene precipitation is observed for a toluene-based VAPEX test. In addition, precipitated asphaltenes from pentane are solid and those from propane remain liquid and mobile. Solid precipitation can further reduce permeability and damage a porous space, and liquid asphaltenes can induce capillary trapping. Zhao et al. experimentally discussed the extraction capacity and solubility of propane in heavy oil [31]. A significant change in oil composition can be observed because of de-asphalting. Using a visual rectangular sand-packed physical model, Mohammadpoor and Torabi experimentally compared the asphaltene precipitation behavior of propane, methane, and a propane-CO_2 mixture [42]. More asphaltene precipitation can be observed around injection points and at an oil–solvent interface. In comparison, a propane injection process can produce more asphaltene precipitation. In addition, CO_2 can be applied as a good alternative solvent or co-solvent to assist thermal recovery [43].

5.4.3 Mathematical modeling for phase behavior of heavy oil–solvent mixtures

Equation of state (EOS) modeling is the most commonly used method to simulate the phase behavior of heavy oil-MMF mixtures, especially for heavy oil–solvent mixtures. Among many cubic EOSs, Peng-Robinson (PR)-EOS has been widely applied [44]:

$$P = \frac{RT}{v-b} - \frac{a(T)}{v(v+b)+b(v-b)} \tag{5.12}$$

where P is pressure, T is temperature, v is a molar volume, R is the universal gas constant, a is a correction term of an intermolecular force, and b is a volume correction term:

$$a(T) = \frac{0.45724R^2T_c^2}{P_c}\alpha(T) \tag{5.13}$$

$$b = \frac{0.0778RT_c}{P_c} \tag{5.14}$$

where α is a function of temperature:

$$\alpha(T) = \left[1 + \left(0.37464 + 1.54226\omega - 0.26992\omega^2\right)\left(1 - \left(\frac{T}{T_c}\right)^{0.5}\right)\right]^2 \tag{5.15}$$

where ω is an acentric factor, T_c is a critical temperature, and P_c is a critical pressure.

To increase the accuracy of estimation, some modified α functions have been proposed. Chen and Yang proposed a α function found to be more accurate for predicting the vapor pressure of heavy hydrocarbon components in heavy oil [45]:

$$\alpha = exp\left\{\begin{array}{c}\left(0.33730 - 0.26346\omega + 0.05297\omega^2\right)\left(1 - T_r\right) \\ +1.01711\ln\left[1 + \left(-0.42587 + 1.16423\omega - 0.10857\omega^2\right)\left(1 - \sqrt{T_r}\right)\right]^2\end{array}\right\} \tag{5.16}$$

where T_r is a reduced temperature.

Another α function can be applied to estimate the vapor pressure of water [11]:

$$\alpha_{H_2O} = \left[c_1 + c_2\left(1 - T_{rw}\right) - c_3\left(1 - T_{rw}^{-1}\right) + c_4\left(1 - T_{rw}^{-2}\right)\right]^2 \tag{5.17}$$

where T_r is the reduced temperature of water, $c_1 = 1$, $c_2 = 0.39222$, $c_3 = 0.07291$, and $c_4 = 0.00706$.

For a correction on the PR-EOS calculated molar volume, the Peneloux volume translation strategy can be applied [46]. It can provide a more accurate prediction for a molar volume [12].

For a heavy oil-MMF mixture, the van der Waals mixing rule is usually applied:

$$a = \sum_i \sum_j x_i x_j a_{ij} \tag{5.18}$$

$$a_{ij} = \left(a_i a_j\right)^{0.5}\left(1 - k_{ij}\right) \tag{5.19}$$

$$b = \sum_i x_i b_i \tag{5.20}$$

where x_i and x_j are the molar fractions of the ith and jth components in the mixture and k_{ij} is a binary interaction parameter (BIP) between the ith and jth components in the mixture.

A BIP usually has a more important role in the phase behavior estimation of a fluid mixture. Li et al. developed BIP correlations for a binary C_3H_8-heavy oil mixture and an n-C_4H_{10}-heavy oil mixture [12]:

$$k_{o3} = -0.4560\frac{T}{T_c} + 0.1817 \tag{5.21}$$

$$k_{o4} = -0.2331\frac{T}{T_c} + 0.1198 \tag{5.22}$$

where k_{o3} is a BIP between C_3H_8 and heavy oil, k_{o4} is a BIP between n-C_4H_{10} and heavy oil, and T_c is the critical temperature of heavy oil.

5.5 Phase behavior of heavy oil–chemical mixture

Besides NCGs and solvents, chemical additives can also be applied to improve the recovery of steam-based recovery processes. For a hybrid thermochemical process, in situ emulsification is one of the most important EOR mechanisms, especially for a hybrid steam surfactant process. In this section, emulsification behavior in a hybrid thermochemical process will be experimentally discussed [17,47].

5.5.1 Oil viscosity with effect of chemical additives

In this section, the viscosity of a heavy oil and chemical additive mixture is experimentally investigated. For a chemical additive, temperature is an important factor for its performance in reservoirs. Considering the effect of high-temperature degradation, the performance of chemical additives in a high-temperature condition is significantly different from that under the room temperature conditions. Therefore, in this section, tests under both room temperature and steam temperature conditions are performed.

A new parameter, a viscosity reduction ratio, is applied to evaluate the performance of chemical additives in the viscosity of heavy crude oil. It can be expressed by the equation:

$$f = \frac{\mu_o - \mu_m}{\mu_o} \times 100\% \tag{5.23}$$

where μ_o is the oil viscosity at a given temperature condition, cp; and μ_m is the viscosity of a heavy oil–chemical additive mixture at a given temperature condition, cp.

5.5.1.1 Tests under room temperature condition (25°C)

In this section, a falling-sphere viscometer is applied to test oil viscosity. The oil sample used is a dead oil sample from the Henan oil field, Sinopec. The heavy oil viscosity is 10,688 cp at 25°C. Two different water-soluble surfactants (viscosity reducers [VRs]) are applied and marked VR1# and VR2#.

The detailed experimental procedure is thus:

- Testing the viscosity of pure heavy crude oil by the viscometer,
- Preparing the brine water with 3 wt% NaCl and a surfactant solution with a predefined mass concentration,
- Mixing the 280-g heavy oil sample and 120-g surfactant solution in the previous step,
- Setting the temperature of the water bath as a predefined temperature condition,
- Using a blender to stir the mixed solution at a rate of 250 rpm for 2 minutes,
- Testing the mixture viscosity,

- ➢ Calculating the viscosity reduction ratio from Eq. (5.23),
- ➢ Repeating the previous steps at least twice, and
- ➢ Calculating the average of all tests to confirm the accuracy of these viscosity tests.

Fig. 5.6 shows the experimental results. For results with a mass concentration of 0.5 wt%, the viscosity reduction ratios of the two surfactants are 89.2% and 86.6%, respectively, and the performance of VR1# is better than that of VR2#. As the mass concentration of a VR increases, the viscosity reduction ratio gradually increases. Once the mass concentration of a chemical additive is higher than 0.5 wt% (critical micelle concentration), the changing tendency of the oil viscosity becomes smooth and slow. The reason is that for a water-soluble VR, the main viscosity-reduction mechanism is to transform a highly viscous heavy crude oil to a low-viscosity oil–water emulsion.

5.5.1.2 Tests under steam temperature conditions (250°C)

Temperature is a sensitive parameter for the properties of chemical additives in porous media. Therefore, for hybrid thermal recovery, the effect of chemical additives on heavy oil properties under steam temperature conditions is more practical. In this section, an enclosed high-temperature high-pressure reactor is applied to test oil viscosity under steam temperature conditions. To reflect the effect of interaction time (or high-temperature endurance time) on the properties of chemical additives, a mixture of heavy oil and a chemical additive will be kept in the reactor for a predefined time. Based on the experimental results under room temperature conditions, the mass concentration of a chemical additive is fixed at 0.5 wt%. Once the endurance time is achieved, the mixture viscosity will be tested.

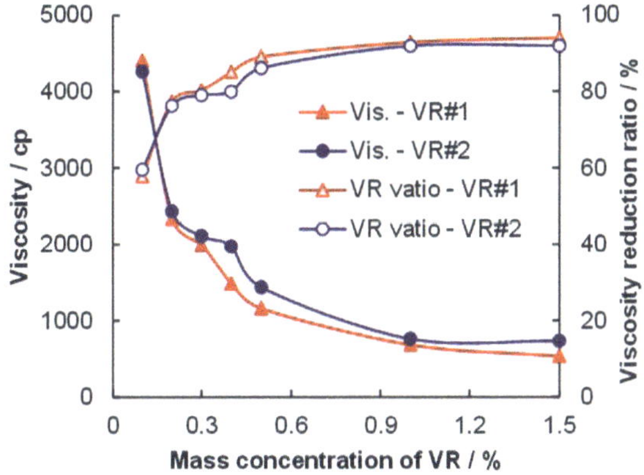

FIGURE 5.6 Experimental results under room temperature conditions. *Vis.*, viscosity; *VR*, viscosity reducer.

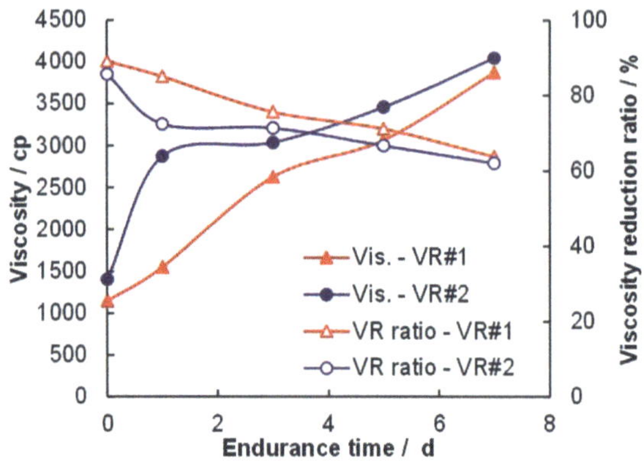

FIGURE 5.7 Experimental results under steam temperature condition. *Vis.*, viscosity; *VR*, viscosity reducer.

Fig. 5.7 shows the experimental results. A high-temperature environment can damage the viscosity reduction of a chemical additive. As temperature increases, the viscosity reduction decreases. For chemical additive VR1#, the viscosity reduction ratio is reduced to 63.8% for an endurance of 7 days. Comparatively, for VR2#, it is 62%. The test results show that the effect of high-temperature degradation is more significant for VR2#. Therefore, combining the test results under room and steam temperature conditions, VR1# is highly recommended. It has higher thermal endurance and viscosity reduction.

5.5.2 Interfacial tension between heavy oil and chemical additive

For the EOR mechanism of in situ emulsification in hybrid thermochemical processes, IFT tension is a dominant factor for emulsification and emulsion stability. In this section, an interfacial rheometer is applied to test the IFT between heavy oil and a chemical additive. During experiments, the mass concentration of a chemical additive is set to 0.5 wt%. The temperature in the high-temperature high-pressure reactor is set to 250°C. The test temperature (water bath temperature) is set to 50°C. The results are shown in Fig. 5.8. When the endurance time increases, the IFT between heavy oil and a chemical additive increases, which increases the difficulty of emulsion forming and damages the stability of the emulsion. Also, the IFT of VR2# is significantly higher than that of VR1#. The IFT increase of VR2# with the endurance time is also higher than that of VR1#. Therefore, for this heavy oil sample, the in situ emulsification behavior can be easily observed in the system of heavy oil and VR1#, and the emulsion stability is higher.

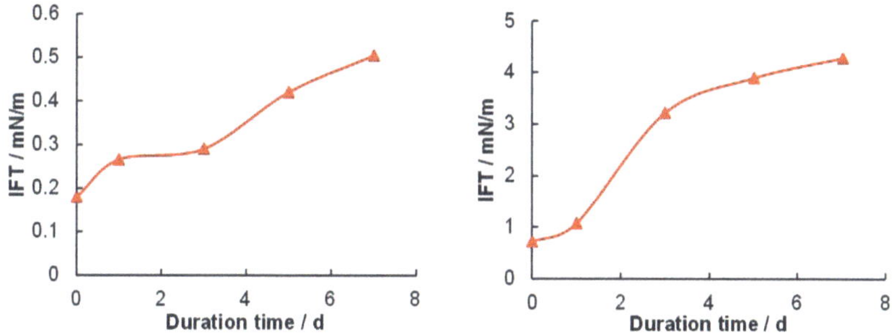

(a) VR1#. (b) VR2#.

FIGURE 5.8 Interfacial tension between heavy oil and chemical additives.

5.5.3 Evaluation method for oil viscosity reduction process

To evaluate the effect of chemical additives on the oil viscosity effectively, a new parameter (i.e., a hybrid viscosity ratio) is proposed:

$$R_i = \frac{\mu_{(T+VR)i}}{\mu_{Ti}} \times 100\% \tag{5.24}$$

where μ_{Ti} is the oil viscosity at temperature T_i, cp; $\mu_{(T+VR)i}$ is the oil viscosity under the coeffect of temperature T_i and a chemical additive, cp; and R_i is a hybrid viscosity ratio, dimensionless. This parameter can be used to evaluate the viscosity reduction degree of chemical additives effectively based on steam injection. A higher viscosity ratio indicates the lesser viscosity reduction ability of a chemical additive.

On the other hand, based on the definition of the viscosity reduction ratio (Eq. 5.23), a parameter of the enhanced viscosity reduction ratio is proposed:

$$f_{(T/VR)i} = \frac{\mu_{Ti} - \mu_{(T+VR)i}}{\mu_{Ti}} \times 100\% \tag{5.25}$$

Fig. 5.9 shows the test results. For pure steam injection, a thermal effect is the dominant for reducing oil viscosity. As the temperature increases, the oil viscosity is gradually reduces. For this heavy oil sample, once the temperature reaches above 60°C, the viscosity reduction ratio by the thermal effect is higher than 90%.

After the consideration of a chemical additive, an extra viscosity reduction mechanism by in situ emulsification is activated. Fig. 5.9A shows that the oil viscosity caused by a thermochemical coeffect is lower than that caused by a pure thermal effect. As the temperature increases, the viscosity reduction ratio by the thermochemical coeffect is significantly higher than only by the thermal effect. Once the temperature is higher than about 100°C, the viscosity

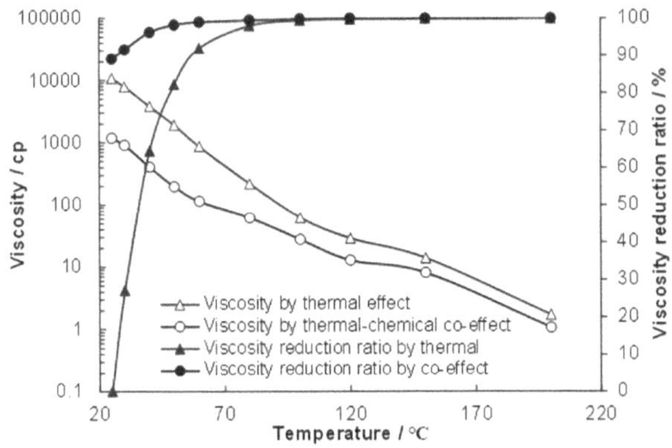

(a) Viscosity and viscosity reduction ratio

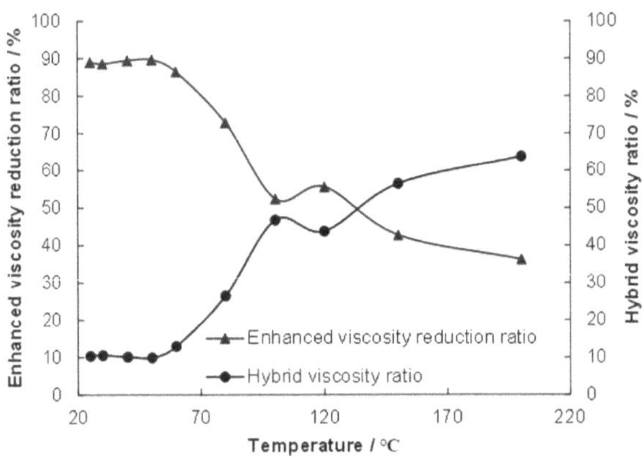

(b) Enhanced viscosity reduction ratio and hybrid viscosity ratio

FIGURE 5.9 Experimental results of oil viscosity under different conditions. *IFT*, interfacial tension; *VR*, viscosity reducer.

reduction ratio difference between the thermal effect and thermochemical coeffect is reduced.

Fig. 5.9B shows the enhanced viscosity reduction ratio and hybrid viscosity ratio. As the temperature increases, the enhanced viscosity reduction ratio is gradually reduced and the hybrid viscosity ratio is increased gradually. As the temperature reaches 100°C, the oil viscosity caused by a thermal effect is only about 61 cp, and thus the extra viscosity reduction capacity by a chemical additive is weakened. The enhanced viscosity reduction ratio has reduced to 36–52%. However, for oil viscosity at 200°C, oil viscosity under thermochemical coeffect conditions is about 63.89% compared with that under pure thermal effect conditions. From the hybrid viscosity ratio, based on the thermal effect, the addition of a chemical additive can further reduce the heavy oil viscosity. Under a low temperature condition, it is more significant, and extra viscosity reduction behavior by a chemical additive can be brought into full play. For a high-temperature condition, although the thermal effect is the dominant mechanism of oil viscosity reduction, a chemical additive can also reduce oil viscosity by a small percentage.

5.5.4 Characteristics of in situ emulsification

5.5.4.1 Experimental method

Heavy oil contains large amounts of resins and asphaltenes. With the combined action of steam and a chemical additive, it can easily be emulsified. In heavy oil reservoirs, the addition of a VR can further transform a highly viscous heavy oil and water–oil emulsion to low-viscosity oil–water emulsion. For a specific hybrid thermochemical process in heavy oil reservoirs, in situ emulsification in porous media has a complicated emulsion formation and flow characteristics. It is highly affected by the combined influences of steam temperature, transport distance, surfactant concentration, oil saturation, and pore structure. In this section, using a sand pack model, the characteristics of in situ emulsification are discussed through the acquisition of temperature and pressure data along the sand pack model.

Fig. 5.10 shows the experimental setup. The whole experimental apparatus includes a fluid injection system, a data acquisition system for temperature and pressure, a sand pack model, and a fluid-gathering system. During the experiments, to reduce heat loss between a fluid injection pipeline and its environment, a heater band is applied. A sand pack model is applied to

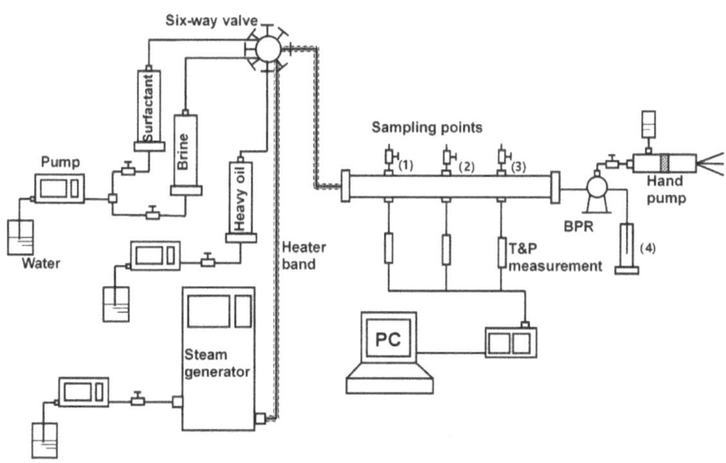

FIGURE 5.10 Experimental setup for in situ emulsification. *T&P*, temperature and pressure.

TABLE 5.8 Experimental condition and parameters of sand pack models.

Test no.	Viscosity reducer concentration (wt.%)	Sand pack model (cm × cm)	Porosity (%)	Permeability (10^{-3} μm^2)	Steam temperature (°C)	Water–oil ratio
1	0.5	L60 × φ3.8	35.6	3520	300	1:1
2	0.5	L60 × φ3.8	32.9	3410	200	1:1
3	0.5	L60 × φ3.8	36.0	6090	300	1:1

simulate a porous medium environment. Along the sand pack model, five pressure measurement points and four temperature measurement points are designed. During experiments, heavy crude oil, steam, and a chemical additive are simultaneously injected into the sand pack model. To analyze in situ emulsification, four fluid sample gathering sites are located along the sand pack model, as shown in Fig. 5.10. For fluid samples collected from the sand pack model, a polarizing microscope is used to observe the characteristics of emulsion formation. By observing the particle size and distribution of emulsion droplets, the characteristics of in situ emulsification in a hybrid thermochemical process can be discussed. Table 5.8 shows the experimental condition and parameters of the sand pack models. The experimental results reveal the effect of steam temperature and permeability on in situ emulsification behavior.

5.5.4.2 Experimental results

Fig. 5.11 presents microscopic images of each fluid sample gathering point after 1.0 pore volume (PV). For the three tests, the emulsion collected is an oil–water emulsion. Compared with temperature, the effect of permeability is more significant. For different sampling points, the distributions of emulsion droplets are significantly different. When a sampling point approaches the outlet of a sand pack model, the size and number of emulsion droplets are significantly altered. Comparing the microscopic images of different tests, a significant difference can be observed for the results of tests 1/2 and 3. The number of emulsion droplets is significantly smaller for test 3 than for tests 1 and 2.

To compare the difference of in situ emulsification behavior accurately under different conditions, the distribution of emulsion sizes along a sand pack

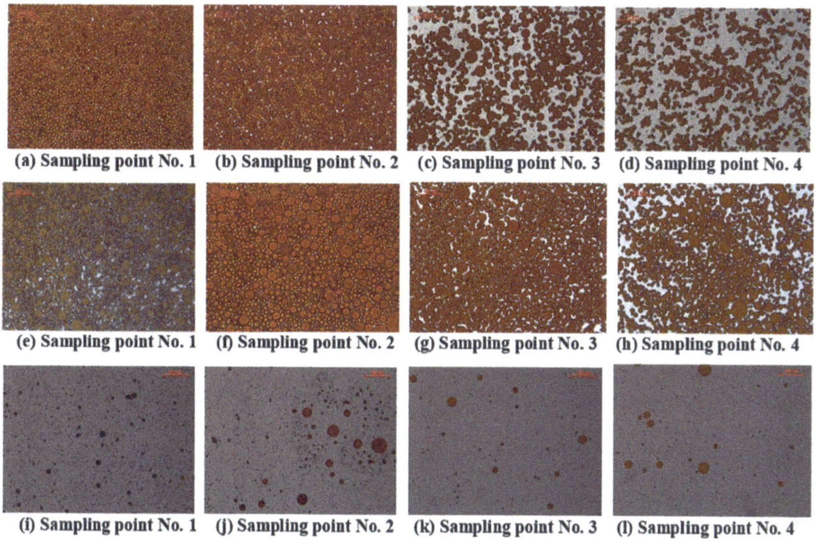

FIGURE 5.11 Microscope observations of emulsions at each sampling point for 1.0 PV injected (magnification times: 200): (A–D) Test 1; (E–H) Test 2; (I–L) Test 3.

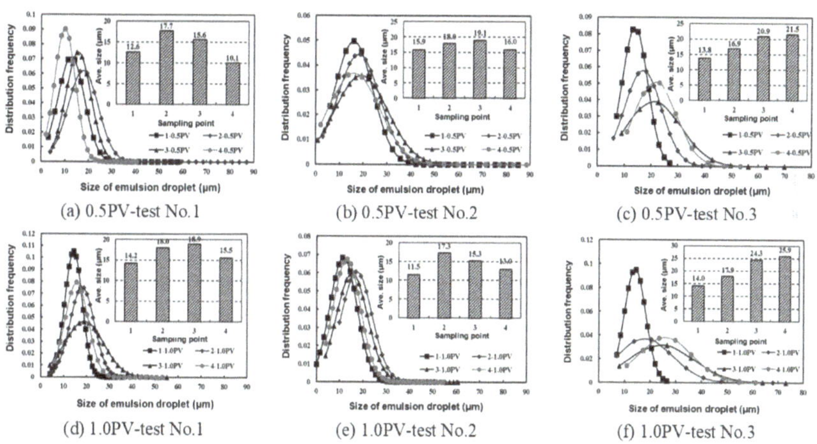

FIGURE 5.12 Particle size distribution and average size at each sampling point.

model is collected, as shown in Fig. 5.12. The average emulsion size is also provided. The results of test 1 are shown in Fig. 5.12A and D. The average size of emulsion droplets along the sand pack model is relatively uniform. Because of the effect of a porous medium environment on in situ emulsification, the size of emulsion droplets at sampling points 2 and 3 is larger than that at sampling points 1 and 4. Fig. 5.12B and E shows the results of test 2. Compared with test 1, the operation temperature of test 2 is lower. The distribution of emulsion sizes along the sand pack model in test 2 is more uniform. When the temperature is reduced, the size of emulsion droplets has a tendency to decrease. The reason is that as the temperature increases, the strength of an interfacial film is reduced and the stability of emulsion is lower. Thus, the emulsion tends to flocculate and coalesce. Overall, considering emulsion transport in a porous medium, the emulsion formed under different temperature conditions differs slightly.

Fig. 5.12C and F shows the results of test 3. Compared with test 1, the permeability of test 3 is higher. This indicates that the pore diameter in the porous medium is higher than that in test 1. The distribution of the sizes of emulsion droplets is wider for test 3, and the difference between sampling points is also larger. In addition, as permeability increases, the average size of emulsion droplets increases. This is because the mechanical shearing effect is lower at a larger pore size environment. Because of a large pore space, the emulsion can easily transport through a pore medium. Therefore, under high permeability, the tensile and coalescence of emulsion droplets dominate the transport of emulsion in porous media. As permeability reduces, mechanical shearing and the broken effect of emulsion caused by a narrow pore space are enhanced. Thus, the size of emulsion droplets is reduced.

In this chapter, the fluid phase behavior of heavy oil and MMFs is discussed. An improvement in fluid phase behavior is one of the most important EOR mechanisms in hybrid EOR processes in heavy oil reservoirs. For the interaction between heavy oil and different additives, the physical properties of reservoir fluids are significantly altered, including the oil density and viscosity, swelling, and emulsification. For a hybrid steam−air injection, because of the effect of low-temperature oxidation reactions, heavy crude oil can be upgraded in situ. In the next chapter, the interaction between heavy oil and MMFs at the molecular level will be discussed.

References

[1] Behzad R, Peyman P, Alireza F, Mahmood RY, Kamran H, Maryam K, Mohammad M, Ahmad D. A new approach to characterize the performance of heavy oil recovery due to various gas injections. Int J Multiphas Flow 2018;99:273−83.

[2] Mohsenzadeh A, Escrochi M, Afraz MV, Karimi G, Al-Wahaibi Y, Ayatollahi S. Non-hydrocarbon gas injection followed by steam−gas co-injection for heavy oil recovery enhancement from fractured carbonate reservoirs. J Petrol Sci Eng 2016;144:121−30.

[3] Nasr TN, Prowse DR, Frauenfeld T. The use of flue gas with steam in bitumen recovery from oil sands. J Can Petrol Technol 1987;26(3):62−9.

[4] Nasr TN, Pierce GE. Steam-CO_2 recovery processes for bottom water oil reservoirs. J Can Petrol Technol 1995;34(7):42−9.

[5] Wang Y, Ren S, Zhang L, Peng X, Pei S, Cui G, Liu Y. Numerical study of air assisted cyclic steam stimulation process for heavy oil reservoirs: recovery performance and energy efficiency analysis. Fuel 2018;211:471−83.

[6] Yu T. Experimental study on air assisted CSS for medium deep ultra heavy oil reservoir. Special Oil Gas Reservoirs 2013;20(5):77−80 [in Chinese].

[7] Ursenbach MG, Moore RG, Mehta SA. Air injection in heavy oil reservoirs: a process whose time has come (again). J Can Petrol Technol 2010;49(1):48−54.

[8] Ardali M, Barrufet M, Mamora DD, Qiu F. A critical review of hybrid steam/solvent processes for the recovery of heavy oil and bitumen. In: SPE 159257 presented at the SPE annual technical conference and exhibition, San Antonio, Texas, USA, 8−10 October, 2012; 2012.

[9] Bayestehparvin B, Farouq Ali SM, Abedi J. Use of solvents with steam - state-of-the-art and limitations. In: SPE 179829 presented at the SPE EOR conference at oil and gas West Asia, Muscat, Oman, March 21−23, 2016; 2016.

[10] Li X, Berg S, Castellanos-Diaz O, Wiegmann A, Verlaan M. Solvent-dependent recovery characteristic and asphaltene deposition during solvent extraction of heavy oil. Fuel 2020;263:116716.

[11] Chen Z, Zhao Z, Yang D. Quantification of phase behavior for solvent/heavy-oil/water systems at high pressures and elevated temperatures with dynamic volume Analysis. SPE J 2020;25(6):2915−31.

[12] Li H, Yang D. Phase behaviour of C3H8-n-C4H10-heavy oil systems at high pressures and elevated temperatures. In: SPE 157744 presented at the SPE heavy oil conference, Calgary, Alberta, Canada, June 12−14, 2012; 2012.

[13] Zhao X, Feng Y, Liao G, Liu W. Visualizing in-situ emulsification in porous media during surfactant flooding: a microfluidic study. J Colloid Interface Sci 2020;578:629−40.

[14] Velayati A, Nouri A. Emulsification and emulsion flow in thermal recovery operations with a focus on SAGD operations: a critical review. Fuel 2020;267:117141.

[15] Bennion DB, Chan M, Sarioglu G, Courtnage D, Wansleeben J, Hirata T. The in situ formation of bitumen-water stable emulsions in porous media during thermal stimulation. In: PETSOC-93-46 presented at the annual technical meeting, Calgary, Alberta, Canada, May 9−12, 1993; 1993.

[16] Chen T, Yuan JY, Serres AJ, Rancier DG. Emulsification of water into bitumen in co-injection experiments. J Can Petrol Technol 1999;38(13):1−6.

[17] Lu C. The improvement effects of gas and chemical agent during steam injection process for heavy oil reservoirs [Ph.D. dissertation]. Beijing: China University of Petroleum; 2015.

[18] Wang X, Wang J, Qiao M. Horizontal well, nitrogen and viscosity reducer assisted steam huff and puff technology: taking super heavy oil in shallow and thin beds, Chunfeng Oilfield, Junggar Basin, NW China, as an example. Petrol Explor Dev 2013;40(1):104−10.

[19] Dong X, Liu H, Hou J, Zhang Z, Chen Z. Multi-thermal fluid assisted gravity drainage process: a new improved-oil-recovery technique for thick heavy oil reservoir. J Petrol Sci Eng 2015;133:1−11.

[20] Jia X, Dong X, Xu J, Chen Z. Multiphase fluid flow and reaction in heterogeneous porous media for enhanced heavy oil production. John Wiley & Sons Ltd.; 2018. p. 319−51.

[21] Chung FTH, Jones RA, Nguyen HT. Measurements and correlations of the physical properties of CO2-heavy-crude-oil mixtures. SPE Reservoir Eng 1988;3(3):822−8.

[22] Gao J, Okuno R, Li HA. An experimental study of multiphase behavior for n-butane/bitumen/water mixtures. SPE J 2017;22(3):783−98.

[23] Gao J, Okuno R, Li HA. A phase-behavior study for *n*-hexane/bitumen and *n*-octane/bitumen mixtures. SPE J 2018;23(1):128–44.

[24] Khansari Z, Kapadia P, Mahinpey N, et al. A new reaction model for low temperature oxidation of heavy oil: experiments and numerical modeling. Energy 2014;64:419–28.

[25] Babu DR, Cormack DE. Effect of low temperature oxidation on the composition of Athabasca bitumen. Fuel 1984;63:858–61.

[26] Mamura DD. New findings in low temperature oxidation of crude oil. In: Paper presented at SPE Asian Pacific oil and gas conference in Kuala Lumpur Malaysia, March 20–22, 1995; 1995.

[27] Monin JC, Audibert A. Thermal cracking of heavy-oil/mineral matrix systems. In: Paper SPE 16269 presented at the 1987 SPE international symposium on oilfield chemistry held in San Antonio Texas, Feb. 4–6, 1987; 1987.

[28] Masao H, Douglas WB, John KD, et al. Thermal cracking models for Athabasca oil sands oil. In: Paper SPE 7549 presented at the 53rd Annual fall technical conference and exhibition of the society of petroleum engineers of AIME held in Houston Texas, October 1–3, 1978; 1978.

[29] Hyne. Aquathermolysis-a synopsis work on the chemical reaction between water (steam) and heavy oil sands during simulated stimulation. Synopsis Report No. 3. 1986. p. 845–9.

[30] Wang C, Liu H, Pang Z, Wang J, Chen C, Wang C, Wu Z. Enhance oil recovery for steam flooding Low temperature oxidative decomposition of heavy oil with air injection. Energy Fuel 2015;29(10):6242–9.

[31] Zhao F, Liu Y, Zhao X, Tan L, Geng Z. Characteristics and mechanisms of solvent extraction of heavy oils from porous media. Chem Technol Fuels Oils 2015;51:33–40.

[32] Badamchi-Zadeh A, Yarranton HW, Svrcek WY, Maini BB. Phase behaviour and physical property measurements for VAPEX solvents: part I. Propane and Athabasca bitumen. J Can Petrol Technol 2009;48(1):54–61.

[33] Kariznovi M, Nourozieh H, Guan JG, Adebi J. Measurement and modeling of density and viscosity for mixtures of Athabasca bitumen and heavy n-alkane. Fuel 2013;112:83–95.

[34] Wu Y, Liu X, Du, Zhou X, Wang L, Li J, Li Y, Li X, Li Y. Scaled physical experiments on drainage mechanisms of solvent-expanded SAGD in super-heavy oil reservoirs. Petrol Explor Dev 2020;47(4):765–71.

[35] Arrhenius SA. Über die Dissociation der in Wasser gelösten Stoffe. Z Phys Chem 1887;1:631–48.

[36] Kendall J, Monroe K. The viscosity of liquids II. The viscosity-composition curve for ideal liquid mixtures. Am Chem J 1917;9:1787–802.

[37] Cragoe CS. Changes in the viscosity of liquids with temperature, pressure and composition. Proc World Pet Cong Lond 1933;2:529–41.

[38] Lederer EL. Viscosity of mixtures with and without diluents. Proc World Pet Cong Lond 1933;2:526–8.

[39] Shu WR. A viscosity correlation for mixtures of heavy oil, bitumen, and petroleum fractions. SPE J 1984;24:277–382.

[40] Pourabdollah K, Mokhtari B. The VAPEX process, from beginning up to date. Fuel 2013;107:1–33.

[41] Dong X, Liu H, Chen Z, Lu N, Wu K, Zhang Q. Enhanced oil recovery techniques for heavy oil and oil sands reservoirs after steam injection. Appl Energy 2019;239:1190–211.

[42] Mohammadpoor M, Torabi F. Effect of solvent type and drainage height on asphaltene precipitation for the solvent percolating gravity drainage mechanism in the vapor extraction process. Spec Top Rev Porous Media Int J 2014;5(2):107–17.

[43] Bayat AE, Junin R, Hejri S, Fazeli A, Afsari K. Application of CO_2-based vapor extraction process for high pressure and temperature heavy oil reservoirs. J Petrol Sci Eng 2015;135:280–90.

[44] Peng DY, Robinson DB. A new two-constant equation of state. Ind Eng Chem Fundam 1976;15(1):59–64.

[45] Chen Z, Yang D. Optimization of the reduced temperature associated with Peng-Robinson equation of state and Soave-Redlich-Kwong equation of state to improve vapor pressure prediction for heavy hydrocarbon compounds. J Chem Eng Data 2017;62(10):3488–500.

[46] Peneloux A, Rauzy E, Freze R. A consistent correction for Redlich-Kwong-Soave volumes. Fluid Phase Equil 1982;8(1):7–23.

[47] Lu C, Liu H, Zhao W, Lu K, Liu Y, Tian J, Tan X. Experimental investigation of in-situ emulsion formation to improve viscous-oil recovery in steam-injection process assisted by viscosity reducer. SPE J 2017;22(1):130–7.

Chapter 6

Molecular dynamic simulation for hybrid enhanced oil recovery processes

6.1 Molecular dynamic simulation for adsorption configurations of heavy crude oil

6.1.1 Introduction

An occurrence state of heavy crude oil in a porous medium has an important role in setting up a development strategy for heavy oil reservoirs. In heavy crude oil, because of the performance of supermolecular structure non-hydrocarbon matter and high oil viscosity characteristics, heterogeneous behavior within the oil can be observed. In addition, as the complexity of a colloid system formed between different molecules in heavy crude oil increases, the heterogeneous behavior of heavy oil molecules in porous media is further aggravated. Near a pore wall surface, heavy oil molecules are usually distributed by an intertwined colloid configuration. In this complicated structure, asphaltene molecules are located at the center, and saturates and aromatic molecules appear as a dispersion phase. Therefore, for the effective development of heavy oil reservoirs, figuring out the structure of a colloid system and stripping the asphaltene molecules from a pore surface are key issues. Specifically, interaction between asphaltene molecules and a pore surface is the nature of these problems. Mullins proposed a Yen−Mullin model to describe the aggregation behavior of asphaltene molecules in porous media; it was observed that π-π stacking was the main accumulation form of asphaltene molecules near a pore surface [1−3]. On the other hand, polymerization is also an important mechanism of asphaltene aggregation [4]. However, the van der Waals force is the root source of an interaction between asphaltene molecules [5]. Moreover, the methods of molecular dynamic simulation (MDS) and dissipative particle dynamics have been used to understand the aggregation structure of heavy oil molecules in different types of

Hybrid Enhanced Oil Recovery Processes for Heavy Oil Reservoirs.
https://doi.org/10.1016/B978-0-12-823954-4.00003-5

pore surfaces. The effects of a boundary condition and shear action can be also simulated [6–8]. To characterize the aggregation behavior of heavy oil molecules accurately around a pore surface is the key to understanding an occurrence state of heavy crude oil in porous media.

For heavy oil reservoirs after steam injection, under the effect of lengthy steam stimulation, an occurrence state of heavy oil in porous media changes significantly from its original state, as discussed in Section 2.9. The occurrence state can have an important influence on the flow behavior of heavy oil in porous media. As the temperature increases, a complicated colloid system of heavy oil takes on an ordered distribution, and thus oil viscosity is reduced. Simultaneously, the rheology of heavy oil is significantly altered. At the molecular level, it is mainly caused by the interplay of heavy oil, steam, and a pore surface [9–12]. For hybrid enhanced oil recovery (EOR) processes, the addition of different additives can further reduce oil viscosity and improve oil mobility to alter the retention and starting behavior of remaining oil significantly in porous media after pure steam injection. Some attempts have been made to study the effect of multicomponent and multiphase fluids on an occurrence state of heavy crude oil, but the mechanism at the molecular level is still unclear. MDS is an effective method to address this issue. It can be used to figure out the dominant factor for a configuration and accumulation morphology of a colloid system [13–15]. Moreover, MDS can be applied to calculate changes in the potential of a mean force and free energy during molecule accumulation [16–19]. In addition, for a hybrid EOR process, MDS can provide a visualized tool to illustrate the interaction behavior between additives and heavy oil molecules [20–22].

6.1.2 Molecular dynamic simulation model development

In this section, an open access MDS software is used to develop the molecular model of heavy crude oil. Considering the complicated composition of heavy oil, both a pure asphaltene model and a full saturates, aromatics, resin, and asphaltene (SARA) model will be developed to discuss interactions between heavy oil and a pore surface. During MDS runs, specific molecule structures of different oil components can have a dominant role, especially for asphaltene molecules. Many different asphaltene molecular structures have been proposed to represent and investigate the accumulation behavior of heavy oil molecules in pores [23]. In this chapter, a commonly used improved Y-M asphaltene model will be applied [24,25]. Fig. 6.1A–D shows the molecular structures of SARA components in heavy crude oil. For a pore surface, a surface hydroxylated α-SiO_2 supercell model is applied to represent a water-wet sandstone porous medium environment, as shown in Fig. 6.1E.

Based on these molecular structures, MDS models under different conditions can be developed to study the adsorption configuration of heavy oil

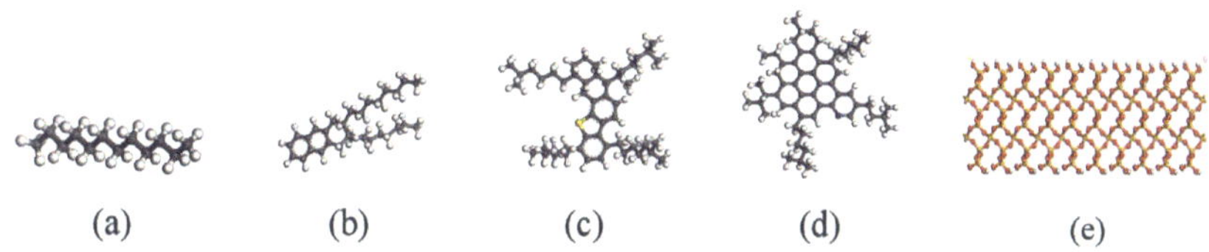

FIGURE 6.1 Molecular structure of (A) saturates, (B) aromatics, (C) resin, (D) asphaltene, and (E) quartz plane. White, gray, blue, red, yellow, and dark yellow balls represent H, C, N, O, Si, and S atoms, respectively.

components on a pore surface. All MDS runs in this chapter are conducted in an NVT (constant-temperature, constant-volume) ensemble. The size of a simulation box is 4.84487 nm × 3.51673 nm × 6.17210 nm. Temperature and pressure are maintained at 333 K and 1 MPa, respectively. The Optimized Potentials for Liquid Simulations All-Atoms (OPLS-AA) force field is applied to describe the molecular force field, and periodic boundary conditions are applied. The simulation time is 10 ns. From this model, a series of MDS runs are performed to discuss the adsorption configuration and interaction behavior between heavy oil molecules and a pore surface.

6.1.3 Simulation results

For a pure asphaltene model, eight asphaltene molecules (Fig. 6.1D) are involved in this model. For a full SARA model, the ratio of SARA components is 4:3:1:1 and the total molecule number is 72. Fig. 6.2 shows the simulation results of the pure asphaltene model and the full SARA model, respectively. Fig. 6.2A illustrates that a face-to-face adsorption configuration can be observed between asphaltene molecules and a quartz surface. The adsorption configuration of offset stacking and face-to-face stacking can be observed between different asphaltene molecules. For the results of the full SARA model (Fig. 6.2B), however, a different adsorption configuration can be observed. It is caused by the complicated interaction between fluid–fluid

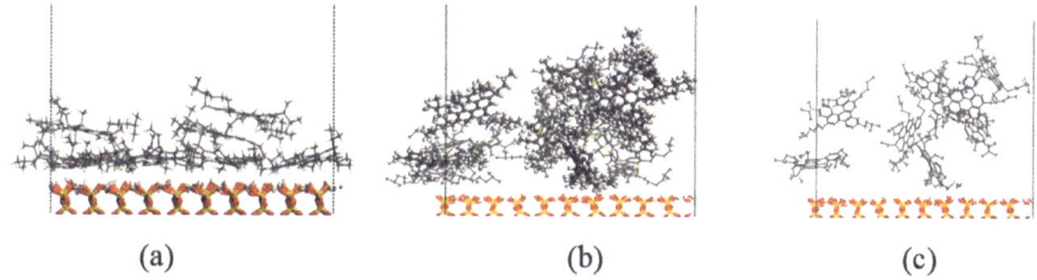

FIGURE 6.2 Simulation results at 333 K: (A) pure asphaltene model; (B) full SARA components model; (C) and asphaltene molecules in a full saturates, aromatics, resin, and asphaltene model.

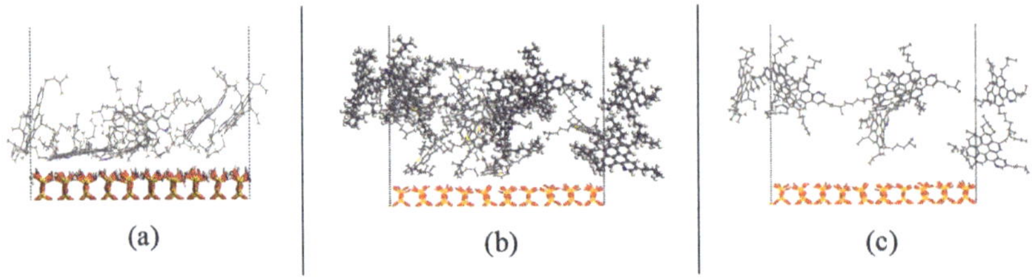

FIGURE 6.3 Simulation results at 393 K: (A) pure asphaltene model; (B) full saturates, aromatics, resin, and asphaltene (SARA) components model; (C) asphaltene molecules in a full SARA model.

molecules and fluid–solid molecules. Fig. 6.2C shows that both face-to-face stacking and offset stacking can be observed between asphaltene molecules and a quartz surface, because some adsorption sites are occupied by aromatics and resin molecules.

In addition, considering the effect of temperature on an adsorption configuration of heavy oil molecules, two MDS runs are performed at 393 K, as shown in Fig. 6.3. Compared with the results in Fig. 6.2, the adsorption configuration of heavy oil molecules has significantly changed. For the pure asphaltene model, the adsorption configurations of face-to-face stacking, offset stacking, and T-shaped stacking can be observed between asphaltene molecules and a quartz surface. Also, the distance between asphaltene molecules and the quartz surface is increased. For the results of the full SARA model (Fig. 6.3C), only the T-shaped adsorption configuration can be observed, and the distance is also increased.

Based on the results in Figs. 6.2 and 6.3, an increase in the operation temperature can increase molecular activity, and thus interaction between fluid molecules and a pore surface is weakened. Once the temperature is increased to a certain value, the fluid molecules (asphaltene) can be fully separated from a pore surface. In the macroscopic view, increasing the temperature in an unswept reservoir zone (i.e., thermal sweep efficiency) should be an EOR direction for heavy oil reservoirs after steam-based recovery.

6.2 Molecular dynamic simulation for heavy oil–water mixtures

In this section, using the same method, a series of simulation runs are performed for interaction behavior for a heavy oil–water mixture. Three commonly used rock types and two typical temperature conditions are employed. Specifically, a surface of methyl quartz (SMQ) is employed to simulate a strong oil-wet surface, a surface of hydroxylated quartz (SHQ) is employed to simulate a weak oil-wet surface, and a calcite plane (CAL) is employed to simulate a water-wet surface [26–28]. Here, 333 K (60°C) represents the temperature of the heavy oil–water mixture in the original condition or post–steam injection condition (residual heat condition), whereas 483 K (210°C) represents the temperature of the heavy oil–water mixture in a steam flooding condition. All MDS runs in this section are performed in 1 ns.

6.2.1 Adsorption configuration and contact angle

Fig. 6.4 gives the equilibrium adsorption configurations of oil droplets at 333 K in three different simulated planes. As shown, the heavy oil droplets are spread out over the SHQ and SMQ planes, and they can form a certain contact angle. This indicates that a quartz plane with different surface properties can have a strong adsorption effect on the oil droplets. Moreover, the heavy oil drops cannot wet the CAL plane, and the oil molecules are far from the calcite plane. The same phenomena can be observed at high temperature conditions (483 K), as shown in Fig. 6.4D–F.

On the other hand, because of a high molecular weight, the distribution of asphaltene molecules around a pore surface can have a great effect on the adsorption configuration of oil droplets. Fig. 6.5 gives an enlarged view of the asphaltene equilibrium adsorption configuration on a quartz plane. For an SHQ plane with weak oil-wet, T-shaped stacking on the surface can be observed. For an SMQ plane with strong oil-wet, π-π stacking can be found. Simultaneously, tight adsorption behavior between asphaltenes and a pore surface is developed, whereas for the CAL plane, a complicated hydrogen bond system of water–calcite can be observed.

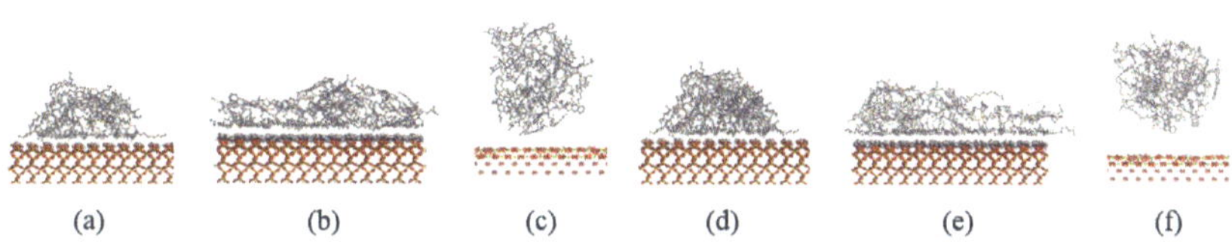

FIGURE 6.4 Equilibrium adsorption configurations of oil–water mixtures on simulated surfaces: (A) surface of hydroxylated quartz (SHQ) plane at 333 K; (B) surface of methyl quartz (SMQ) plane at 333 K; (C) calcite plane (CAL) plane at 333 K; (D) SHQ plane at 483 K; (E) SMQ plane at 483 K; (F) CAL plane at 483 K. Water molecules are hidden for clarity.

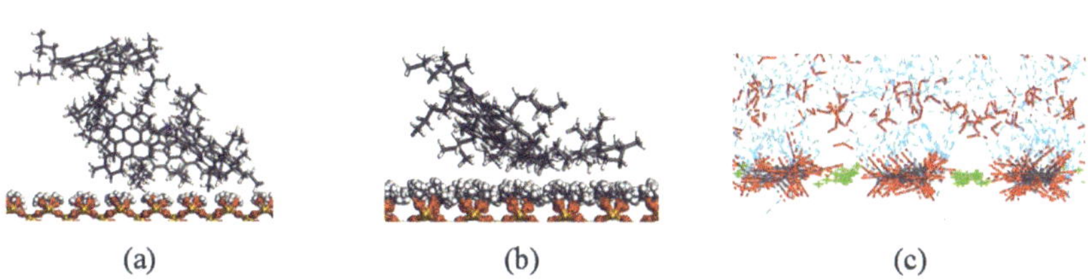

FIGURE 6.5 Typical asphaltene adsorption configurations on different pore surfaces: (A) surface of hydroxylated quartz plane; (B) surface of methyl quartz plane; and (C) water–calcite plane (CAL) hydrogen bond (blue line) system on CAL plane.

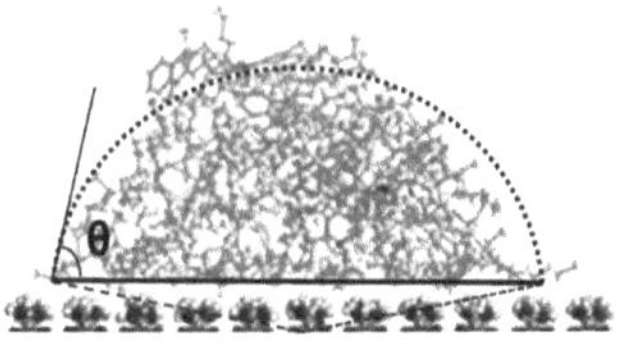

FIGURE 6.6 Schematic illustration for calculating contact angle.

TABLE 6.1 Contact angles of oil droplets on three simulated surfaces.

	Simulated surface		
Temperature (K)	Surface of hydroxylated quartz (degrees)	Surface of methyl quartz (degrees)	Calcite plane
333	68	48	Nonwet
483	79	45	Nonwet

To describe the behavior of an oil−water mixture quantitatively on different pore surfaces, the contact angles of heavy oil droplets can be calculated. Fig. 6.6 gives a schematic for calculating the contact angle.

The results are shown in Table 6.1. The contact angle of oil droplets on a strong oil-wet (SMQ) surface is the lowest. The methyl group on the SMQ plane is favorable for the adsorption of heavy oil molecules. In addition, temperature has little influence on the contact angle of an oil−water system on an SMQ surface, and the contact angle is about 45 degrees at 483 K. Comparatively, the SHQ plane is water-wet. As shown, the oil droplets can have a greater contact angle on the SHQ plane than the SMQ plane. By increasing the temperature, the contact angle increases from 68 to 79 degrees. This indicates that water molecules will reoccupy the origin adsorption sites of heavy oil on the SHQ plane. For the CAL plane, because of a strong hydrogen bond interaction between water molecules and the CAL plane, the oil phase is the nonwet phase on a CAL plane. Moreover, temperature has little effect on the contact angle, and oil droplets remain nonwetting on the calcite plane.

6.2.2 Interaction energy

6.2.2.1 Center of mass of oil droplets from pore surface

To understand the adsorption behavior of heavy oil on a pore surface, the center of mass (COM) of oil droplets from the pore surface is evaluated. Fig. 6.7 gives the distance between the COM of oil droplets and the simulated surface. As shown in Fig. 6.5, oil droplets can adsorb and spread out on an oil-wet quartz plane. This indicates that the COM of oil droplets from a quartz

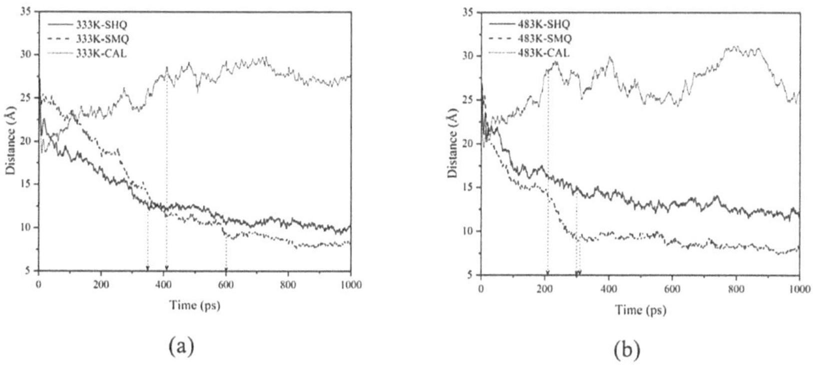

FIGURE 6.7 Distance between center of mass of oil droplets and pore surface: (A) 333 K and (B) 483 K. *CAL*, calcite plane; *SHQ*, surface of hydroxylated quartz; *SMQ*, surface of methyl quartz.

plane can gradually decrease with time. As shown in Fig. 6.7A, the COM of oil droplets on an SHQ plane can be stable at 12.5 Å when the time reaches about 350 ps, and that on an SMQ plane is stabilized around 9.0 Å when the time reaches 600 ps. The lower equilibrium height of the COM on the SMQ plane indicates that oil droplets are more closely attached to the surface. Also, at a higher temperature, the equilibrium height of the COMs will move to 14.0 and 8.5 Å for the SHQ and SMQ planes, respectively. Moreover, the time required to stabilize is reduced to 300 and 310 ps, respectively. From the difference in the required equilibrium time at different conditions, it can be concluded that a strong oil-wet system can easily be stabilized at a high temperature. The adsorption behavior of a heavy oil-water mixture on a quartz surface is highly affected by its hydrophobic property.

In addition, because of the water-wet property of a CAL plane, the distance of COM is higher than for the other two planes. Fig. 6.7 shows that the equilibrium distance of COM is about 28.0 Å for the two different temperature conditions, whereas the required equilibrium time is shortened from 390 to 210 ps as the temperature rises.

6.2.2.2 *Interaction energy*

The interaction energy between fluids and a simulated surface is analyzed in this section (Fig. 6.8). For the interaction between oil droplets and simulated surfaces, the van der Waals force is the dominant interaction energy. In comparison, for that of a water-surface system, the electrostatic (ET) force is usually not neglected. As shown in Fig. 6.8, the interaction energy of oil droplets on an SMQ plane is much higher than on the other two planes, and the effect of temperature is not significant. It is caused by the stable hydrophobic sites on an SMQ plane, and the oil droplets can spread over an SMQ plane with a small contact angle to form a stable adsorption layer.

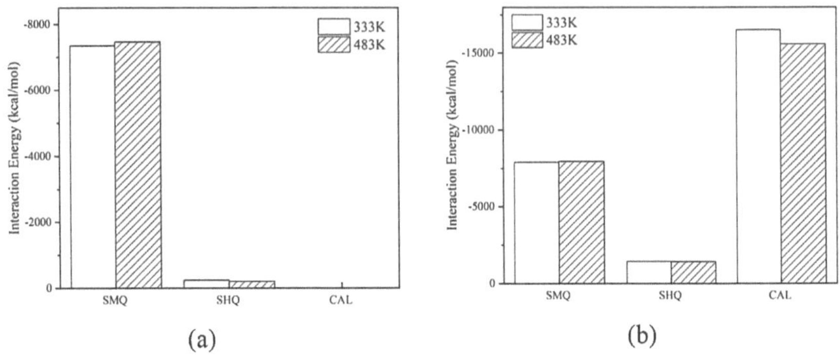

FIGURE 6.8 Total interaction energy between fluids and pore surfaces under different temperature conditions: (A) oil-surface and (B) water-surface. *CAL*, calcite plane; *SHQ*, surface of hydroxylated quartz; *SMQ*, surface of methyl quartz.

Second, for an SHQ plane with a weak oil-wet property, oil–SHQ interaction energy is lower and water–SHQ interaction energy is dominant. Simultaneously, ET interaction energy can contribute a large proportion to the total interaction energy. Fig. 6.8 shows that as the temperature increases to 483 K, oil–SHQ interaction energy can significantly decrease but the water–SHQ interaction energy is rarely affected. This is why oil droplets can form a large contact angle on an SHQ plane and a contact angle rises with temperature.

Third, because of the strong water-wet property of a CAL plane, oil droplets are far from the surface. Therefore, the water molecules are attached on the CAL surface and a strong hydrogen bond can be observed. ET interaction energy is the dominant interaction energy. In addition, Fig. 6.8B shows that as the temperature increases, the total interaction energy of the water–CAL plane slightly decreases. This is why the fluctuation behavior of oil droplet COM occurs during the equilibrium stage.

A heavy oil–water mixture can form a large contact angle on an SHQ plane with weak oil-wet adsorption sites, and the hydrophobicity of the SHQ plane decreases at a higher temperature. For an SMQ plane, the oil-wet property is enhanced and oil droplets can spread over the surface with a small contact angle. For a CAL plane, it is a water-wet surface with strong hydrophilicity. Furthermore, the MDS simulation results show that the wettability of SMQ and CAL planes are insensitive to temperature.

6.3 Molecular dynamic simulation for a hybrid thermal–noncondensable gas process

For hybrid EOR processes, the injection of different additives can further enhance interactions between fluids and a pore surface [29,30]. Therefore, during hybrid EOR processes, the distribution of different fluid molecules can be significantly changed compared with their original distribution. In this section, using MDS, a series of simulation runs are performed for the interaction behavior of heavy oil molecules and nitrogen molecules. The basic MDS model for heavy oil is the same as the full SARA model in Section 6.1. The initial configuration is shown in Fig. 6.9A. The number of nitrogen molecules in a dense layer and a nitrogen layer is set to 2000 and 500, respectively. The original density is set to 0.5 g/cm^3. Based on this model, a series of simulation runs is performed for an adsorption configuration and interaction energy between different molecules.

6.3.1 Adsorption configuration

Fig. 6.9 shows the simulation results of adsorption configuration under different temperature conditions. After the 10-ns simulation is completed, this model can be considered to achieve the equilibrium state. Compared with the simulation results of pure SARA molecules in Figs. 6.2 and 6.3, the adsorption configurations of different heavy oil molecules in Fig. 6.9 do not present a significant difference. For the adsorption configuration between asphaltene molecules and SiO_2 molecules, it is the same as that of the pure SARA model. This indicates that the injection of nitrogen cannot significantly alter the adsorption configuration of fluid molecules. This observation is because of the low solubility of nitrogen in heavy oil; nitrogen molecules cannot enter the complicated SARA molecule system. Therefore, for a hybrid thermal−noncondensable gas process, the injection of nitrogen cannot unlock remaining oil saturation in a swept zone, and temperature is still the dominating factor for the adsorption configuration of fluid molecules. For this process, an increase in thermal sweep efficiency is the main EOR mechanism.

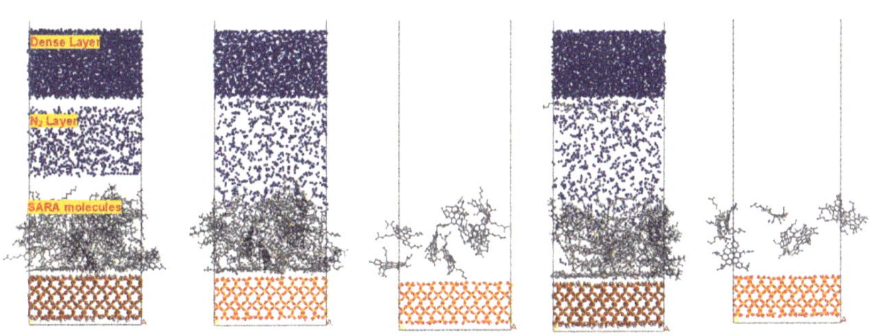

(a) (b) (c) (d) (e)

FIGURE 6.9 Simulation model and adsorption configuration in saturates, aromatics, resin, and asphaltene (SARA) + N_2 model: (A) original configuration; (B) adsorption configuration after equilibrium at 333 K; (C) asphaltene molecules after equilibrium at 333 K; (D) adsorption configuration after equilibrium at 393 K; (E) asphaltene molecules after equilibrium at 393 K.

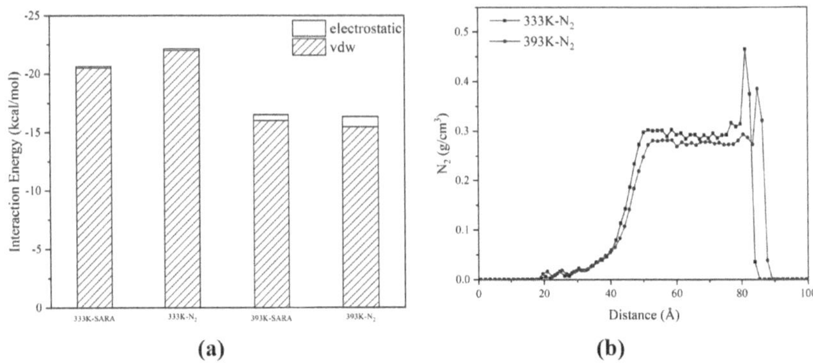

FIGURE 6.10 Simulation results at different temperature conditions: (A) interaction energy and (B) N_2 density profile. *vdw*, van der Waals.

6.3.2 Interaction energy

This section discusses changes in interaction energy between different molecules. Fig. 6.10A shows the interaction energy between fluid molecules and a pore surface in both the pure SARA model and SARA + N_2 model. As shown, the addition of nitrogen cannot significantly alter the interaction energy between fluid molecules and the pore surface. This is why the adsorption configuration is not changed in this SARA + N_2 model. On the other hand, Fig. 6.10A shows that the higher the temperature, the smaller the difference between them. It is caused by the high molecular activity of nitrogen at a higher temperature. Fig. 6.10B shows a density profile of N_2 molecules at different simulation temperatures. As the temperature increases, the distance between N_2 molecules and a pore surface increases.

6.4 Molecular dynamic simulation for a hybrid thermal-solvent process

In this section, using the same method, a series of simulation runs is performed for the interaction behavior of heavy oil molecules and solvent molecules in hybrid thermal-solvent processes. Three commonly used solvent types, CO_2, C_3H_8 and C_4H_{10}, are simulated. The basic MDS model for heavy oil is still the same as the full SARA model in Section 6.1. The initial molecular configurations are shown in Fig. 6.11A, F, and K. The number of solvent molecules in a dense layer and a solvent layer is set to 1000 and 500, respectively. Based on this model, a series of simulation runs is performed for an adsorption configuration and interaction energy between different molecules.

6.4.1 Adsorption configuration

Fig. 6.11 shows the simulation results of adsorption configurations in different solvent environments at different temperature conditions. First, for CO_2

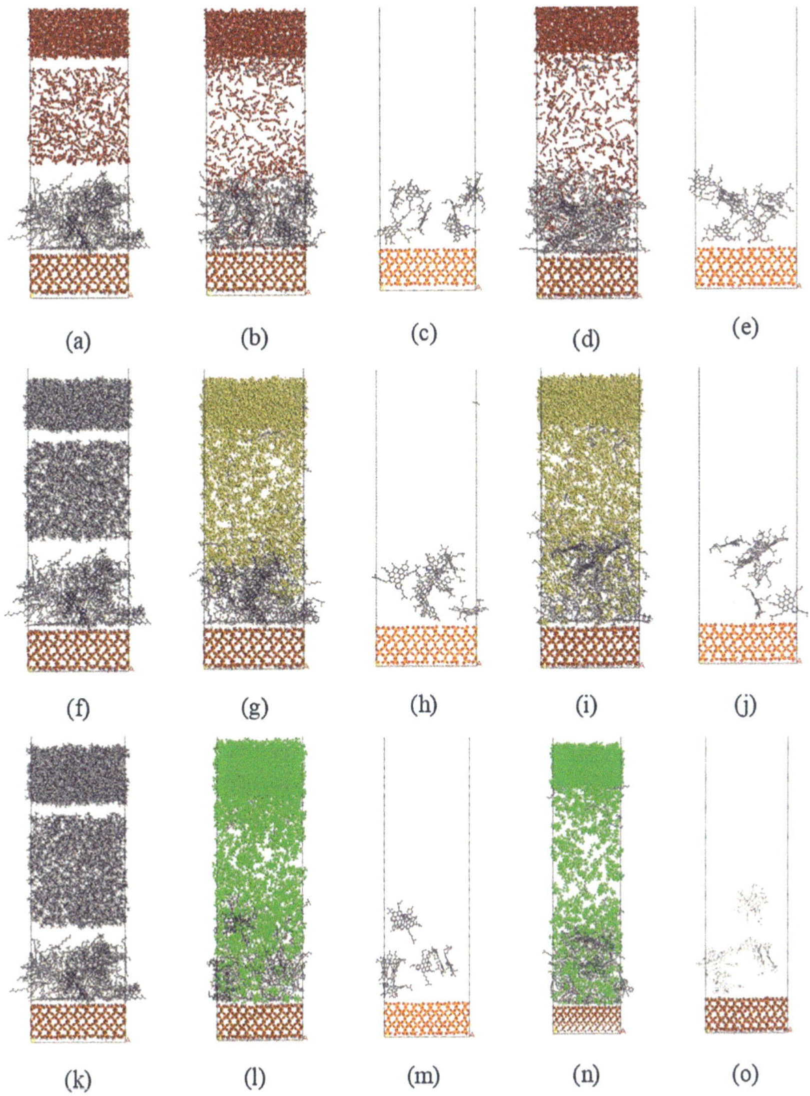

FIGURE 6.11 Simulation model and adsorption configuration in three different saturates, aromatics, resin, and asphaltene plus solvent models: (A) original configuration of model A; (B) adsorption configuration of model A after equilibrium at 333 K; (C) asphaltene molecules of model A after equilibrium at 333 K; (D) adsorption configuration of model A after equilibrium at 393 K; (E) asphaltene molecules of model A after equilibrium at 393 K; (F) original configuration of model B; (G) adsorption configuration of model B after equilibrium at 333 K; (H) asphaltene molecules of model B after equilibrium at 333 K; (I) adsorption configuration of model B after equilibrium at 393 K; (J) asphaltene molecules of model B after equilibrium at 393 K; (K) original configuration of model C; (L) adsorption configuration of model C after equilibrium at 333 K; (M) asphaltene molecules of model C after equilibrium at 333 K; (N) adsorption configuration of model C after equilibrium at 393 K; (O) asphaltene molecules of model C after equilibrium at 393 K.

(model A), the critical temperature is 304 K, which indicates that the two temperature conditions investigated in this section have reached the supercritical temperature condition. As shown in Fig. 6.11B–E, by comparing with the results of the SARA + N_2 model, more CO_2 molecules can enter the oil layer and more CO_2 molecules can interact with the quartz surface. This indicates a higher adsorption selectivity of CO_2 compared with N_2. Therefore, compared with the results of the full SARA model, some adsorption sites can be occupied by CO_2 molecules, and the heavy oil molecules will move far from the quartz surface. Simultaneously, the adsorption configuration between asphaltene molecules and SiO_2 molecules has changed significantly and is dominated by offset stacking and T-shaped stacking. Table 6.2 shows the number of solvent molecules in the oil layer; 16 CO_2 molecules are released from the oil layer by the increase in temperature from 333 to 393 K. This is caused by asphaltene precipitation, and oil aggregates become denser with a rise in temperature.

Second, for C_3H_8 (model B), in Fig. 6.11G–J, C_3H_8 molecules are marked yellow in this figure. The same as for the results of CO_2 molecules, C_3H_8 molecules can enter the oil layer to alter the adsorption configuration between oil molecules and a quartz surface. Especially for the adsorption configuration between asphaltene molecules and SiO_2 molecules, it shows offset stacking and T-shaped stacking at 333 K. However, for 393 K, only T-shaped stacking can be observed. For adsorption configuration between different asphaltene molecules, face-to-face stacking is dominant. Table 6.2 shows that a great difference can be observed for the number of C_3H_8 molecules in the oil layer, as a result of a change in temperature. At high temperature conditions, C_3H_8 molecules can easily interact with oil molecules. Thus, the aggregates become loose and move from the quartz plane.

Third, for C_4H_{10} (model C), compared with CO_2 and C_3H_8, the molecular weight is higher and the dissolving capacity of C_4H_{10} in heavy oil is high.

TABLE 6.2 Number of solvent molecules in the oil layer.

	Number of molecules in oil layer	
Solvent	**333 K**	**393 K**
CO_2	166	150
C_3H_8	174	290[a]
nC_4H_{10}	265[a]	304[a]

[a]*Because the CH_3 groups of C_3H_8 and nC_4H_{10} molecules are still in the solvent layer, the number molecules are only up to the boundary of resin and asphaltene molecules, owing to the high interaction energy between them and solvent molecules.*

Therefore, Table 6.2 shows that more C_4H_{10} molecules can enter the oil layer than CO_2 and C_3H_8. About half of C_4H_{10} molecules in the solvent layer have entered the oil layer. Also, the results of adsorption configurations in Fig. 6.6L–O show that only T-shaped stacking between asphaltene molecules and SiO_2 molecules can be observed under the two temperature conditions. The asphaltene molecules are farther from the pore surface than those for CO_2 and C_3H_8.

6.4.2 Interaction energy

In this section, interaction energy between asphaltene and a pore surface is discussed. As shown in Fig. 6.12, interaction energy highly depends on the solvent type and temperature condition. For a CO_2 atmosphere, interaction energy is slightly increased from −21.48 to −22.39 kcal/mol as the temperature increases. This indicates that CO_2 molecules are released from the oil layer by an increase in temperature. Thus, for a C_3H_8 atmosphere, it has the highest interaction energy among the three different solvents. According to Fig. 6.11 (h), some asphaltene molecules can adsorb onto a quartz plane with π-π stacking. As the temperature rises, the interaction energy drops sharply to −22.55 kcal/mol. This indicates that the aggregates of asphaltene molecules become loose. For an nC_4H_{10} atm, the interaction energy at 393 K is the lowest, indicating that nC_4H_{10} molecules can fully dissolve into the heavy oil layer and reoccupy adsorption sites originally occupied by heavy oil molecules.

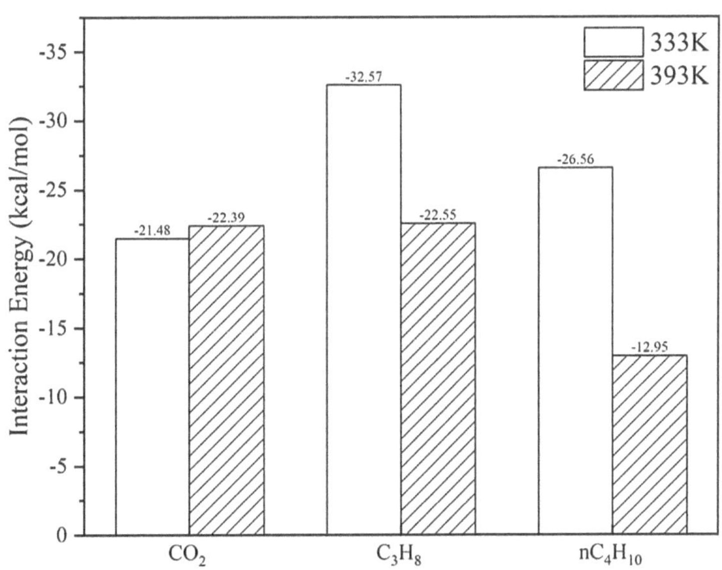

FIGURE 6.12 Interaction energy of asphaltene on quartz at different temperatures in solvent atmosphere.

6.4.3 Extraction behavior of solvent

Compared with other additives, a solvent has a mechanism to extract light oil components. In this section, the extraction and diffusion behaviors of solvents are discussed using MDS. Specifically, for extraction behavior, by analyzing the mass fractions of saturates and aromatics, the proportion of light components extracted from an oil layer can be obtained, which is caused by extraction by a solvent. Moreover, the diffusion coefficients of different solvents can also be calculated.

The mass fractions of light components extracted from an oil layer can be calculated by the formula:

$$m_w = \frac{m_{\text{extracted}}}{m_{\text{total}}} \times 100\% \tag{6.1}$$

where m_w is the mass fraction, wt%; $m_{\text{extracted}}$ is the weight of saturates or aromatics extracted from the oil layer, g; and m_{total} is the origin weight of saturates or aromatics in the oil layer, g.

A diffusion coefficient (D) of a solvent can also be calculated from the equation:

$$MSD = \frac{1}{6} \lim_{t \to \infty} \frac{d}{dt} \langle |R_i(t) - R_i(0)|^2 \rangle \tag{6.2}$$

where MSD is the mean squared displacement.

In Eq. (6.2), the factor 1/6 on the right-hand side is caused by the three dimensions (x, y, z) and two directions [31].

Simulation results are shown in Tables 6.3 and 6.4 and Fig. 6.13. First, for CO_2, Table 6.3 shows that as the temperature increases, CO_2 can extract more light oil components from the heavy oil layer. Moreover, as mentioned in Section 6.4.1, high temperature can enhance the precipitation of asphaltenes in a CO_2 atmosphere. Most CO_2 molecules outside the oil layer can contribute to a higher diffusion coefficient.

TABLE 6.3 Saturates and aromatics fractions extracted from an oil layer in a solvent atmosphere[a].

	m_w, wt.%		
Temperature (K)	CO_2	C_3H_8	nC_4H_{10}
333	13.05	33.65	23.50
393	22.38	59.81	33.53

[a]*Because the CH_3 groups of C_3H_8 and nC_4H_{10} molecules are still in the solvent layer, the number of molecules in this table is only up to the boundary of resins and asphaltenes molecules, owing to the high interaction energy between them and solvent molecules.*

TABLE 6.4 Solvent molecule diffusion coefficient, D ($\times 10^{-8}$ m^2/s).

Solvent	333 K	393 K
CO_2	1.603	4.028
C_3H_8	0.822	1.876
nC_4H_{10}	1.149	1.566

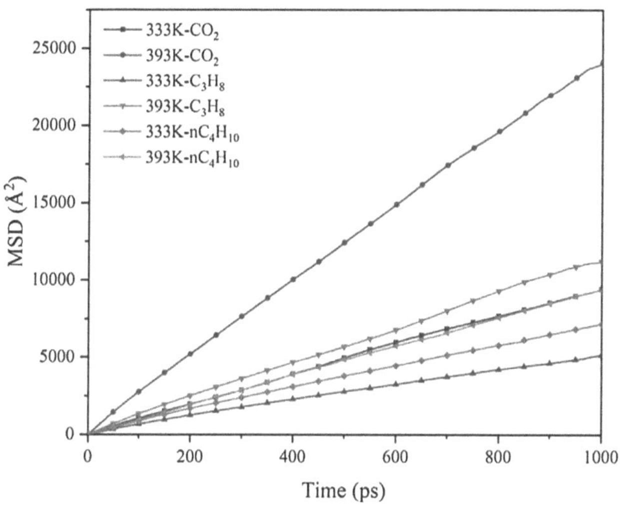

FIGURE 6.13 Mean squared displacement (MSD) of three solvents at 333 and 393 K.

Comparatively, for C_3H_8 and nC_4H_{10}, in Table 6.3, their extraction ability gradually increases with a rise in temperature. These hydrocarbon solvents can extract more light oil components than CO_2. Comparing the extraction behavior of C_3H_8 and nC_4H_{10}, C_3H_8 performs better than nC_4H_{10}, and the diffusion coefficient of nC_4H_{10} at 393 K is lower that of C_3H_8 (Table 6.4). This is because of the strong diffusion capacity of nC_4H_{10} in the heavy oil layer (Fig. 6.11 and Table 6.2), and a large amount of nC_4H_{10} molecules can enter the oil layer, which can benefit swelling of the entire heavy oil layer. Moreover, the heavy oil components can migrate to the top side of the heavy oil layer, and saturates and aromatics also partly move to the top of the heavy oil layer and form a relatively loose aggregate. Thus, the oil—solvents interface is blurred and the mass fraction of light oil components extracted from the oil layer are reduced. This behavior can help control the precipitation of asphaltenes. It also suggests that nC_4H_{10} is a good solvent in a hybrid thermal-solvent process.

Solvent molecules tend to accumulate on a surface of silica and an oil—solvent interface. The extraction and swelling behavior of CO_2 are limited with increasing temperature. C_3H_8 can extract maximum light oil components, whereas nC_4H_{10} can both extract light fractions and control the precipitation of asphaltenes, demonstrating good extraction ability and diffusing capacity. This indicates that nC_4H_{10} can perform better than the other two solvents at the microscopic level.

6.5 Molecular dynamic simulation for a hybrid thermal-chemical process

Different from these two hybrid processes, the mechanisms of a hybrid thermochemical process include emulsification, wettability alteration, and interface tension reduction. From a microscopic point of view, the most important issue for a hybrid thermochemical process is to change the interaction behavior among oil, water, and a pore wall [32—35]. In this section, MDS is applied to this issue. The monomer molecule structure of a chemical agent applied in this section is shown in Fig. 6.14; it is composed of three anionic and nonionic surfactants assembled by a low degree of polymerization and can be applied as a viscosity reducer (VR) for heavy oil. In this section, the degree of polymerization is set to 3. All MDS are last for 1 ns at 333 and 393 K, respectively.

6.5.1 Adsorption configuration

Fig. 6.15 shows an equilibrium adsorption configuration. The VR molecules can spontaneously penetrate a heavy oil layer and significantly change the original adsorption configuration of an asphaltene. The VR molecules can interpenetrate and wind with the asphaltene aggregates, which helps form loose VR—water—oil aggregates. Because of the interaction between VR molecules and a solid wall, the asphaltene molecules are displaced from the quartz plane. In this way, the VR molecules will help the process of depolymerization and viscosity reduction for heavy crude oil.

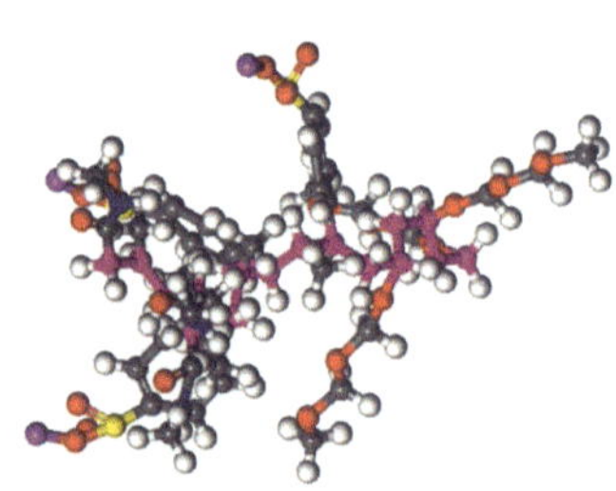

FIGURE 6.14 Monomer molecule structure of a viscosity reducer ($C_5H_{11}NO$—$C_9H_{11}O_3NaS$—$C_5H_{12}O_3$).

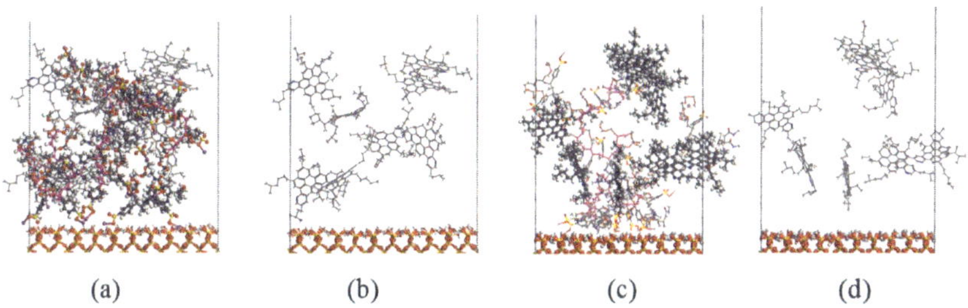

FIGURE 6.15 Adsorption configurations: (A) viscosity reducer and asphaltene adsorption configuration after equilibrium at 333 K; (B) asphaltene molecules after equilibrium at 333 K; (C) adsorption configuration after equilibrium at 393 K; (D) asphaltene molecules after equilibrium at 393 K. Other fractions are hidden for clarity.

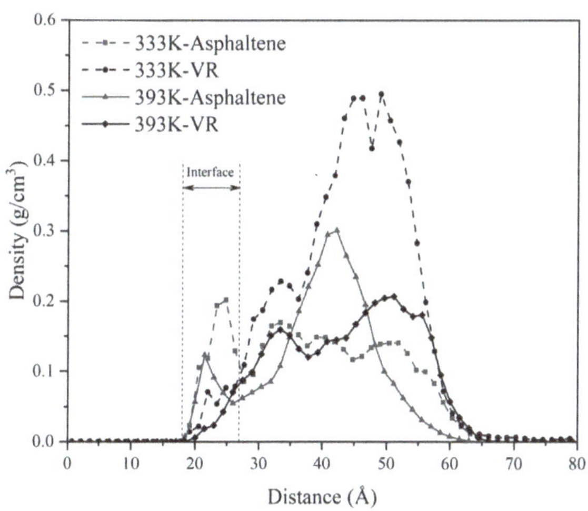

FIGURE 6.16 Density profiles of asphaltene and viscosity reducer (VR) molecules under different temperature conditions.

Density profiles of asphaltene and VR molecules under different temperature conditions are provided in Fig. 6.16. A small proportion of VR molecules can penetrate a deep side of an oil layer and directly interact with a quartz plane at 333 K, whereas as the temperature increases to 393 K, the VR molecules are evenly distributed in the heavy oil layer and thus a peak value of asphaltene stacking at the oil–quartz interface decreases and the peak width is shortened.

6.5.2 Interaction energy

Fig. 6.17 shows the interaction energy of asphaltene molecules and other molecules. Among the three different types of interaction behavior, interaction

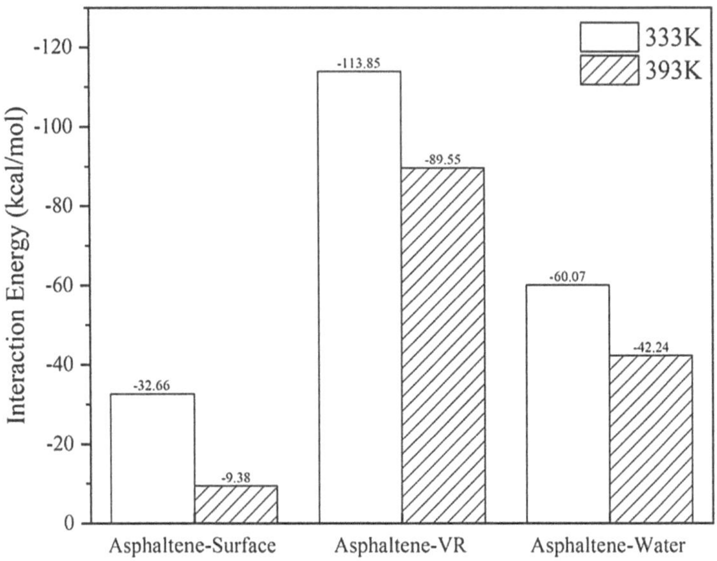

FIGURE 6.17 Interaction energy between asphaltene molecules and other molecules under different temperature conditions. *VR*, viscosity reducer.

energy between asphaltene and VR molecules is always the highest under different temperature conditions and dominates interaction behavior. As the temperature increases, all three different interaction energies decrease because of higher molecular activity at a high temperature. Simulation results in Fig. 6.17 further illustrate that adding a VR solution can alter the interaction behavior of heavy crude oil in porous media and significantly help recover oil.

6.6 Discussion

For the effective development of heavy oil reservoirs, oil viscosity is one of the most important parameters. From a molecular point of view, the molecular structure and molecular weight of different components have a dominant role in oil viscosity. Fluid viscosity can show an exponential dependence on molecule weight. If the molar fraction of large molecules is high, the fluid viscosity is high.

Table 6.5 shows the average molecular weight and molar fraction of SARA components in some typical heavy crude oils. In the SARA components, the average molecular weight of asphaltene is the highest and that of saturates is lowest. From the molar fractions of different heavy oils, the molecular weight increases with the molecular polarity. With an increase in the molar fractions of resin and asphaltene, the molecular weight of heavy oil molecules increases. Thus, the polarity and oil viscosity increase. For molecular polarity, a dipole moment is usually applied to evaluate the strength of polarity quantitatively.

TABLE 6.5 Molecular weights and molar fractions of saturates, aromatics, resin, and asphaltene components in some heavy oils.

	Saturates		Aromatics		Resin		Asphaltene		
No.	Molecular weight (g/mol)	Molar fraction (mol %)	Molecular weight (g/mol)	Molar fraction (mol %)	Molecular weight (g/mol)	Molar fraction (mol %)	Molecular weight (g/mol)	Molar fraction (mol %)	Oil viscosity (cp@80°C)
1	517	35	683	32	1280	26	1835	7	295
2	708	25	869	29	2333	37	3780	10	1844
3	591	23	870	29	2360	42	2461	5	3429
4	528	21	920	29	2434	39	4743	10	5532
5	467	26	666	27	1279	37	3148	10	8146
6	526	21	735	25	2108	46	2631	8	14,015
7	497	21	666	27	1444	38	5777	14	20,283
8	559	18	841	25	1911	46	3232	12	46,990

Also, the higher the dipole moment, the higher the molecular polarity. Similarly, for SARA components in heavy crude oil, the dipole moment of asphaltene is the highest and that of saturates is the lowest. This is caused by the more complicated molecular structure of asphaltene compared with the other three components.

For molecular structures and compositions of different components in heavy crude oil, the molecules of saturates and aromatics usually do not contain metallic elements. However, for resin and asphaltene, their metallic elements can be commonly observed, including nickel, vanadium, calcium, magnesium, and iron. The elements nickel and vanadium are the most important biomarkers in petroleum. During the formation of petroleum, they can interact with large molecules in terms of porphyrin and nonporphyrin chelates. Because the proportions of nickel and vanadium in resin and asphaltene are higher, under the effect of a chelating process, the system viscosity will be high. On the other hand, the chelating process is also influenced by the proportion of heteroatoms. Therefore, for heavy crude oil, the higher the molar fraction of resin and asphaltene, the higher the elemental proportion of sulfur, nitrogen, nickel, and vanadium, and the higher the oil viscosity.

Based on a supermolecular structure model of asphaltene, a molecular structure of asphaltene such as supermolecules can be classified as a unit piece, crystal-like association, micelle, supermicelle, clusters, floccule, and liquid crystal. In an MDS process, a unit piece is usually the basic structure of a supermolecular structure. For asphaltene, a π-π complexation effect is the dominant interaction between the unit pieces of the asphaltene structure. In addition, the effects of a bridged link among heteroatoms, metallocene complexation, and hydrogen bond are included.

Based on the discussion in this chapter, for the MDS of heavy crude oil, because of complicated intermolecular interactions caused by resin and asphaltene in heavy oil, a special microscopic structure can form. Also, the special molecular structure of resin and asphaltene can induce intermolecular noncovalent bonding. Thus, the difference levels of a supermolecular structure can be formed for suspension in liquid hydrocarbons, and they can mutually combine to form a network structure. Under the effect of van der Waals force, resin molecules with a low molecular weight and low molecular polarity can interact with asphaltene molecules and will adsorb in the vicinity of an asphaltene nucleus to form a spatial repulsion layer. Moreover, the associated colloid of resins and asphaltene will be surrounded by aromatics and further transit to aliphatic hydrocarbons. Asphaltene and resin can be connected by a hydrogen bond to form a certain amount of a micelle system. Therefore, to develop heavy oil reservoirs effectively, destroying or weakening the intermolecular interaction of large molecules of resin and asphaltene will be the most important task. A hybrid EOR process is an important solution for this task. In the next chapter, microscale physical simulations for hybrid EOR processes will be discussed.

References

[1] Mullins CO, Sabbah H, Eyssautier J, et al. Advances in asphaltene science and the Yen-Mullins model. Energy Fuels 2012;26:3986–4003.

[2] Mullins CO, Seifert JD, Zuo YJ, et al. Clusters of asphaltene nanoaggregates observed in oilfield reservoirs. Energy Fuels 2013;27(4):1752–61.

[3] Mullins CO. The asphaltenes. Annu Rev Anal Chem 2011;4:393–418.

[4] Agrawal AM, Yarranton HW. An asphaltene association model analogous to linear polymerization. Ind Eng Chem Res 2001;40(21):4664–72.

[5] Rogel E. Simulation of interactions in asphaltene aggregates. Energy Fuels 2000;14(3):566–74.

[6] Ruiz-Morales Y, Mullins CO. Coarse-grained molecular simulations to investigate asphaltenes at the oil–water interface. Energy Fuels 2015;29(3):1597–609.

[7] Ruiz-Morales Y, Alvarez-Ramírez F. Mesoscale dissipative particle dynamics to investigate oil asphaltenes and sodium naphthenates at the oil–water interface. Energy Fuels 2021;35(11):9294–311.

[8] Xu J, Xue S, Zhang J, et al. Molecular design of the amphiphilic polymer as a viscosity reducer for heavy crude oil: from mesoscopic to atomic scale. Energy Fuels 2021;35(2):1152–64.

[9] Mahdavi E, Zebarjad FS, Ayatollahi S, Taghikhani V. Experimental investigation on the effect of asphaltene types on the interfacial tension of CO_2-hydrocarbon systems. Energy Fuels 2015;29:7941–7.

[10] Coelho R, Ovalles C, Hascakir B. Clay-asphaltene interaction during hybrid solvent-steam injection into Bitumen reservoirs. SPE 18072 presented at the SPE Canada heavy oil technical conference, Calgary, Alberta, Canada, 7-9 June, 2016.

[11] Hoepfner PM, Limsakoune V, Chuenmeechao V, Maqbool T, Fogler SH. A fundamental study of asphaltene deposition. Energy Fuels 2017;27(2):725–35.

[12] Takanohashi T, Sato S, Tanaka R. Molecular modeling of asphaltene aggregates and computer simulation of their relaxation behaviors. Am Chem Soc Div Fuel Chem 2001;46(2):357–8.

[13] Hosseini-Dastgerdi Z, Tabatabaei-Nejad SAR, Khodapanah E, et al. A b comprehensive study on mechanism of formation and techniques to diagnose asphaltene structure; molecular and aggregates: a review. Asia Pac J Chem Eng 2015;10:1–14.

[14] Mansoori AG, Riazi RM, Vafaie-Sefti M. An analysis of methods for determination of onsets of asphaltene phase separations. J Petrol Sci Eng 2004;42(2–4):145–56.

[15] Yaseen S, Mansoori GA. Molecular dynamics studies of interaction between asphaltenes and solvents. J Petrol Sci Eng 2017;156:118–24.

[16] Silva SH, Sodero A, Bouyssiere B, et al. Molecular dynamics study of nanoaggregation in asphaltene mixtures: effects of the N, O, and S hcteroatoms. Energy Fuels 2016;30:5656–64.

[17] Silva SH, Alfarra A, Korb J, et al. The role of metalloporphyrins on the physical-chemical properties of petroleum fluids. Fuel 2017;188:374–81.

[18] Headen TF, Boek ES, Jackson G, et al. Simulation of asphaltene aggregation through molecular dynamics: insights and limitations. Energy Fuels 2017;31:1108–25.

[19] Fang T, Wang M, Li J, et al. Study on the asphaltene precipitation in CO_2 flooding: a perspective from molecular dynamics simulation. Ind Eng Chem Res 2018;57:1071–7.

[20] Fang T, Zhang Y, Yan Y, et al. Molecular insight into the oil extraction and transport in CO_2 flooding with reservoir depressurization. Int J Heat Mass Tran 2020;148:119051–8.

[21] Tang X, Xiao S, Lei Q, et al. Molecular dynamics simulation of surfactant flooding driven oil-detachment in nano-silica channels. J Phys Chem B 2019;123(1):277–88.

[22] Tang J, Qu Z, Luo J, et al. Molecular dynamics simulations of the oil-detachment from the hydroxylated silica surface: effects of surfactants, electrostatic interactions, and water flows on the water molecular channel formation. J Phys Chem B 2018;122(6):1905–18.

[23] Rashid Z, Devi Wilfred C, Gnanasundaram N, et al. A comprehensive review on the recent advances on the petroleum asphaltene aggregation. J Petrol Sci Eng 2019;176:249–68.

[24] Ahmadi M, Chen Z. Insight into the interfacial behavior of surfactants and asphaltenes: molecular dynamics simulation study. Energy Fuels 2020;34(11):13536–51.

[25] Mullins CO. The modified yen model. Energy Fuels 2010;24:2179–207.

[26] Ji D, Liu G, Zhang X, et al. Molecular dynamics study on the adsorption of heavy oil drops on a silica surface with different hydrophobicity. Energy Fuels 2020;34:7019–28.

[27] Fang T, Zhang Y, Ding B, et al. Static and dynamic behavior of CO_2 enhanced oil recovery in nanoslits: effects of mineral type and oil components. Int J Heat Mass Tran 2020;153:119583.

[28] Hu M, et al. Molecular dynamics simulation of asphaltene deposition during CO_2 miscible flooding. Petrol Sci Technol 2011;29(12):1274–84.

[29] Fang T, Wang M, Gao Y. Enhanced oil recovery with CO_2/N_2 slug in low permeability reservoir: molecular dynamics simulation. Chem Eng Sci 2019;197:204–11.

[30] Liu B, Wang C, Zhang J, et al. Displacement mechanism of oil in shale inorganic nanopores by supercritical carbon dioxide from molecular dynamics simulations. Energy Fuels 2017;31(1):738–46.

[31] Benkabou F, Aourag H, Becker JP, Certier M. Molecular dynamics study of zinc-blende GaN, AlN and InN. Mol Simulat 2000;23(4–5):327–41.

[32] Zhou L, Yan Y, Li S, et al. Molecular dynamic simulation study on formation of water channel in oil film detachment process controlled by surfactant polarity. Chem Phys Lett 2021;771(16):138502.

[33] Aminian A, ZareNezhad B. Oil-detachment from the calcium carbonate surfaces via the actions of surfactant, nanoparticle and low salinity brine: an insight from molecular dynamic simulation. Chem Eng Sci 2019;202(20):373–82.

[34] Li P, Asim Khan M, Xia M, et al. Efficient preparation and molecular dynamic (MD) simulations of gemini surfactant modified layered montmorillonite to potentially remove emerging organic contaminants from wastewater. Ceram Int 2019;45(8):10782–91.

[35] Ivanova AA, Cheremisin NA, Barifcani A, et al. Molecular dynamics study of the effect of sodium and chloride ions on water-surfactant-hydrocarbon interfaces. Chem Phys 2021;548(1):111243.

Chapter 7

Microscale experiments for hybrid enhanced oil recovery processes

7.1 Microscale experimental methods

Microscale experiments are important physical simulation techniques in petroleum engineering. Through microscale experiments, pore-scale fluid–solid interactions and fluid-exchanging mechanisms can be observed intuitively [1–3]. Visualized models are commonly used experimental models in microscale experiments. They include visualized etch chips and visualized glass bead micromodels. Both can be applied to study a microscale physical process in porous media.

7.1.1 Visualized etch chips

Visualized etch chips have been widely applied to simulations of recovery processes for conventional and unconventional petroleum reservoirs [4–6]. Experiments in a visualized etch chip are usually called tests on a chip, as shown in Fig. 7.1.

Considering the advantages of small-scale pores in etch chips, more accurate fluid flow behavior can easily be observed by a high-speed microscope compared with traditional experiments. On the other hand, compared with an

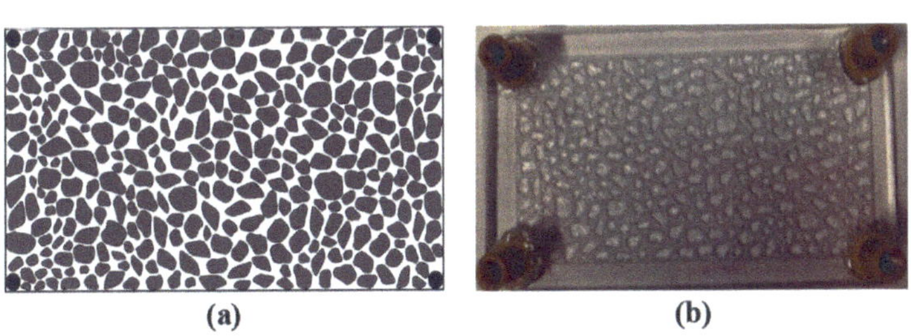

FIGURE 7.1 Typical visualized etch chip. (A) Chip design. (B) Etch chip.

Hybrid Enhanced Oil Recovery Processes for Heavy Oil Reservoirs.
https://doi.org/10.1016/B978-0-12-823954-4.00002-3

actual fluid flow environment in a reservoir, four limitations can be found in etch chips:

(1) In visualized etch chips, pores and throats are usually simulated by different categories of grooves on glass [7–9]. Therefore, they can be considered to be two-dimensional models and differ significantly from an actual pore structure of a reservoir. Tests on an etch chip cannot truly reflect fluid flow characteristics in porous media.
(2) Because of the characteristics of visualization, etch chips are usually manufactured using glass. Therefore, they have low-pressure capability. Only tests under low-pressure conditions are feasible. The maximum allowable pressure is usually below 1 MPa.
(3) The temperature resistance of etch chips is low; thus, tests under high-temperature conditions usually are not recommended.
(4) For experiments using an etch chip, a high-speed microscope is necessary. Thus, considering the requirement of an operation environment via a microscope, the environmental temperature in each experiment should be low. In addition, experiments under room temperature conditions are recommended.

7.1.2 Visualized glass bead micromodels

To explore microscale mechanisms of thermal recovery processes, another type of visualized model is developed [10,11]. Different from etch chips, this type of visualized model is developed using two high-pressure quartz glass sheets. Between the glass sheets, one or two layers of glass beads fill to simulate a porous medium, as shown in Fig. 7.2. After glass beads completely fill the space between glass sheets, the two glass sheets are spliced with a high-temperature sealant to prevent fluid leakage during experiments. Also, a

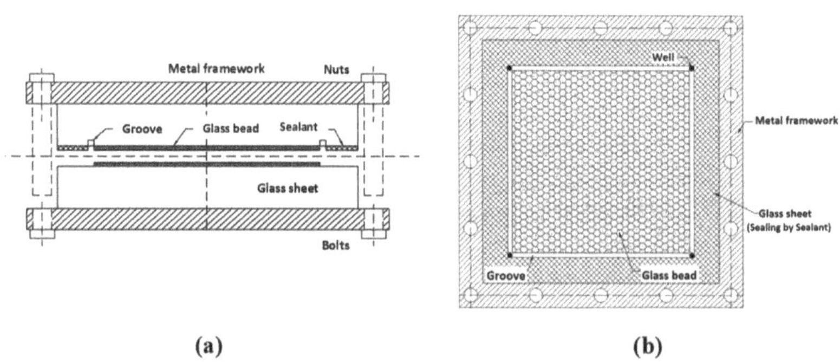

FIGURE 7.2 Visualized glass bead micromodel. (A) Side view. (B) Top view.

hollow metal framework is used to clamp this model [12]. Moreover, around the four corners of this micromodel, four predefined holes are set to simulate injection or production wells, as shown in Fig. 7.2B. Finally, compared with visualized etch chips, this micromodel has a higher pressure capacity, and the maximum allowable pressure can reach 3 MPa.

During experiments, this model can be connected to a steam generator, a fluid container, a microsyringe pump, a high-speed microscope (image acquisition system), and a liquid acquisition device to simulate different recovery processes.

7.1.3 Experimental methods

According to different experimental purposes, two experimental methods exist: static and dynamic. Through a series of static and dynamic experiments, fluid states and microscale flow behaviors in reservoirs can be well-illustrated.

7.1.3.1 Static experiments for a fluid occurrence state in porous media

The purpose of static experiments is to study a fluid state in porous media [13,14]. For a given petroleum reservoir, a fluid occurrence state in the reservoir has an important role in evaluating a reservoir type and oil reserves (e.g., a saturated or unsaturated state, and a gas or liquid state). Especially for unconventional tight and shale reservoirs, because of their small pore sizes, a fluid occurrence state is significantly different from that in conventional reservoirs. Similarly, for heavy oil reservoirs, because of interactions among heavy oil, formation water, and rock surfaces, fluid distribution characteristics are tremendously altered from their original occurrence states in these reservoirs. Based on static experiments under different conditions, a fluid occurrence state can be well-evaluated.

7.1.3.2 Dynamic experiments for fluid flow behavior in porous media

Different from static experiments, the purpose of dynamic experiments is to explore fluid flow behavior in pores under different conditions [15,16]. Especially for various recovery processes for heavy oil reservoirs (pure steam injection and hybrid steam-additive processes), a series of complicated reactions of heavy crude oil and hybrid additives in porous media (e.g., steam condensation, emulsification, and flow redirection) makes it difficult to evaluate pore-scale fluid flow behavior accurately. In this situation, through a set of data and image acquisition processes, microscale experiments can provide a visualized technique to identify and characterize details of fluid flow in porous media.

7.2 Microscale experiments on pure steam injection

7.2.1 Experimental method

Using a visualized glass bead micromodel proposed in Section 7.1.2, a steam injection process in heavy oil reservoirs is simulated [17]. The size of this micromodel is 20 cm × 20 cm. Silica glass beads with size 40 mesh are selected to simulate a porous medium. During experiments, a high-speed microscope is applied to record the imaging results; the collected images are used to evaluate oil saturation distribution and the recovery performance of steam flooding. The experimental setup is shown in Fig. 7.3. The entire experimental setup includes fluid injection, a model, and data acquisition.

The detailed experimental procedure is:

- ➢ connecting experimental apparatuses,
- ➢ testing gas tightness to check for leaks,
- ➢ injecting simulated formation water into the micromodel,
- ➢ injecting heavy oil to simulate initial oil saturation distribution,
- ➢ setting the temperature to the required temperature condition (reservoir temperature) for more than 24 h,

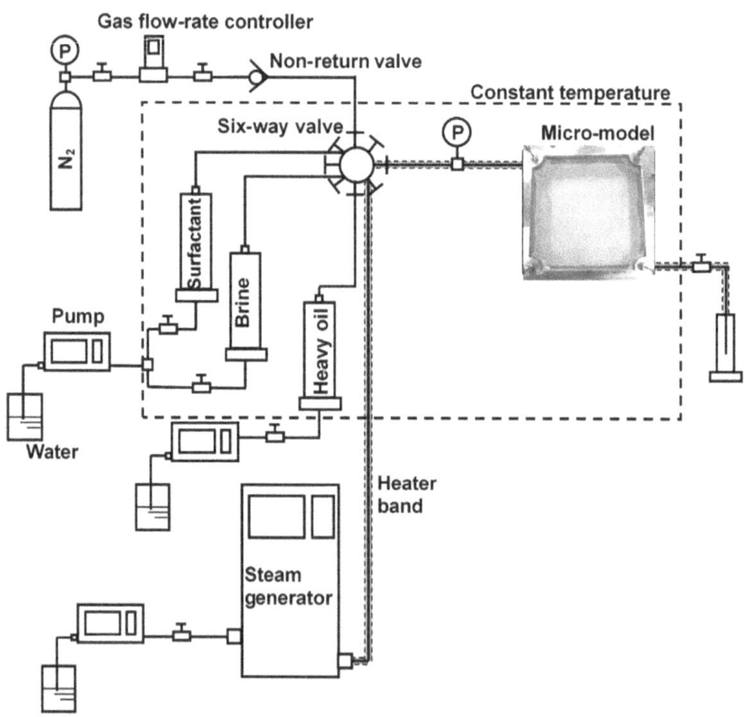

FIGURE 7.3 Experimental setup of visualized microscale experiments.

- ≻ setting the steam injection rate to the required rate to simulate steam flooding until the water cut reaches above 95%,
- ≻ acquiring data acquisition for images during the experiments, and
- ≻ collecting fluid samples after experiments.

7.2.2 Experimental results

7.2.2.1 Oil saturating process

Fig. 7.4 shows an oil saturating process for this micromodel. As heavy crude oil is injected into the micromodel, a uniform oil front can be observed and no obvious breakthrough phenomena are found because of the higher viscosity of the displacing fluid compared with that of the displaced fluid. On the other hand, the uniform oil front indicates the homogeneous properties of this micromodel. After the micromodel is fully saturated with heavy oil, both the inlet and outlet valves are closed, and it is set to the required temperature for about 24 h. Then, it is ready for steam injection.

7.2.2.2 Steam injection

Fig. 7.5 shows the results of steam injection. As the high-temperature steam is injected, the thermally swept volume gradually increases. Because of the large difference in viscosity, an obvious phenomenon of viscous fingering can be

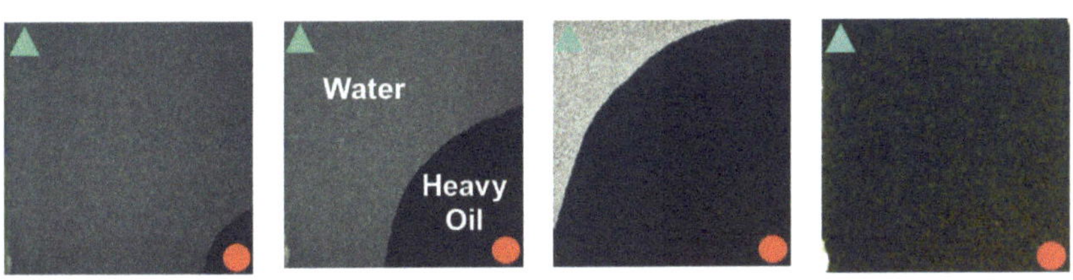

FIGURE 7.4 Oil saturating process.

(a) (b) (c)

FIGURE 7.5 Steam injection process. (A) 0.4 PV. (B) 0.8 PV. (C) 1.2 PV. *PV*, Pore volume.

observed and the injected steam preferentially moves along the main streamlines between the injector and producer. Moreover, a fractal structure can be observed. As the steam front reaches the model outlet, the expansion of steam sweep efficiency is hindered. In reality, the formation of a main steam injection path in a reservoir inhibits the flow of steam, and a large amount of remaining oil can be observed along the two sides of the steam injection path. For pure steam injection, a steam swept volume mainly concentrates around the main stream line, and the expansion of sweep efficiency is limited.

The imaging results in Fig. 7.5 show that two different types of remaining oil can be observed after steam injection, including the corner-trapped remaining oil and the bypassed remaining oil. The formation of the corner-trapped remaining oil highly depends on the wettability and pore structure of the model; the bypassed remaining oil is mainly caused by the microscopic heterogeneity of a micromodel. Because of the low steam injection rate, the injected steam and the steam condensate must bypass the flowing paths with high flow resistance.

7.3 Microscale experiments on a hybrid thermal–noncondensable gas process

Using the same experimental setup as in Fig. 7.3, the recovery performance of a hybrid thermal–noncondensable gas (NCG) process can be experimentally studied [18]. The NCG component used in this section is nitrogen. Because of the low thermal conductivity and large compressibility of nitrogen, it is highly recommended to co-inject it with steam to enhance thermal recovery. By observing the transportation behavior of fluids in porous media, the effect of nitrogen on the recovery of a hybrid thermal-NCG process can be studied.

7.3.1 Recovery enhancement of hybrid thermal–noncondensable gas process

Fig. 7.6 shows images collected during steam-nitrogen co-injection in this micromodel. At the end of pure steam injection, in which nitrogen is injected with steam, because of the interaction of nitrogen and pore surfaces, truncation of the continuous nitrogen stream can be observed. Thus, the original fluid flow in the porous medium is significantly affected by small nitrogen bubbles. For such nitrogen gas bubbles, some can move with oil and water and are finally produced from the outlet. On the other hand, some dispersed nitrogen bubbles can be trapped and gather in relatively large pores, and they may stay in these pores for a certain amount of time. As shown in Fig. 7.6A, the trapped gas bubbles can show the important performance of temporary plugging. Fig. 7.6 demonstrates that the width of fluid flow path A is initially larger than that of fluid flow path B. Fluid flow path A represents relatively large pores, and path B represents relatively small pores in a porous medium. Therefore,

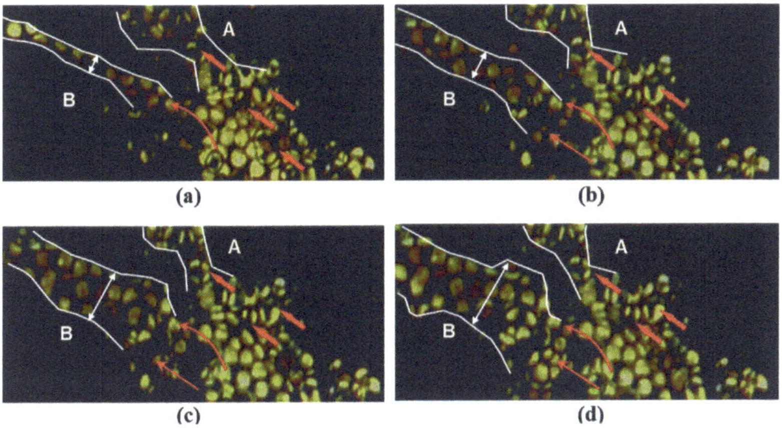

FIGURE 7.6 Temporary plugging behavior of noncondensable gases in pores.

with nitrogen injection, the injected gas primarily accumulates in the relatively large pores. As the gas bubbles gather around path A, the subsequent injected fluid (steam and nitrogen) will redirect to fluid flow path B. Therefore, an obvious change in the width of fluid flow path B can be observed (Fig. 7.6), and the swept area around fluid flow path A remains in the initial state. Comparatively, the sweep efficiency of this micromodel can be significantly improved under the effect of nitrogen and steam co-injection. In addition, the results of a pressure drop show that the co-injection of steam and nitrogen can raise the pressure drop from 400 to 660 kPa. Compared with pure steam injection, recovery increases by about 10.7%.

The imaging results also show that nitrogen can have an indirect influence on emulsification in porous media. On the one hand, the injected nitrogen can further help the movement of emulsion droplets in porous media. On the other hand, the temporary plugging behavior of nitrogen is helpful for redirecting subsequent fluid injection. Thus, it can significantly increase sweep efficiency.

7.3.2 Antiwater coning behavior of a hybrid thermal–noncondensable gas process

Another important application of NCGs used in heavy oil reservoirs is to prevent water coning. For heavy oil reservoirs with a bottom/boundary aquifer, in a later stage of a thermal recovery, water coning tremendously inhibits improvements in recovery. Considering the unique characteristics of NCGs, they have been applied to prevent water coning in heavy oil reservoirs with aquifers. After injection into a reservoir, NCGs will primarily enter low-resistance water-channel paths. Under the effect of dispersed gas bubbles, water coning paths can be effectively plugged, and thus the recovery will be significantly improved. In this section, using the same experimental setup as in Fig. 7.3, antiwater coning of nitrogen in heavy oil reservoirs is discussed [19].

Fig. 7.7A–C shows images of water coning. As the process continues, obvious water coning can be observed. In addition, because of the effect of a pressure drop along the wellbore, water coning is usually first observed around a heel-end section of the wellbore. As production from the producer continues, the degree of the water coning along the wellbore gradually increases. Once water coning is observed in a toe-end section of the horizontal well, production is terminated and nitrogen injection is activated to prevent water coning.

The results in Fig. 7.7D–E demonstrate that after nitrogen is injected into the model, nitrogen will preferentially enter the water channeling paths. Thus, the bottom water will be pushed back to the aquifer. It can be also observed that nitrogen exists as a dispersed state around the producer. However, in the water channeling paths, the nitrogen is distributed as a continuous gas slug. Fig. 7.7F shows the results of a reproduction process of the producer. During reproduction, a distribution range of the continuous gas slug is reduced. However, the dispersed gas state region does not present a significant change.

From the experimental results, the mechanisms of antiwater coning by nitrogen can be summarized thus:

(1) After nitrogen is injected into a water zone, a dispersed state in the gas phase can easily be formed. Although the dispersed gas state can occupy a large amount of pore space and have a low stability, it can still plug the pores and prevent water coning.

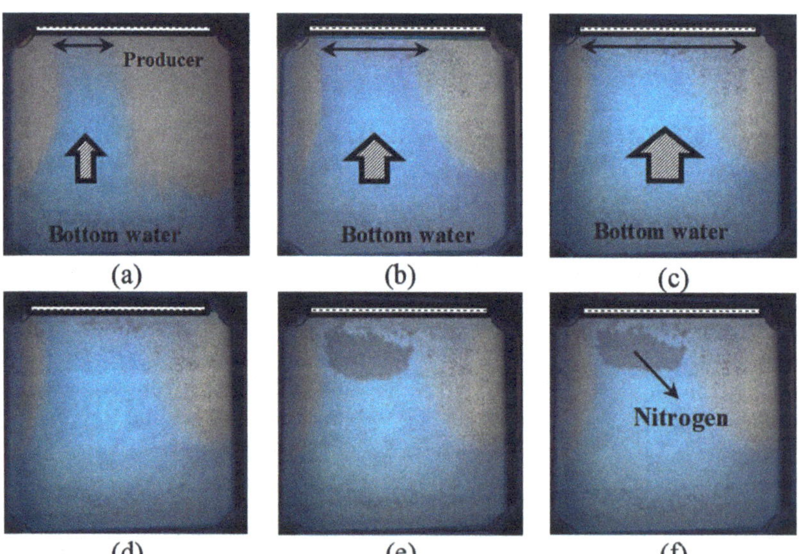

FIGURE 7.7 Antiwater coning of nitrogen. (A–C) Water coning process; (D–E) nitrogen injection process for antiwater coning; (F) reproduction process.

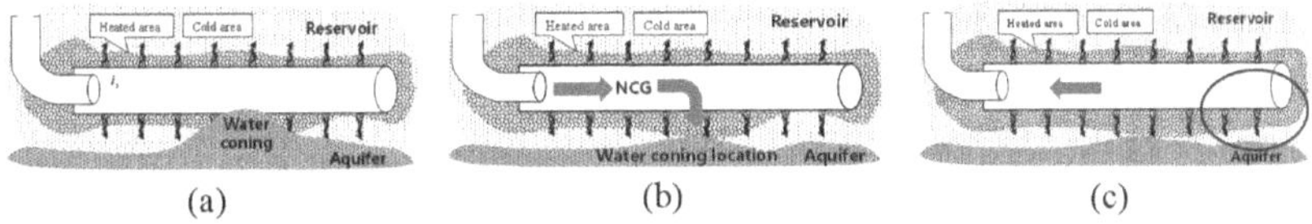

FIGURE 7.8 Schematic of antiwater coning by noncondensable gas (NCG) in a horizontal well. (A) Water coning; (B) NCG injection; (C) horizontal well reopened.

(2) The movability of dispersed nitrogen is significantly reduced and it can effectively prevent water coning.
(3) During movement, large dispersed nitrogen can separate to form a series of small gas bubbles.
(4) The liquid-membrane geometry of dispersed nitrogen can result in a small change during movement in a porous medium. It does not have an important influence on antiwater coning.

Fig. 7.8 illustrates the process of antiwater coning by NCG. Considering a long horizontal wellbore, water coning can be observed in a high-permeability segment along the horizontal well (Fig. 7.8A). In an actual steam injection operation in the field, because of the great heterogeneity in permeability along a horizontal wellbore, it can be commonly observed. High permeability indicates low flow resistance, and the possibility of water coning is higher. After water coning is observed in a horizontal well, and if no treatment is performed, the water coning wellbore segment will continuously expand until the whole wellbore is affected. NCG injection is a potential controlling method for water coning.

For antiwater coning by NCG, the injected NCG will preferably enter a reservoir through water coning wellbore segments (Fig. 7.8B). This is caused by the low flow resistance of the water coning segments. Then, in the reservoir area around the water coning segments, once NCG is injected, the bottom water will be successfully pushed back to an aquifer zone. On the other hand, a gas–water–oil three-phase flow zone can be created. For a three-phase flow process, flow resistance is significantly enhanced. Thus, once a horizontal well is reopened, because of the large flow resistance, the previous water coning wellbore segments will be well-controlled. Moreover, the recovery of the horizontal well is improved.

7.4 Microscale experiments on behavior of heavy oil–CO_2 system in porous media

Different from the NCG of N_2, because of the characteristics of CO_2 dissolution in heavy oil, CO_2 can have the flow behavior of gas and liquid phases. It is usually considered to be a solvent. In this section, a visualized experimental method is proposed to discuss the flow behavior of a heavy oil–CO_2 system in a porous medium.

7.4.1 Experimental method

To investigate the flow behavior of multicomponent multiphases in a porous medium, a visualized experimental method is proposed [20]. The experimental setup is shown in Fig. 7.9. A sand pack model is first used to form complicated multicomponent fluid flow in the medium. Quartz sands with a mesh of 80–120 are applied to simulate the environment of the porous medium. During experiments, heavy oil, steam, and CO_2 are injected into the sand pack model at the same time. Simultaneously, the injection parameters of different fluids (e.g., an injection rate, steam quality, and oil viscosity) can be correspondingly altered based on an experimental design. Then, a visualized window is applied to observe the flow state of a fluid mixture, and a high-speed microscope is applied to record imaging results. The visualized size of this window is 5 cm × 20 cm. With this visualized window and a high-speed microscope, the fluid state in the porous medium can be well-described.

The detailed experimental procedure is:

- ➢ sand pack filling with quartz sand,
- ➢ connecting experimental apparatuses,
- ➢ gas tightness testing to check for leaks,
- ➢ porosity and permeability testing,
- ➢ setting the temperature to the required temperature condition (reservoir temperature) for more than 24 h,

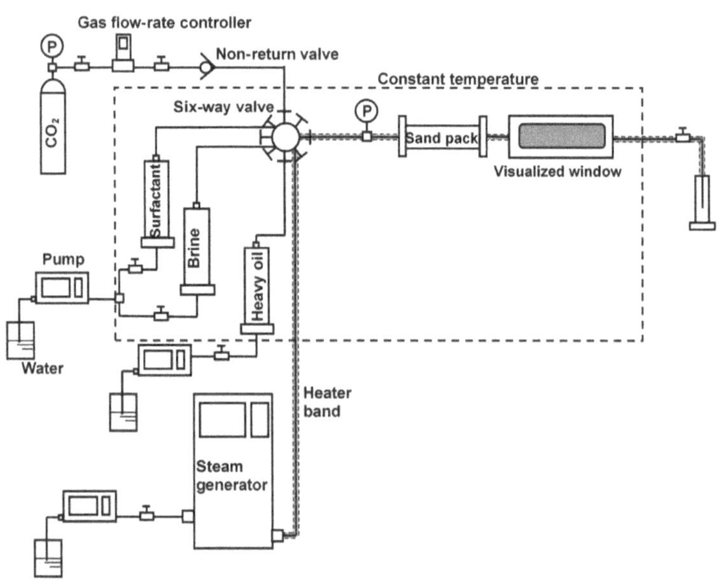

FIGURE 7.9 Experimental setup of visualized experiments for fluid flow state.

- ➢ injecting heavy oil, steam, and CO_2 simultaneously according to a required ratio,
- ➢ data acquisition for images during the experiments, and
- ➢ collecting fluid samples after experiments.

7.4.2 Experimental results

In this section, five different experiments are performed. The detailed experimental parameters are shown in Table 7.1. Based on these experiments, the effect of CO_2 on heavy oil—water flow in a porous medium can be investigated. The experimental results are shown in Fig. 7.10.

For the flow state of a water—CO_2 system in a porous medium, Fig. 7.10A and B shows that during fluid injection, CO_2 is dispersed into gas bubbles with different sizes. Compared with the flow of water, flow resistance in the gas—water two-phase process is higher, and the gas bubbles can plug pore throats well. Considering this behavior of CO_2, it can also be a potential additive to prevent water coning in heavy oil reservoirs with a bottom aquifer. The injection of CO_2 can lead to plugging of the gas phase and also enhanced

TABLE 7.1 Experimental parameters.

No.	Porosity/ %	Permeability/ μm^2	Temperature/ °C	Pressure/ MPa	Water cut/%	Fluid
1	36.6	4.3	25	1	/	Water—CO_2
2	35.9	3.9	50	1	/	Water—CO_2
3	36.2	4.2	25	1	/	Heavy oil—CO_2
4	35.8	4.2	50	1	/	Heavy oil—CO_2
5	36.8	4.1	50	5	20	Heavy oil-steam—CO_2

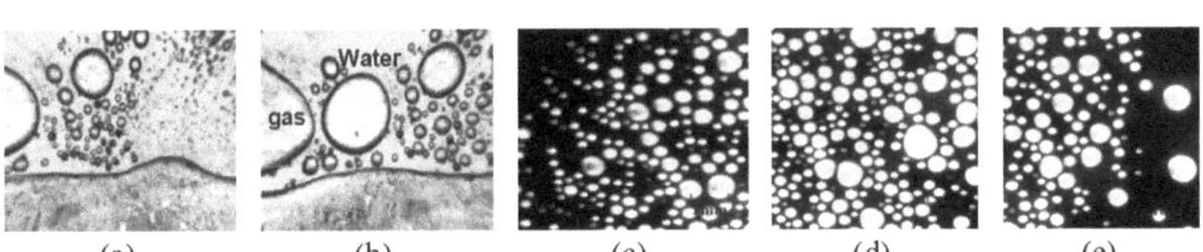

FIGURE 7.10 Experimental results. (A) Experiment 1; (B) experiment 2; (C) experiment 3; (D) experiment 4; (E) experiment 5.

recovery of the liquid CO_2 phase. On the other hand, during experiments in increasing the CO_2 injection rate the size of CO_2 bubbles increased because of shear interaction between CO_2 bubbles and water.

Then, for the flow state of a heavy oil—CO_2 system in a porous medium, Fig. 7.10C and D shows that CO_2 can stay in a heavy oil environment by small gas bubbles with different sizes (below the critical saturation pressure of CO_2). Compared with pure heavy oil, the system has a lower viscosity. This behavior is called foamy oil. When the temperature is raised, the size of gas bubbles is increased and the stability of gas bubbles is reduced. During the flow of gas bubbles in porous media, with an increase in the gas bubble size, gas bubbles can be merged or broken. Furthermore, an increase in temperature indicates a reduction in oil viscosity. Thus, the interaction force between heavy oil and gas bubbles is reduced and the stability is reduced. Fig. 7.10E illustrates that with the addition of the water phase, the number and distribution frequency of large size CO_2 gas bubbles increase. This is because increasing the water cut can reduce the oil viscosity, and confinement of the liquid phase on the gas phase is reduced. Moreover, the addition of the water phase can result in the formation of water—oil emulsion droplets. It can plug steam channeling paths. Furthermore, with an increase in the water cut, a transition from a water—oil emulsion to oil—water emulsion can be observed.

7.5 Microscale experiments on hybrid thermochemical process

In this section, using the same experimental setup as in Fig. 7.3, the recovery performance of different hybrid thermochemical processes is experimentally discussed.

7.5.1 Experiment of nitrogen-foam injection process

Nitrogen-foam injection is a typical hybrid thermochemical process applied to improve sweep efficiency at a later stage of steam injection. In this process, nitrogen and a foaming agent are co-injected into a formation to form a stable foam system. It can effectively plug steam channeling paths and improve steam injection profiles [17,21].

In this section, using the same visualized glass bead micromodel as in Section 7.1.2, steam and then nitrogen-foam injections are combined and simulated. The experimental results are shown in Fig. 7.11. Because of the effect of the microscopic heterogeneity of pore structures, a large amount of bypassed remaining oil and pore-trapped remaining oil can be observed. When a nitrogen-foam system is injected, a previous steam flow zone will be occupied by nitrogen-foam, and thus flow resistance is enhanced. Therefore,

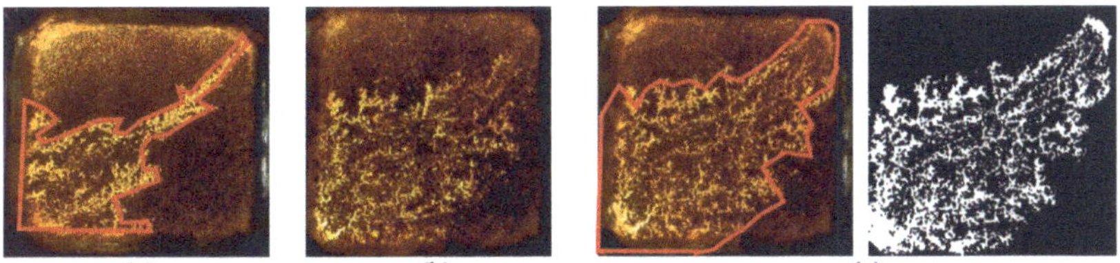

FIGURE 7.11 Results of nitrogen foam injection. (A) End of steam flooding; (B) image of nitrogen-foam injection at 15 min; (C) original image and binarization image for nitrogen-foam injection at 30 min.

the subsequent injected steam will redirect the previous fluid flow paths, and the sweep efficiency will be enhanced. For the imaging results of visualized experiments, a binarization method can be used to evaluate the sweep efficiency of different displacement processes quantitatively. Here, this method is applied to analyze the difference of sweep efficiency of pure steam injection and nitrogen-foam injection. Because of the effect of microscopic heterogeneity, the injection of a nitrogen-foam system can enhance sweep efficiency by 7.41% based on steam injection. On the other hand, when the effect of microscopic heterogeneity is no longer considered, the affected zone of injected fluids (Fig. 7.11) is considered to be a swept region. In comparison, nitrogen-foam injection can enhance sweep efficiency by more than 22.37%.

7.5.2 Experiment of hybrid steam-surfactant process

In this section, using the same visualized microscale model as before, a combination of first steam injection and then surfactant solution injection is simulated. Different from NCG-foam injection, the enhanced oil recovery (EOR) mechanisms of surfactant solution injection include reducing interfacial tension between the oil and water phases and reducing oil viscosity by in situ emulsion. Compared with emulsification in pure steam injection, emulsification behavior in a hybrid steam–surfactant process is significantly different. The former is caused by steam condensation and the latter is caused by a change in interfacial tension between the oil and water phases [12,22,23].

Fig. 7.12 shows the experimental results. For steam injection, the swept area is located in the main streamline and a large amount of scattered remaining oil and concentrated continuous remaining oil can be observed (Fig. 7.12A). Overall, the sweep efficiency of steam flooding is low. Then, at the end of steam flooding, a surfactant solution is injected (Fig. 7.12B and C). Compared with steam flooding, the sweep efficiency is not significantly increased. However, higher sweep efficiency can be observed in the swept region because of the low interfacial tension between heavy oil and water. Thus, an oil film attached on a pore surface can easily be dropped. Then, at the

FIGURE 7.12 Results of hybrid steam–surfactant. (A) End of steam flooding process; (B) surfactant solution injection process at 10 min; (C) surfactant solution injection process at 20 min; (D) end of subsequent steam flooding.

end of surfactant solution injection, steam injection is reactivated (Fig. 7.12D). As shown, under the coeffects of steam and surfactant, the scattered remaining oil around the main streamline is effectively displaced. Simultaneously, the swept area is improved. Based on surfactant solution injection, recovery is increased by about 15.87%.

7.5.3 Experiment of high-temperature gel injection process

Compared with NCG–foam and surfactant injection processes, a high-temperature gel (HTG) system has higher plugging ability. It can be applied to improve the recovery of heavy oil reservoirs with serious steam channeling behavior. For an HTG system, during injection, it is a chemical agent–based solution system, and thus it can be smoothly injected into a reservoir. Then, after an amount of the HTG system is injected and the required gelling time is achieved, the plugging ability of the HTG system will be increased. Thus, it can effectively plug steam channeling paths and improve steam injection profiles [24].

In this section, using the same visualized microscale model as before, a steam injection process and then an HTG injection process are simulated [25]. Fig. 7.13 shows the experimental results. Like the steam injection process in the previous section, at the end of steam flooding, a steam breakthrough path can be observed and a large amount of remaining oil remains in the reservoir

FIGURE 7.13 Results of high-temperature gel (HTG) injection process. (A) Steam flooding; (B) end of steam flooding; (C) HTG injection; (D) subsequent steam flooding.

(Fig. 7.13B). Thereafter, the HTG solution system is injected (Fig. 7.13C). As shown, during HTG system injection, the chemical agents injected flow along the steam channeling paths. Once the designed HTG volume is achieved, the injection process is terminated. Injection of the HTG system will effectively plug the channeling paths, and thus the displacing fluid will be redirected. It can help displace remaining oil in small pores. Compared with the end of steam flooding, sweep efficiency is increased by about 7.8% after HTG injection. Subsequently, steam flooding is restarted, as shown in Fig. 7.13D. With the performance of the HTG system, the previous remaining oil will be effectively unlocked. On the other hand, the solidified HTG can still move and enter the deep reservoir. Compared with the sweep efficiency in Fig. 7.11C, the sweep efficiency at the end of the subsequent steam flooding process is increased by about 29%. Compared with the results of the NCG–foam system and the surfactant system (Figs. 7.11 and 7.12), the improvement in sweep efficiency by the HTG system is the most significant.

7.6 Microscale experiments on movement of emulsion droplets in hybrid processes

Based on the experimental results in the previous sections, emulsification is the most commonly observed phenomenon in both pure steam injection and hybrid EOR. In this section, based on a visualized microscale model, the movement of emulsion droplets in a hybrid steam–surfactant process is discussed [18,26].

7.6.1 Formation of emulsion in porous media

Fig. 7.14A and B shows the oil-water emulsion observed during a hybrid steam–surfactant process. Because of low interfacial tension between surfactant and heavy oil, the addition of surfactant can gradually disperse oil droplets in the porous medium. After a certain interaction between surfactant and oil droplets, the droplets can be totally stripped from the pore surface. Next, the stripped small oil droplets can migrate by oil–water emulsion. On the other hand, under the effect of flocculation and coalescence, these small

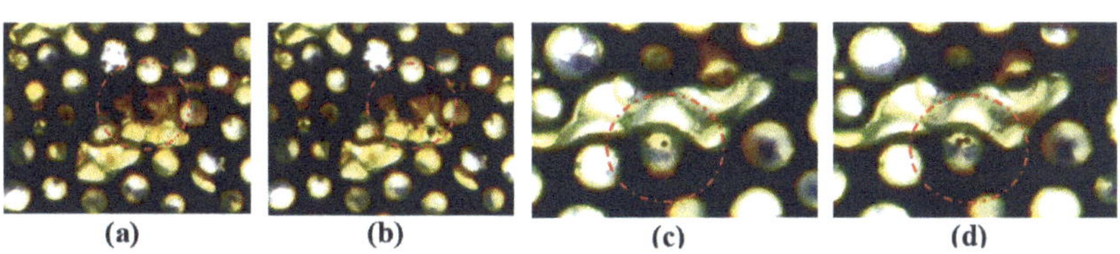

FIGURE 7.14 Oil–water emulsion in a hybrid steam-surfactant process.

droplets can merge in the porous medium and then form a large oil droplet (Fig. 7.14C and D). Moreover, because of the effect of pore surface cutoff and fluid disturbance, a large emulsion can form a series of small emulsions.

7.6.2 Pore blockage of emulsion

Fig. 7.15A and B indicates the blockage behavior of a single emulsion in a porous medium. During fluid flow in the porous medium, a single oil–water emulsion droplet can be stuck in a pore throat. Thus, a fluid flow path will be plugged by emulsion. Also, as fluid injection continues, under the effect of a pressure difference, the emulsion droplet begins to deform and attempts to go through a pore. This behavior of emulsion in a porous medium is called a deformed blockage effect. It is caused by a lower displacement pressure drop along an emulsion flow direction, and the pressure drop has not reached a critical threshold pressure drop. Once the displacement pressure difference across a pore throat is higher than the threshold pressure drop of emulsion droplets, the drops can deform and successfully go through pore throats. This process is called deformed movement.

On the other hand, as multiple emulsion droplets pass through pore throats, mutual extrusion and collision will dominate the process. In such a small space, the accumulation of multiple emulsion droplets can be observed. The accumulation can be classified as bridging accumulation and annular accumulation, as shown in Fig. 7.15C and D. Both accumulation behaviors can reduce effective fluid flow space. During a hybrid steam–surfactant process, the continuous generation, movement, and accumulation of emulsion droplets in a porous medium can lead to a reduction in fluid flow space and an increase in flow resistance. The accumulation of emulsion droplets can form a series of emulsion clusters that can further increase fluid flow resistance and improve sweep efficiency.

In hybrid EOR processes for heavy oil reservoirs, especially for a hybrid thermochemical process, fluid flow is dominated by the movement of emulsion droplets. Because of the complicated interplay between emulsions and pore surfaces, the formation, deformation, and accumulation of emulsion droplets in porous media can significantly affect EOR. For the flow behavior of emulsion droplets in porous media, more details can be found in Section 5.5.4.

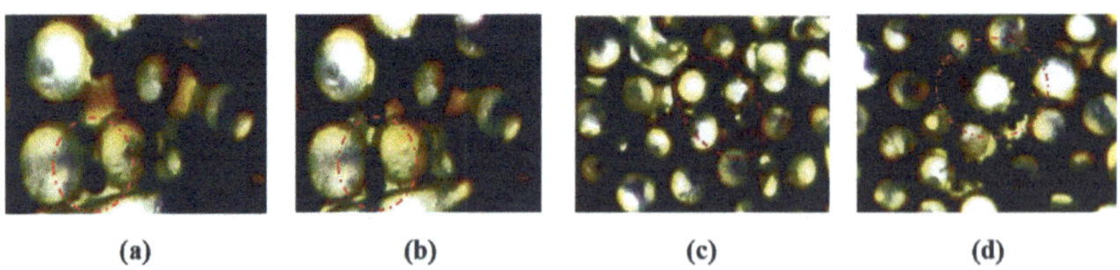

FIGURE 7.15 Oil–water emulsion in a hybrid steam-surfactant process.

7.7 Pore-scale enhanced oil recovery mechanisms of hybrid enhanced oil recovery processes

Based on the previous discussion, it can be concluded that from a microscale point of view, a hybrid EOR process can effectively improve pore-scale displacement efficiency, increase microscale sweep efficiency, and thus enhance heavy oil recovery. In this section, the pore-scale EOR mechanisms of different hybrid EOR processes are summarized.

First, in a hybrid thermal−NCG process, the injection of NCG can take full advantage of NCG and steam. Especially for N_2, because of the low dissolution of N_2 in heavy oil, the injection of N_2 can enter a deep reservoir and increase a heat-swept area. On the other hand, from the experimental observation in Section 7.3, after the injection of N_2, it can be distributed in a porous medium by gas bubbles with different sizes through cutoff behavior. These small nitrogen gas bubbles can help move heavy oil. Moreover, through temporary plugging caused by trapped gas bubbles, sweep efficiency can be increased. This hybrid EOR process can be a potential technique to enhance heavy oil recovery in post−cyclic steam stimulation oil reservoirs. In addition, for heavy oil reservoirs with serious water coning, because of the effect of dispersed gas bubbles, the injection of nitrogen can help prevent water coning.

Second, for a hybrid thermal-solvent process, different from NCG in heavy oil, a solvent can easily dissolve in heavy crude oil. Therefore, compared with pure steam injection, heavy oil viscosity in a hybrid thermal-solvent process is lower. On the other hand, because of the diffusion and extraction behavior of solvent in porous media, heavy oil recovery can be significantly improved based on a steam injection process. In this chapter, based on the microscale experimental observation of a heavy oil−CO_2 system, CO_2 can be present simultaneously in gas and liquid phases (below the saturation pressure of CO_2), and the liquid CO_2 phase can act like a solvent in porous media. For the gas CO_2 phase, it can take full advantage of foamy oil to reduce oil viscosity. In addition, one of the most important issues for a hybrid thermal-solvent process is asphaltene deposition behavior.

Third, for a hybrid thermal-chemical process, compared with hybrid thermal-NCG and thermal-solvent processes, it is applied to plug steam channeling paths and increase sweep efficiency. In this chapter, three different microscale experiments are performed to simulate the recovery of nitrogen-foam injection, surfactant solution injection, and HTG injection. For a nitrogen-foam system, based on steam injection, sweep efficiency in heavy oil reservoirs can be increased by about 22.37%. In comparison, an HTG system has higher plugging ability than an NCG-foam system; from experimental results, sweep efficiency can be increased by about 29% after the steam injection profile control of HTG injection. However, for a surfactant solution, except for the plugging mechanism caused by emulsification, it can reduce heavy oil viscosity. From the microscale experimental results, the sweep efficiency of a surfactant solution injection process increased by only about 3.87%. Compared with the two other hybrid thermal-chemical processes, the increase in sweep efficiency is lower. In this process, however, the scattered remaining oil around a main streamline can be effectively displaced.

In this chapter, a series of microscale experiments is performed to address EOR mechanisms of different hybrid thermal recovery processes. The microscale experimental results show that three different hybrid EOR processes can enhance heavy oil recovery from different aspects. A hybrid thermal–NCG process can improve the recovery performance of steam injection by increasing sweep efficiency and temporary plugging. A hybrid thermal-solvent process can further reduce oil viscosity and extract light oil components. A hybrid thermal-chemical process can effectively plug steam channeling paths and improve steam injection profiles. Although the microscale experiments can well describe fluid flow in pores, considering the low-pressure limitations of a microscale visualized experimental method, macroscale experiments are closer to actual reservoir conditions. In next chapter, a series of macroscale experiments will be performed to explain the reservoir-scale recovery performance of hybrid EOR processes in heavy oil reservoirs.

References

[1] Cui J, Babadagli T. Use of new generation chemicals and nano materials in heavy-oil recovery: visual analysis through micro fluidics experiments. Colloid Surface Physicochem Eng Aspect 2017;529:346–55.

[2] Emadi A, Sohrabi M, Jamiolahmady M, Ireland S. Visualization of oil recovery by CO2-foam injection; effect of oil viscosity and gas type. SPE 152996 presented at the SPE improved oil recovery symposium, Tulsa, Oklahoma, USA, April 14–18, 2012.

[3] Ortiz-Arango JD, Kantzas A. Visualization of viscous coupling effects in heavy oil reservoirs. SPE 117675 presented at the international thermal operations and heavy oil symposium, Calgary, Alberta, Canada, October 20–23, 2008.

[4] Yu F, Jiang H, Xu F, Zhen F, Wang J, Cheng B, Su H, Li J. A multi-scale experimental study of hydrophobically-modified polyacrylamide flood and surfactant-polymer flood on enhanced heavy oil recovery. J Petrol Sci Eng 2019;182:106258.

[5] Argüelles-Vivas FJ, Babadagli T. Pore-scale investigations on the dynamics of gravity-driven steam-displacement process for heavy-oil recovery and development of residual oil saturation: a 2D visual analysis. SPE J 2016;21(6):1943–59.

[6] Ali T, Trivedi J. Evaluation of foam generated with the hydrocarbon solvent for extra-heavy oil recovery from fractured porous media: pore-scale visualization. J Petrol Sci Eng 2017;157:1170–8.

[7] Lele P, Syed AH, Riordan J, Mosavat N, Guerrero A, Fadaei H, Sinton D. Deformation of microdroplets in crude oil for rapid screening of enhanced oil recovery additives. J Petrol Sci Eng 2018:298–304.

[8] Maurice A, Theisen J, Gabriel J-CP. Microfluidic lab-on-chip advances for liquid–liquid extraction process studies. Curr Opin Colloid Interface Sci 2020;46:20–35.

[9] Li J, Su H, Jiang H, Yu F, Liang T, Zhao Y, Gao Y, Hossein H. Application of microfluidic models in oil and gas field development. Petrol Sci Bull 2018;3(3):284–301.

[10] Lago M, Huerta M, Gomes R. Visualization study during depletion experiments of Venezuelan heavy oils using glass micromodels. J Can Petrol Technol 2002;41(1).

[11] Mohammadzadeh O, Chatzis I. Pore-level investigation of heavy oil recovery using steam Assisted gravity drainage (SAGD). IPTC 13403 presented at the international petroleum technology conference, Doha, Qatar, December 7–9, 2009.

[12] Lyu X, Liu H, Pang Z, Sun Z. Visualized study of thermochemistry assisted steam flooding to improve oil recovery in heavy oil reservoir with glass micromodels. Fuel 2018;218:118–26.

[13] Liu Y, Dong X, Chen Z, Hou Y, Luo Q, Chen Y. A novel experimental investigation on the occurrence state of fluids in microscale pores of tight reservoirs. J Petrol Sci Eng 2021;196:107656.

[14] Li J, Liu Y, Gao Y, Cheng B, Meng F, Xu H. Effects of microscopic pore structure heterogeneity on the distribution and morphology of remaining oil. Petrol Explor Dev 2018;45(6):1043–52.

[15] Wang C, Jiang H, Ma M, Xu F, Yu F, Zhao Y, Li J. Study of the variation of pore-scale residual oil flow based on a microfluidic model. Petrol Sci Bull 2020;5(3):376–91.

[16] Wu Y, Fang S, Dai C, Sun Y, Fang J, Liu Y, He L. Investigation on bubble snap-off in 3-D pore-throat micro-structures. J Ind Eng Chem 2017;54:69–74.

[17] Wu Z. Stimulation mechanisms and application of air assisted steam for heavy oil reservoirs. PhD dissertation. Beijing: China University of Petroleum; 2018.

[18] Lu C. The improvement effects of gas and chemical agent during steam injection process for heavy oil reservoirs. PhD dissertation. Beijing: China University of Petroleum; 2015.

[19] Wang Y, Liu H, Chen Z, Wu Z, Pang Z, Dong X, Chen F. A visualized investigation on the mechanisms of anti-water coning process using nitrogen injection in horizontal wells. J Petrol Sci Eng 2018;166:636–49.

[20] Wang L. Technology application and oil displacement mechanisms of in-situ combustion after steam injection in heavy oil reservoirs. PhD dissertation. Beijing: China University of Petroleum; 2018.

[21] Wu Z, Liu H, Pang Z, Wu C, Gao M. Visualization experiment of nitrogen foam assisted steam-flooding in heavy-oil reservoir. Special Oil Gas Reservoirs 2016;23(5):126–9.

[22] Wang Y, Liu H, Zhang Q, Chen Z, Wang J, Dong X, Chen F. Pore-scale experimental study on EOR mechanisms of combining thermal and chemical flooding in heavy oil reservoirs. J Petrol Sci Eng 2020;185:106649.

[23] Wang Y, Liu H, Wang J, Dong X, Chen F. Formulation development and visualized investigation of temperature-resistant and salt-tolerant surfactant-polymer flooding to enhance oil recovery. J Petrol Sci Eng 2019;174:584–98.

[24] Wang C, Liu H, Wang J, Hong C, Dong X, Meng Q, Liu Y. A novel high-temperature gel to control the steam channeling in heavy oil reservoir. SPE 170049 presented at the SPE heavy oil conference-Canada, Calgary, Alberta, Canada, 10–12 June, 2014.

[25] Wu Z, Liu H, Pang Z, Wu Y, Wang X, Gao M. Visualized study on blocking characteristics and EOR mechanisms of high-temperature gel during steam injection process, IPTC 18622 presented at the international petroleum technology conference, Bangkok, Thailand, November 14–16, 2016.

[26] Lu C, Zhao W, Liu Y, Dong X. Pore-scale transport mechanisms and macroscopic displacement effects of in-situ oil-in-water emulsions in porous media. J Energy Resour Technol 2018;140(10):102904.

Chapter 8

Macroscale experiments for hybrid enhanced oil recovery processes

8.1 One-dimensional sand pack displacement experiments

One-dimensional (1D) sand pack displacement experiments are the most commonly used method for studying the displacement efficiency of different recovery processes in heavy oil reservoirs. Because of the unconsolidated characteristics of reservoir rocks, directly obtaining a rock sample from a reservoir is challenging. In this section, a series of 1D sand pack displacement experiments are performed to analyze the displacement efficiency of different hybrid enhanced oil recovery (EOR) processes in heavy oil reservoirs [1,2].

8.1.1 Experimental method

A physical model of a 1D sand pack tube will be applied in this section; it is usually considered a uniform test model. Therefore, the sweep efficiency of a sand pack tube displacement experiment is 100% and only the displacement efficiency can be tested. Fig. 8.1 shows an experimental setup. The sand pack model is 60 cm long and 3.8 cm in diameter. During experiments, it will be filled with unconsolidated quartz sand with a uniform mesh. The maximum affordable pressure of an ISCO pump is about 70 MPa, and the injection rate can be controlled between 0.001 and 60 mL/min.

The detailed experimental procedure is as follows:

- ➢ Sand pack preparation using unconsolidated quartz sand;
- ➢ Connecting the experimental apparatuses;
- ➢ A gas tightness test to check for leaks;
- ➢ Formation injection for a certain volume amount;
- ➢ Porosity and permeability tests;
- ➢ Heavy oil injection to establish the initial oil saturation;
- ➢ Setting the temperature of an oven to be the reservoir condition that stays for more than 24 h, and simulating an aging process;

Hybrid Enhanced Oil Recovery Processes for Heavy Oil Reservoirs.
https://doi.org/10.1016/B978-0-12-823954-4.00004-7

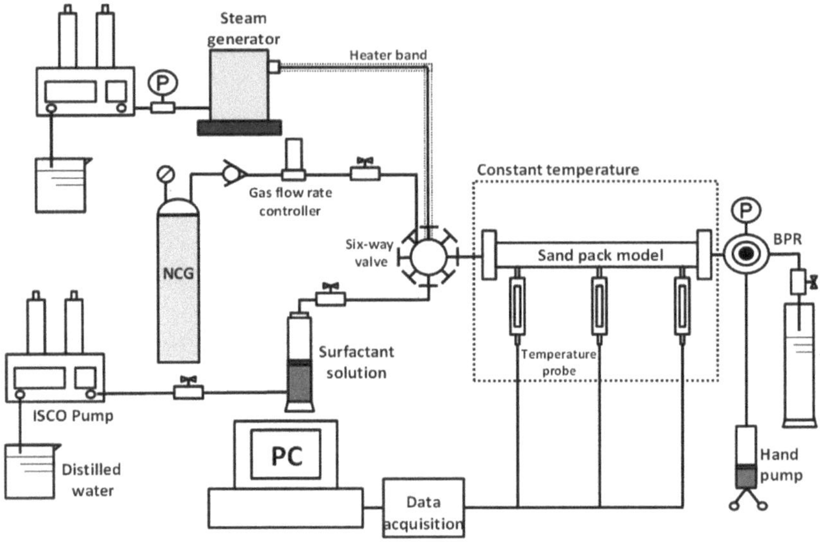

FIGURE 8.1 Experimental setup for sand pack tube displacement experiments. *BPR,* Back pressure regulator.

- ➢ Steam or hybrid fluid injection to simulate a displacement process;
- ➢ Recording the experimental data of liquid production, temperature, and pressure; and
- ➢ Once the water cut at the outlet has reached 98%, an experiment is terminated.

8.1.2 Experimental results of steam−noncondensable gas process

Using the experimental setup in Fig. 8.1, displacement experiments for different hybrid thermal−noncondensable gas (NCG) processes are performed in this section, including steam−CO_2, steam−N_2, and steam−flue gas co-injection. To analyze the effect of oil viscosity on displacement efficiency, two different oil samples are applied; their physical properties are shown in Table 8.1. Based on these two different heavy oil samples, 13 groups of displacement experiments are conducted. The physical properties of the 13 sand pack tubes are shown in Table 8.1.

Table 8.2 shows the specific displacement method, fluid injection parameters, and experimental results of all 13 experiments. Both CO_2 and flue gas can improve the recovery of pure steam flooding. Table 8.1 illustrates that for experiments 1−4, they apply the same oil sample and quartz sand mesh (i.e., the same permeability condition). This indicates that the results are

TABLE 8.1 Physical properties of sand pack tubes in displacement experiments.

No.	Oil sample	Quartz sand mesh	Porosity (%)	Permeability (10^{-3} μm^2)	Initial oil saturation (%)	Residual water saturation (%)
1	Oil I	70–100	34.76	8721	87.54	12.46
2	Oil I	70–100	36.76	9622	88.02	11.98
3	Oil I	70–100	36.82	9184	86.77	13.23
4	Oil I	70–100	36.54	9355	87.49	12.51
5	Oil I	80–120	28.38	4410	90.14	9.86
6	Oil I	80–120	28.67	4901	88.55	11.45
7	Oil I	80–120	28.52	3711	89.72	10.28
8	Oil II	70–100	35.88	8952	90.16	9.84
9	Oil II	70–100	35.29	9132	89.57	10.43
10	Oil II	70–100	36.25	10,571	90.63	9.37
11	Oil II	80–120	28.78	2857	87.21	12.79
12	Oil II	80–120	27.95	3031	89.60	10.40
13	Oil II	80–120	28.35	2886	88.03	11.97

comparable. The co-injection of steam and CO_2 can increase the displacement efficiency of steam flooding by 8.1%; the displacement efficiency of the co-injection of steam and flue gas increases by 3.01%. Therefore, increasing molar fractions of CO_2 in flue gas can improve the recovery of the hybrid steam–flue gas process. In the results of experiments 1 and 4, the displacement efficiency of the hybrid steam-N_2 process is lower by 13.1% than that of the pure steam flooding process. It is caused by a lower solubility of N_2 in heavy oil, and, therefore, the addition of N_2 enhances the occurrence of steam breakthrough. As heavy oil viscosity increases, the solubility of N_2 in heavy oil decreases and a reduction in displacement efficiency caused by N_2 co-injection increases. In addition, by comparing the experimental results of different oil samples, as the oil viscosity increases, the displacement efficiency decreases. In addition, for experimental results with different permeability conditions (cases 1 and 5), as the permeability decreases, the displacement efficiency is correspondingly reduced.

TABLE 8.2 Experimental results of displacement experiments.

No.	Oil sample	Displacement method	Steam injection rate (mL/min)	Steam/gas ratio	Displacement efficiency (%)
1	Oil I	Steam flooding	3	3:1	69.40
2	Oil I	Steam + CO_2 co-injection	3	3:1	77.50
3	Oil I	Steam + flue gas co-injection	3	3:1	72.41
4	Oil I	Steam + N_2 co-injection	3	3:1	56.30
5	Oil I	Steam flooding	3	3:1	51.37
6	Oil I	Steam + CO_2 co-injection	3	3:1	59.33
7	Oil I	Steam + flue gas co-injection	3	3:1	54.24
8	Oil II	Steam flooding	3	3:1	58.82
9	Oil II	Steam + CO_2 co-injection	3	3:1	66.27
10	Oil II	Steam + flue gas co-injection	3	3:1	62.27
11	Oil II	Steam flooding	3	3:1	47.80
12	Oil II	Steam + CO_2 co-injection	3	3:1	58.51
13	Oil II	Steam + flue gas co-injection	3	3:1	53.15

Fig. 8.2 shows the results of a drop in pressure and water cut for the 13 experiments. Fig. 8.2A demonstrates that for the results in cases 1–4, the displacement pressure drop in case 4 (steam plus N_2 co-injection) is the highest among the four processes, and that case 2 (steam plus CO_2 co-injection) is the smallest. Compared with pure steam injection, the pressure drop in steam–CO_2 co-injection is reduced by 15%–40% because of the improvement of CO_2 on oil properties. On the other hand, because of the high solubility of CO_2 in heavy oil, its pressure drop curve is the most stable and the fluctuation is the smallest. The higher the operation pressure, the most obvious the improvement in recovery by CO_2. For steam–flue gas co-injection (cases 3, 7, 10, and 13), the pressure drop is between pure steam injection and

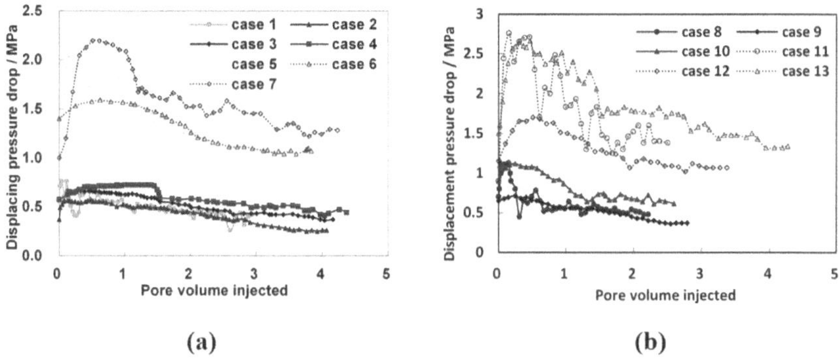

FIGURE 8.2 Results of displacement pressure drop: (A) cases 1–7; (B) cases 8–13.

steam–N_2 co-injection. According to the results of the three hybrid processes, increasing molar fractions of CO_2 can significantly increase the recovery of steam injection. In addition, for the results with different oil samples, because of the high oil viscosity, the pressure drop of oil sample II is the highest.

8.1.3 Experimental results of steam–chemical process

Based on the experimental setup in Fig. 8.1, a sand pack tube model is applied in this section to study the displacement efficiency of a steam-chemical co-injection process [3,4]. The chemical additive used in this section is a viscosity reducer (VR). During experiments, the VR will be co-injected with steam by a water solution (mass concentration: 0.5 wt%). The oil samples used in the experiments are from the Shengli oil field, Sinopec.

8.1.3.1 Ratio of oil viscosity reduction

Chemically reducing the heavy oil viscosity is the main EOR mechanism of a VR. The ratio of an oil viscosity reduction can be applied to evaluate its performance. Before sand pack displacement experiments, two different sets of experiments are performed to analyze the viscosity of a heavy oil–VR solution mixture under different temperature conditions, including tests in an environment with a rising temperature (an injection process) and tests in an environment with a decreasing temperature (a production process). The experimental results are shown in Fig. 8.3. The viscosity of the heavy oil–VR solution mixture is obviously lower than the heavy oil viscosity. However, the changing behavior of the environmental temperature can have an important effect on the viscosity of the mixture system. The mixture viscosity at a rising temperature is lower than that at a decreasing temperature. This indicates the temperature-irreversible behavior of the VR.

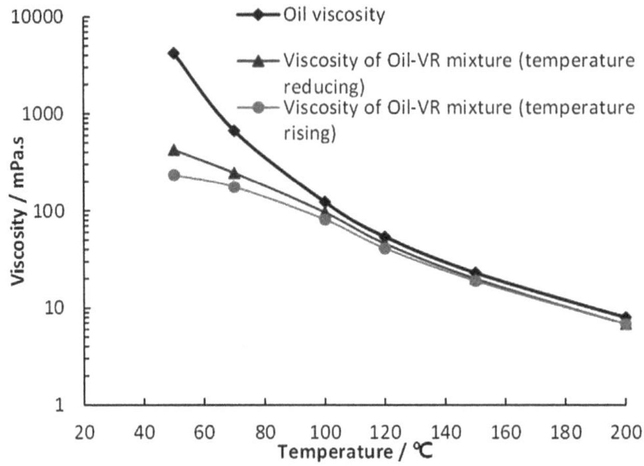

FIGURE 8.3 Mixture viscosities under different conditions. *VR*, viscosity reducer.

8.1.3.2 Displacement experiments

Based on these experimental results, two sets of displacement experiments with different heavy oil samples are performed to study the displacement efficiency of hybrid thermal-chemical processes in heavy oil reservoirs. In this section, both pure steam injection (method A) and first steam injection and then steam−chemical co-injection (method B) are performed. Table 8.3 shows the basic parameters of the sand pack displacement experiments. The two heavy oil samples used in the experiments are dead oil samples from the Henan oil field, Sinopec. For the steam flooding experiments, the steam injection rate is set to 3 mL/min. For the experiments of first steam flooding and then steam-chemical co-injection, to make the results comparable, the injection rates of both steam and a chemical agent are set to 1.5 mL/min.

TABLE 8.3 Basic parameters of sand pack displacement experiments.

						Injection rate (mL/min)	
No.	Heavy oil sample	Oil viscosity (mPa·s@25°C)	Porosity (%)	Permeability (μm^2)	Method	Steam	Viscosity reducer
1	I	10,688	33.6	1306	A	3	\
2	I	10,688	1334	31.9	B	3/1.5	1.5
3	II	50,910	4744	32.8	A	3	\
4	II	50,910	4701	35.9	B	3/1.5	1.5

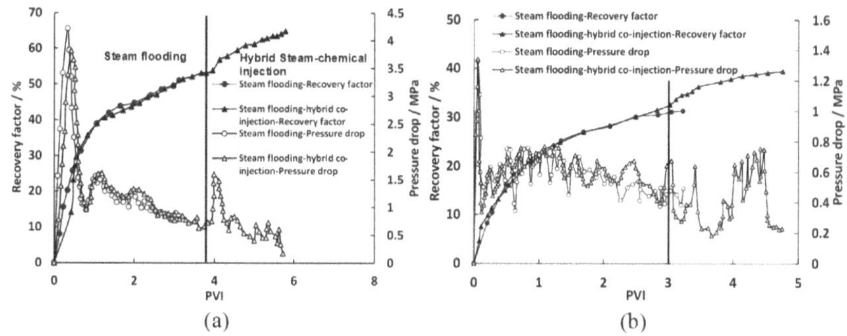

FIGURE 8.4 Results of hybrid steam—chemical co-injection: (A) results of oil sample 1 and (B) results of oil sample 2. *PVI*, Pore volume injected.

Fig. 8.4 shows the experimental results. Based on steam flooding, hybrid steam—chemical co-injection can significantly increase displacement efficiency. Because of the higher viscosity of heavy oil sample 1, displacement efficiency is higher, even though the permeability of sand pack models for heavy oil sample 2 is higher. The difference in incremental displacement efficiency caused by hybrid steam—chemical co-injection between oil samples 1 and 2 is 3.17%. On the other hand, Fig. 8.4 shows that once hybrid steam—chemical co-injection is performed, an obvious increase in pressure drop can be found. caused by the plugging of emulsification between heavy oil and the chemical agent. Moreover, as fluid injection continues, the pressure drop is gradually reduced.

8.2 Similarity criterion in three-dimensional experiments

Physical simulations, numerical simulations, and field pilot tests are three commonly used methods for evaluating recovery performance in oil reservoirs. Compared with a field pilot test, physical and numerical simulations have the advantages of lower costs, higher efficiency, and more repeatability and predictability. With advances in computer technology, numerical simulation has been widely applied in the whole development process of an oil reservoir, but physical simulation is still necessary. It can provide validations and empirical correlations for numerical simulation and is also the most important process for describing complicated phenomena and basic mechanisms. In petroleum reservoirs, a physical model can be classified into two types: an elemental model and a scaling model. A scaling model can be used to simulate some expected phenomena in a field-scale reservoir model. Also, through reasonable explanations for experimental results, the performance of a field-scale reservoir model can be predicted.

For a scaling model, the most important issue is the similarity principle. Especially for an entire three-dimensional experimental process, including experimental operation, data treatment and result explanation, the similarity theory can have an important role. Geometrical, emotional, and movement similarities are three basic rules during the similarity design of a scaling model.

For the similarity criterion in different recovery processes, a formula analysis and dimensional analysis are two commonly used methods. A formula analysis is based on an analysis of a mathematical formula group for a physical process, and a dimensional analysis is based on a description of variables related to the physical process. Therefore, the similarity criterion determined by the formula analysis usually has a clear physical meaning. Once there is a mathematical formula group for a specific recovery process, the similarity criterion can be easily obtained and is the recommended method. In some cases, however, if a relationship between different variables is unclear, the dimensional analysis is the only method to obtain the similarity criterion.

8.2.1 Pure steam injection processes

8.2.1.1 Cyclic steam stimulation (CSS)

From the high-pressure similarity principle proposed by Pujol and Boberg, the requirements for a scaling model of a cyclic steam stimulation (CSS) process can be summarized as [5]:

- ➢ The geometrical similarity must be achieved. A field-scale reservoir model and its corresponding laboratory-scale model should have the same length–width ratio, height–length ratio, reservoir dip, and heterogeneity.
- ➢ The parameters related to reservoir fluids, rocks, and heat and mass transfer should be the same between the field-scale model and the laboratory-scale model.
- ➢ The field-scale model and laboratory-scale model should have the same initial and boundary conditions.

8.2.1.2 Steam flooding

Steam flooding is an important follow-up recovery process after CSS in heavy oil reservoirs. Since a scaling physical simulation method was proposed for steam injection, many attempts have been made to discuss recovery performance and various effects in steam flooding in heavy oil reservoirs (e.g., a well pattern, additives, an aquifer, the slug size, and well spacing) [6–9]. Table 8.4 shows the similarity criterion numbers used for steam injection in heavy oil reservoirs.

For a complicated reservoir system, achieving all of the similarity numbers is usually impossible. Therefore, it is common to realize the most important similarity numbers; the secondary similarity numbers can appropriately relax the restrictions.

TABLE 8.4 Similarity criterion numbers for steam injection.

No.	Scaling parameter	Similarity[a]
1	Time	$\pi_1 = \frac{\alpha t}{L^2}$
2	Permeability	$\pi_2 = \frac{K\rho_o g t}{\varphi(1-S_{wc}-S_{or})\mu_o L}$
3	Steam quality	$\pi_3 = \frac{xL_v}{C_w \Delta T}$
4	Steam injection rate	$\pi_4 = \frac{i_s t}{\varphi(1-S_{wc}-S_{or})\rho_w L^3}$
5	Pressure drop	$\pi_5 = \frac{\Delta p}{\rho_o g L}$
6	Compressibility	$\pi_6 = \frac{Kt}{\mu_o C_t \varphi L^2}$

[a]*Where α is the thermal conductivity, 10^{-3} m^2/h; t is time, d; L is the length, m; K is permeability, $10^{-3}\mu m^2$; ρ_o is the oil density, g/cm^3; g is the gravitational acceleration, m/(s·s); ϕ is porosity; S_{wc} is the irreducible water saturation; S_{or} is the residual oil saturation; μ_o is the oil viscosity, mPa·s; x is the steam quality; L_v is the latent heat of saturated steam, kJ/kg; C_w is the specific heat of saturated water, kJ/(kg·°C); ΔT is the temperature difference between steam temperature and reservoir temperature, °C; i_s is the mass steam injection rate, t/d; ρ_w is the water density, g/cm^3; Δp is a pressure drop, MPa; and C_t is the total compressibility, MPa^{-1}.*

8.2.1.3 Steam-assisted gravity drainage process

The similarity criterion in a steam-assisted gravity drainage (SAGD) process in heavy oil reservoirs is based on the theory proposed by Butler [10,11], as shown in Table 8.5.

TABLE 8.5 Similarity criterion numbers of steam-assisted gravity drainage process.

No.	Scaling parameter	Similarity	Physical meaning
1	Well horizontal length	L	Geometrical similarity
2	Key parameter of steam chamber control in steam-assisted gravity drainage	$B_3 = \sqrt{\frac{Kgh}{\alpha\varphi\Delta S_o m v_s}}$	Nondimensional parameter
3	Production time	$t' = \frac{Kgt}{\varphi\Delta S_o v_s h}$	Nondimensional time
4	Steam injection rate	$q = \sqrt{\frac{2\varphi K g\alpha h}{m v_s}}$	Ratio of steam injection rate

8.2.2 Hybrid enhanced oil recovery processes

For hybrid EOR processes, because of their complicated EOR mechanisms, the similarity criterion is significantly different from that for pure steam injection.

First, for a hybrid thermal–NCG process, fluid mixing is the most important mechanism. Considering a density difference between gas and oil–water, the terms of gravity and a viscosity ratio are two variables that should be met. On the other hand, the operation pressure and steam temperature in a scaling model should also be consistent with those in the prototype model, because pressure and temperature can have an important effect on the properties of NCGs, especially the solubility of gas in oil.

Second, for a hybrid thermal-solvent process, its most important EOR mechanism is the improvement of oil properties by solvent (e.g., viscosity reduction by dissolution, asphaltene deposition, and diffusion) [12], because this mechanism is highly related to an operation condition (e.g., temperature, pressure, and a rock type). Therefore, to be fully close to a prototype model, these conditions in a scaling model should approach the prototype conditions as much as possible. On the other hand, for the diffusion mechanism of solvent, considering the size of a scaling model, the similarity of a diffusion coefficient is usually relaxed. However, an unstable interface between solvent and oil is considered during a scaling process [13]. For the operation parameters of steam, including steam quality, steam injection rate, and total steam injection volume, the similarity criterion in pure steam injection is still valid.

Third, for a hybrid thermal-chemical process, many nonscaling experiments (including polymer, emulsion, and foam) have been performed, and thus their results cannot be directly extended to the field scale. They are applied to the investigation of displacement mechanisms. For the similarity criterion in a hybrid thermal-chemical process, besides the similarity numbers in pure steam injection, the mechanisms of chemical agents (e.g., a non-Newtonian effect, emulsification, adsorption, and dispersion) should also be satisfied. However, in a specific scaling model or within one physical simulation process, to simulate all effects effectively is difficult or even impossible. According to an experimental purpose or condition, some scaling numbers are relaxed. Specifically, during a 3D experimental process, if a rigorous similarity design is followed, the reservoir permeability is usually above $100 \times 10^{-3}\ \mu m^2$. Thus, the movement similarity in porous media is destroyed. Therefore, another constraint condition based on a Reynolds number or Forchheimer coefficient is required.

For the similarity criterion, the rock and fluid properties in a scaling model are the same as those in a protype model. The operation conditions, including temperature and pressure, should be the same. Such a treatment can effectively simulate the thermal expansion of rock, distillation of oil, and even the interaction between fluids and rock. For other parameters or materials, scaling or nonscaling depends on the specific experimental purpose and condition.

8.3 Three-dimensional experiment on performance of steam injection processes

Theoretically, oil recovery is the product of displacement efficiency and sweep efficiency. Therefore, to improve the recovery of a specific heavy oil reservoir, we can increase the displacement efficiency or sweep efficiency. Based on 1D sand pack displacement experiments in Section 8.1 and 2D visualized displacement experiments in Chapter 7, 3D experiments are an effective method to illustrate the recovery of a specific process. Many different 3D high-temperature, high-pressure physical models have been proposed to study the performance of thermal recovery [9,14–16]. In this section, a series of pure steam 3D experiments are performed to discuss changes in reservoir conditions after steam injection. It is the basis for subsequent hybrid EOR processes [17].

8.3.1 Experimental method

A cubic physical model is applied in this section to conducting 3D experiments, as shown in Fig. 8.5. The inner space of this model is 40 cm × 40 cm × 40 cm. The maximum affordable pressure condition is 10 MPa. The inner surface is covered with a layer of heat insulation material. This model can be used to simulate the performance of different thermal recovery processes in heavy oil reservoirs, including pure steam injection (CSS, steam flooding, and SAGD) and hybrid EOR. During experiments, a series of temperature and pressure transducers are installed in this model to test changes in temperature and pressure. Fig. 8.6 shows the experimental setup. During experiments, considering the effect of heat loss, a heater band is applied to reduce heat loss between injected steam and the environment. On the other hand, for 3D experiments, the permeability of a scaling model is usually an important parameter. Therefore, to determine the permeability of 3D model accurately, a sand pack tube is recommended to test the permeability of quartz sand or glass beads with different meshes.

(a)

(b)

FIGURE 8.5 Three-dimensional physical model: (A) outer image and (B) inner space.

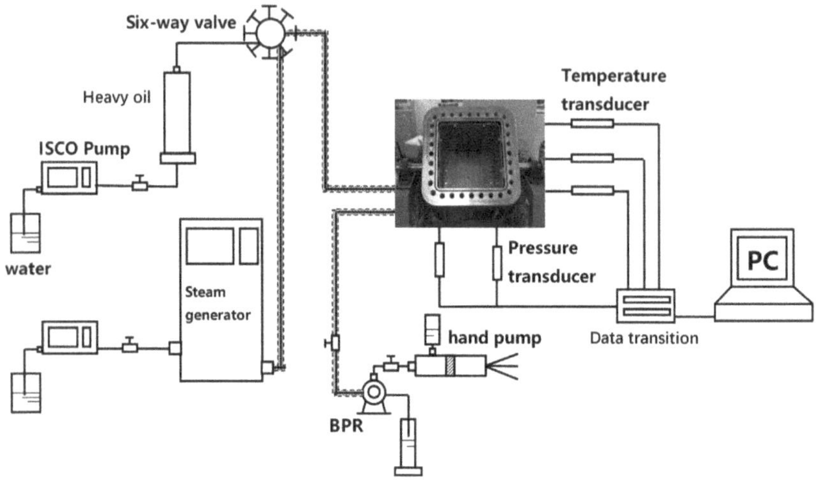

FIGURE 8.6 Experimental setup.

Before experiments, according to the similarity criterion, the model and operation parameters are first determined. The detailed experimental procedure is:

- ➢ Test the permeability of quartz sand with different meshes with a sand pack tube (3.8 cm × 35 cm) and determine the quartz sand mesh used in the 3D experiments;
- ➢ Install the model inner surface for thermal insulation and avoid steam breakthrough through the model inner surface;
- ➢ Prepare the oil sands using quartz sand, a heavy oil sample, and simulation water;
- ➢ Install a simulation well;
- ➢ Fill the prepared oil sands and set the temperature and pressure transducers;
- ➢ Prepare the model cap rock, including an antifluid-channeling layer, cushion rubber, and model metal cap;
- ➢ Move the 3D model into the air bath system and set temperature as the actual reservoir temperature (35°C);
- ➢ Re-saturate heavy oil and water to offset the saturation void caused by the model-filling process;
- ➢ Fix the model at the reservoir temperature condition for more than 48 h to simulate aging;
- ➢ Inject steam to simulate the recovery of pure steam injection;
- ➢ Record liquid production, temperature distribution, and pressure distribution;
- ➢ Once an obvious steam breakthrough is observed, an advanced hybrid EOR process is simulated.

In this section, two groups of pure steam injection 3D experiments are simulated, including CSS-steam flooding and SAGD [17]. The experimental parameters are shown in Tables 8.6 and 8.7. The well placements of these two experiments are shown in Fig. 8.7.

First, for CSS-steam flooding experiments, the four wells in Fig. 8.7A are alternately performed with CSS; there are 12 CSS cycles of these four wells.

TABLE 8.6 Three-dimensional experimental parameters for cyclic steam stimulation (CSS)-steam flooding process.

Phase	Parameter	Prototype	Model
Basic parameter	Well pattern	Inverted nine-spot	One-fourth of an inverted nine-spot pattern
	Well spacing (m, cm)	50	25
	Reservoir thickness (m, cm)	10	5
	Porosity (%)	30	30
	Initial oil saturation (%)	70	90
	Permeability (10^{-3} μm^2)	1500	57,990
	Oil viscosity @ SC (mPa·s)	15,000	6013
	Steam temperature (°C)	250	250
	Reservoir temperature (°C)	26	26
	Steam quality	0.7	0.7
	Time (a, min)	1	27.25
CSS phase	Steam injection rate (t/d, mL/min)	100	188
	Steam injection time (d, s)	10	45
	Soaking time (d, s)	4	18
Steam flooding phase	Steam injection rate (t/d, mL/min)	16	30

SC, Surface condition.

TABLE 8.7 Three-dimensional experimental parameters for steam-assisted gravity drainage (SAGD) process.

Phase	Parameter	Prototype	Model
Basic parameter	Distance of producer from reservoir bottom (m, cm)	3	1.2
	Distance between injector and producer (m, cm)	5	2
	Perforation density (m^{-1})	300	206
	Wellbore inner radius (m, cm)	0.1	0.6
	Horizontal length (m, cm)	800	320(actual 40 cm)
	Well pair spacing (m, cm)	100	40
	Reservoir thickness (m, cm)	30	12
	Permeability (μm^2)	4	40
	Oil viscosity @ RC (mPa·s)	100×10^4	100×10^4
	Initial oil saturation (%)	70	70
	Steam temperature (°C)	225	225
	Steam quality (%)	80	80
	Reservoir temperature (°C)	7	20
	Reservoir pressure (MPa)	2.0	2.0
SAGD phase	Steam injection pressure (MPa)	2.4	2.4
	BPR pressure (MPa)	1.8–2.0	1.8–2.0
	Time (a, min)	1	59.5
	Steam injection rate (t/d, mL/min)	300	40

RC, Reservoir condition.

This indicates that each well can experience three CSS cycles. After 12 CSS cycles are completed, a steam flooding process is activated. During steam flooding, based on the results from the CSS period, one of the four wells is selected to be the steam injection well; the other three are production wells. Finally, once the steam breakthrough between different wells is observed, an experiment is terminated.

Second, for SAGD experiments, a horizontal well pair is placed at the reservoir bottom, as shown in Fig. 8.7B. During SAGD, steam circulation is first performed to preheat the reservoir for thermal connection between the two

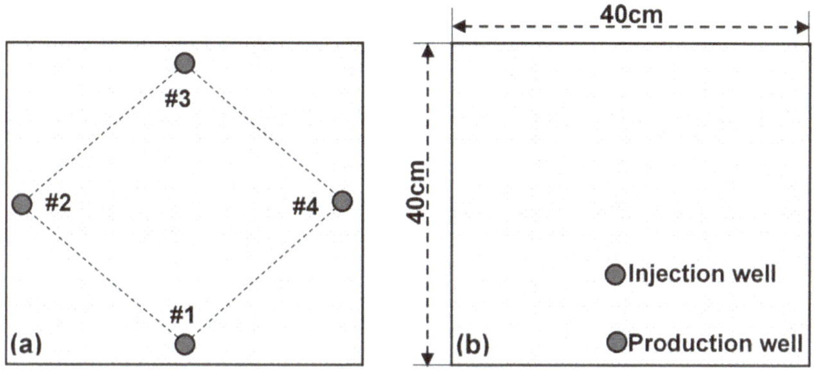

FIGURE 8.7 Well placement in three-dimensional experiments: (A) cyclic steam stimulation–steam flooding; (B) steam-assisted gravity drainage.

horizontal wells. Once the reservoir temperature between the two horizontal wells reaches more than 80°C, the oil drainage phase will be activated. Then, the upper horizontal well starts to inject steam and the bottom horizonal well is converted to production.

8.3.2 Experimental results

8.3.2.1 Cyclic steam stimulation–steam flooding process

First, in the CSS stage, oil recovery after 12 CSS cycles is about 19.1%. The cumulative oil volumes of the four wells are 218.3 mL (1), 93.9 mL (2), 180.3 mL (3), and 174.8 mL (4). Their difference is caused by reservoir heterogeneity. Fig. 8.8 shows the temperature distributions of the top and bottom of the reservoir after CSS. An obvious steam overlap can be observed. The temperature at the reservoir top is higher than that at the reservoir bottom. On the other hand, Fig. 8.8 illustrates that at the end of the CSS operation in well 3, the reservoir temperature around well 2 is much higher than that in the other

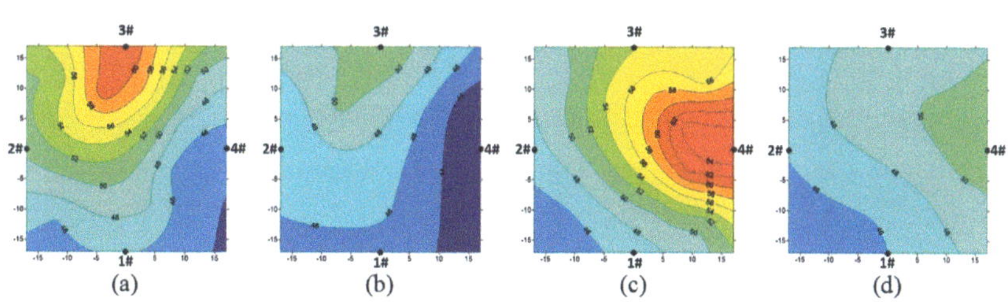

FIGURE 8.8 Temperature distributions after cyclic steam stimulation (CSS): (A) temperature distribution at the top of reservoir after CSS in well 3; (B) temperature distribution at the bottom of reservoir after CSS in well 3; (C) temperature distribution at the top of reservoir after CSS in well 4; (D) temperature distribution at the bottom of reservoir after CSS in well 4.

two wells. The same temperature distribution can be found around well 4. Well 3 is the subsequent CSS well followed by well 2, and well 4 is the subsequent CSS well followed by well 3.

At the end of the 12 CSS cycles, the cumulative oil production of well 1 is the highest. This indicates that the degree of recovery of well 1 is the highest. Therefore, during the steam flooding stage, well 1 is selected as the steam injection well; the other three wells are selected as the production wells. Once an obvious steam breakthrough between well 1 and the other three wells is observed, the corresponding production well will be shut in. Simultaneously, considering the oil redistribution behavior during the shut-in period, three steam flooding operations are performed. One steam flooding operation starts from steam injection and ends in shutting in all wells. Table 8.8 shows the cumulative oil production volumes of each well during each steam flooding. The cumulative oil volume during the steam flooding stage is 866.9 mL. Therefore, at the end of steam flooding, oil recovery has achieved 43.91% and is enhanced by about 24.81%.

Fig. 8.9 shows the temperature distributions of the top and bottom of the reservoir after the third steam flooding operation. The thermal sweep area is significantly increased based on the CSS process, and the temperature at the reservoir top is higher than that at the reservoir bottom. Moreover, because of differences in recovery and well spacing, the time of steam breakthrough at each well is different. During steam flooding, steam breakthrough is first observed between wells1 and 4. Then, to increase the thermal sweep area, well is shut-in, and thus the steam front approaches well 2. Also, the shut-in operation of well 4 increases the expanding rate of the steam front. Once well 2 is shut-in, the steam front moves to well 3. During the entire steam flooding process, the model temperature continuously increases. However, because of low steam sweep efficiency, the temperature in the corner area between 2 and 3 and the corner area between 3 and 4 is always low.

TABLE 8.8 Cumulative oil of each well during steam flooding operation.

Steam flooding operation stage	Steam injection well	Cumulative oil volume (mL)			Total oil (mL)
		2	3	4	
First steam flooding	1	41	252.5	127	420.5
Second steam flooding	1	32	167.5	32.5	232.0
Third steam flooding	1	17	148.9	48.5	214.4
Total	866.9 mL				

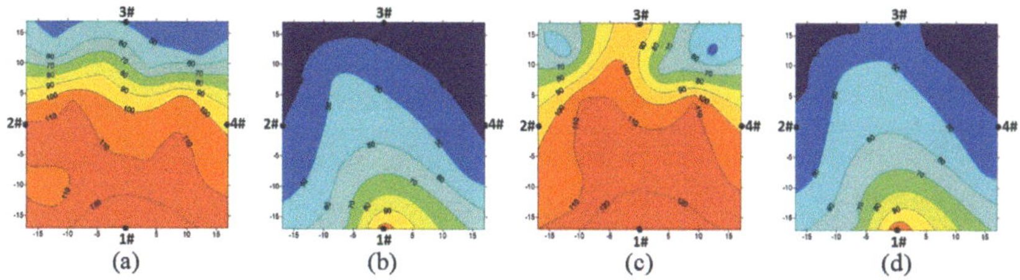

FIGURE 8.9 Temperature distributions after third steam flooding operation: (A) reservoir top temperature as observation of steam breakthrough between 1 and 2; (B) reservoir bottom temperature as observation of steam breakthrough between 1 and 2; (C) reservoir top temperature as observation of steam breakthrough between 1 and 3; (D) reservoir bottom temperature as observation of steam breakthrough between 1 and 3.

At the end of the third steam flooding operation, a large amount of remaining heavy oil remains located in the reservoir. To improve recovery, all wells are closed for about 24 h to simulate redistribution of the reservoir fluids. Then, considering the recovery of the previous steam flooding operation, 3 is selected as the new steam injection well; the other three wells are chosen to be the new production wells. Thus, three new steam flooding operations (fourth through sixth) are performed, and an obvious improvement in oil production can be observed. The current oil recovery factor reaches 57.7%. With the effect of the previously formed steam breakthrough paths, once well 3 starts injecting steam, a steam breakthrough path is first observed in well 1, indicating that the previous recovery process can have an important role in subsequent recovery processes. Operation of well shut-in at a given time can induce the redistribution of oil and water in the reservoir. Then, by converting an injection/production well pattern, sweep efficiency can be improved, which helps an EOR process in heavy oil reservoirs.

8.3.2.2 Steam-assisted gravity drainage process

For SAGD, steam circulation for about 30 min is first performed to preheat the interwell section. Fig. 8.10A shows the temperature distribution after steam circulation. After the preheating phase, the temperature between horizontal wells reaches greater than 100°C. It is supposed that the thermal connection between horizontal wells has reached the goal. Then, the horizontal well is converted to steam injection and the bottom horizontal well is converted to production. As shown in Fig. 8.10, as steam injection continues, the reservoir temperature increases. The steam chamber rises. As the steam injection process lasts for about 130 min, the steam chamber contacts the reservoir top. Then it expands horizontally. This stage is called the plateau stage; it usually contributes most of the oil production during the entire SAGD process. At about 500 min, the steam chamber reaches the reservoir boundary; it is the wind-down stage. The rate of oil production gradually decreases. To improve SAGD recovery, a follow-up EOR process or strategy may be required in this stage.

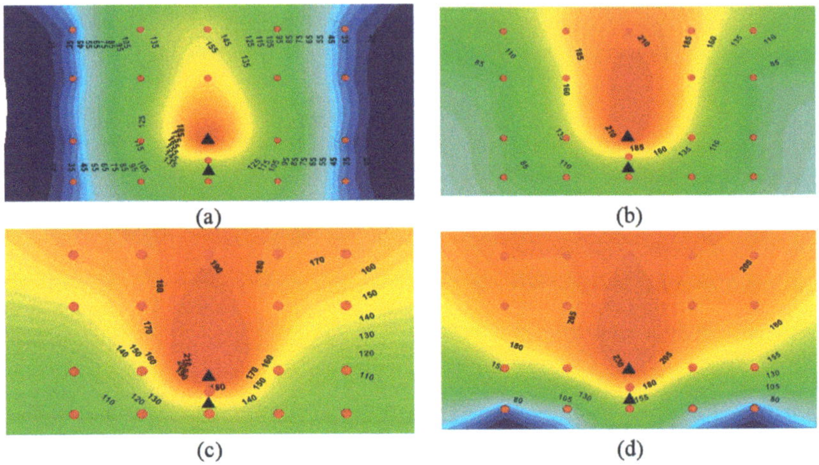

FIGURE 8.10 Temperature distributions at different times of steam-assisted gravity drainage process: (A) end of preheating phase; (B) 130 min; (C) 500 min; (D) 800 min.

Fig. 8.11 gives the experimental results of oil recovery, steam–oil ratio, oil production rate, and water cut. The results of the oil production rate show that the entire SAGD process can be classified into three stages: rising, plateau, and winding down. The final oil recovery is 67.3%. First, in the rising stage, the steam chamber mainly expands vertically and the oil production rate increases rapidly. Then, as steam injection continues, the front of the steam chamber contacts the reservoir top and expands laterally. During this stage, the oil production rate can reach the plateau stage. Thereafter, once the steam chamber reaches the reservoir boundary, it arrives at the winding down stage and oil production rate is reduced.

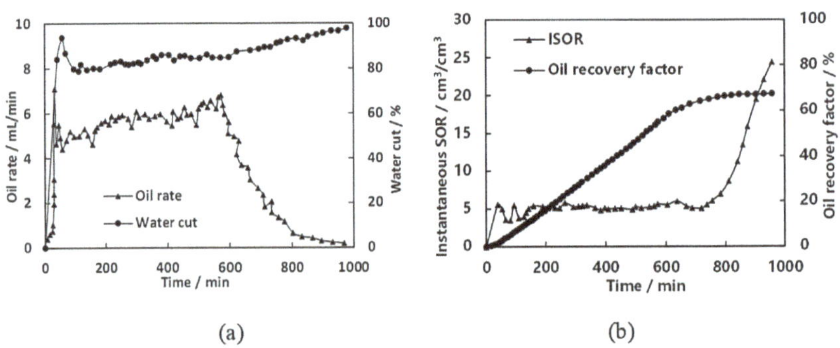

FIGURE 8.11 Three-dimensional experimental results of steam-assisted gravity drainage process: (A) oil rate and water cut; (B) instant steam–oil ratio and oil recovery factor.

8.4 Three-dimensional experiments on performance of hybrid enhanced oil recovery processes

Based on pure steam 3D experiments in Section 8.3, a series of 3D experiments for hybrid EOR processes are performed in this section to study their recovery after pure steam-based recovery, including hybrid thermal–NCG and a hybrid–chemical processes.

8.4.1 Hybrid thermal–noncondensable gas process

8.4.1.1 Recovery performance after cyclic steam stimulation

Using the same 3D experimental method as before, 3D experiments for hybrid steam–N_2 co-injection are performed to study recovery after CSS [19]. Only one vertical well is placed in the center of the 3D physical model (Fig. 8.5). For experimental parameters, the reservoir thickness in the 3D model is 10 cm, the oil viscosity is 6742 mPa·s, and the initial reservoir temperature is 61°C. The other parameters are the same as the parameters in Table 8.6. Table 8.9 gives the detailed operation parameters. As shown, during experiments, five cycles of the CSS process and six of the hybrid steam–N_2 co-injection process

TABLE 8.9 Operation parameters in three-dimensional experiments.

Phase	No.	Steam injection volume (mL)	N_2 injection volume (surface condition)	Gas–steam ratio (surface condition)	Injection time (min)
Cyclic steam stimulation phase	1	120	—	—	20
	2	140	—	—	20
	3	160	—	—	20
	4	180	—	—	20
	5	200	—	—	20
Hybrid steam-N_2 co-injection phase	1	200	100	0.5:1	20
	2	200	200	1:1	20
	3	200	300	1.5:1	20
	4	200	400	2:1	20
	5	200	500	2.5:1	20
	6	200	600	3:1	20

are simulated. Meanwhile, for CSS, the cyclic steam injection volume is increased by 10% for different CSS cycles. For hybrid steam−N_2 injection, the cyclic N_2 injection volume is increased for different cycles.

The experimental results are shown in Fig. 8.12. In the CSS phase, cyclic oil production volume is gradually reduced. The oil production volume in the fifth CSS cycle is only about 19% of the oil production volume in the first CSS cycle. Once it is converted to hybrid steam−N_2 co-injection, an obvious improvement in recovery can be observed and the cyclic oil production volume is increased.

Fig. 8.13 shows temperature distributions at different operation times. Compared with the temperature distribution after CSS, the thermal sweep area in hybrid steam-N_2 co-injection is significantly increased. Because of the low solubility of N_2, it can help to improve heating during thermal recovery.

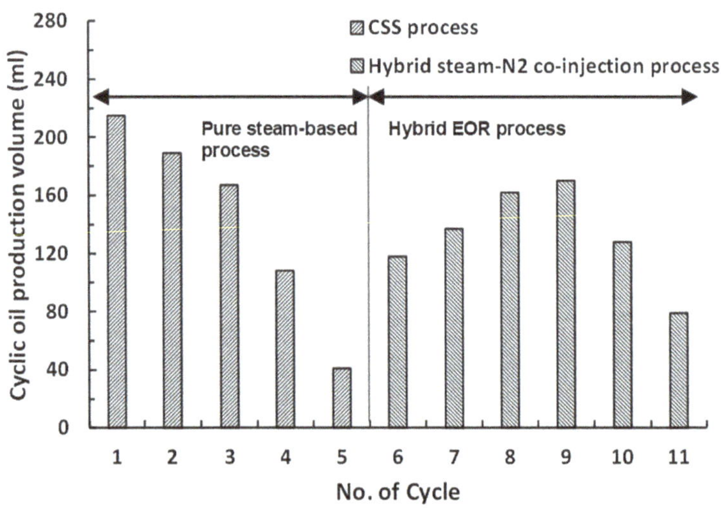

FIGURE 8.12 Cyclic oil production in cyclic steam stimulation (CSS) and hybrid steam−N_2 co-injection. *EOR*, enhanced oil recovery.

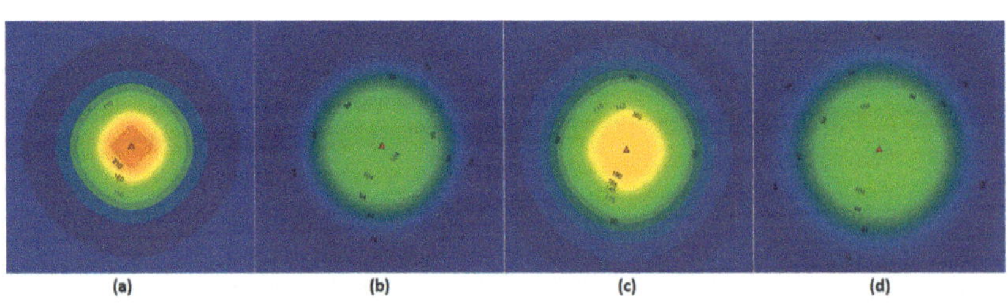

FIGURE 8.13 Temperature distributions: (A) after steam injection in fifth cyclic steam stimulation (CSS) cycle; (B) after production stage in fifth CSS cycle; (C) after fluid injection in sixth hybrid enhanced oil recovery (EOR) process; (D) after production stage in sixth hybrid EOR process.

8.4.1.2 Recovery performance after steam-assisted gravity drainage

Next, 3D experiments for hybrid steam-NCG and multithermal fluid-assisted gravity drainage (MFAGD) are performed to study their recovery after SAGD [1,2,20]. Based on real-time temperature observations, the NCGs in multithermal fluids can reduce heat loss between a reservoir and its cap rock. After it is switched to MFAGD, the temperature at the reservoir top is suddenly reduced. Compared with conventional SAGD, the temperature difference between the injector and the reservoir top has reached 30°C. Fig. 8.14A gives the temperature results in the steam injection well, interwell region, and production well during experiments. The switch time from SAGD to MFAGD is 800 min. Before the start of MFAGD, the steam chamber has reached the reservoir boundary. As shown, after switching to MFAGD, the temperature shows a tendency to increase and then decrease. At 960 min (260 min after the start of MFAGD), the temperatures in the steam injection well, interwell region, and production well are the same. This indicates the occurrence of steam breakthrough between the steam injection well and production well; most of the injected steam is directly produced from the production well. Fig. 8.14B shows the average pressure, which indicates that the changing behavior of temperature and pressure is similar. After steam breakthrough, the average pressure is also suddenly reduced. However, compared with temperature, the reduction in pressure is more significant. In Fig. 8.14, the increase in temperature and pressure from about 889 min is mainly caused by the dissolution of NCGs in heavy oil, especially the CO_2 component in NCGs. It can further increase the swelling of oil. Also, once the saturation condition is achieved (889 min), the subsequent injected NCGs will induce an increase in the reservoir temperature and pressure until the reservoir pressure reaches the injection pressure. Once steam breakthrough occurs, both temperature and pressure are reduced.

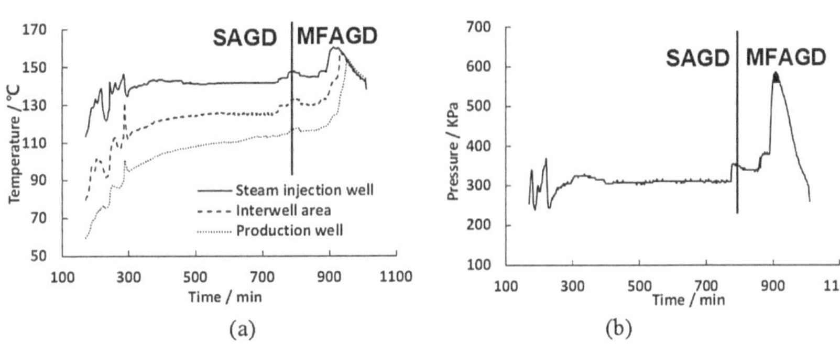

FIGURE 8.14 Temperature and pressure: (A) temperature of different locations; (B) average pressure. *MFAGD*, multithermal fluid-assisted gravity drainage; *SAGD*, steam-assisted gravity drainage.

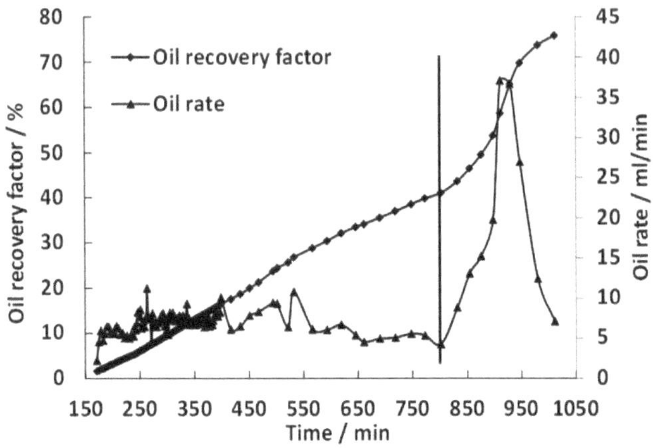

FIGURE 8.15 Recovery rate and water cut curve of drainage.

Fig. 8.15 shows the results of the oil recovery factor and oil rate. As the oil recovery factor of the SAGD phase reaches about 41%, hybrid EOR is activated. The injection of multithermal fluids can reduce the water cut and increase the oil production rate. The subsequent MFAGD process can enhance oil recovery by about 35%, and the final recovery reaches 75.8%. It is caused by the performance of NCG components in multithermal fluids. The lateral expansion of the steam chamber can be significantly improved, and the volume of the steam chamber is also increased.

8.4.2 Hybrid thermal-chemical process

Based on the 3D experimental results of CSS-steam flooding in Section 8.3, a subsequent hybrid thermal-chemical process is performed to study recovery after steam flooding [17,18]. In this process, well 3 will be selected as the injection well; the other three wells (1, 2, and 4) are chosen as the production wells. In this stage, two different hybrid thermal-chemical processes are proposed, including hybrid steam—surfactant co-injection and high-temperature gel (HTG) system injection. Table 8.10 shows the experimental parameters of the hybrid thermal-chemical process. First, hybrid steam-surfactant co-injection lasts about 120 min. The cumulative oil production volume is about 36 mL and the incremental oil recovery is only 1%. The current oil recovery reaches 58.7%. All of these indicate that displacement efficiency within steam breakthrough paths is relatively high, and to enhance heavy oil recovery only by increasing displacement efficiency is ineffective. Lengthy steam flooding can easily induce low sweep efficiency by steam fingering and overlap. Therefore, to plug steam breakthrough paths and redirect fluid flow will be only the effective method for the EOR of this heavy oil reservoir.

TABLE 8.10 Experimental parameters of hybrid thermal–chemical process.

Stage	Parameter	Prototype	Model
Hybrid steam–surfactant co-injection process	Steam injection rate (t/d, mL/min)	13.3	25
	Surfactant solution injection rate (t/d, mL/min)	2.7	5
HTG injection process	HTG solution injection rate (t/d, mL/min)	16	30
	HTG solution volume (PV)	0.1	0.1

HTG, high-temperature gel.

In HTG system injection, well 3 is set to be the injection well; the other three wells are set to be the production wells. Steam flooding is restarted to simulate the reservoir status of steam breakthrough. Once obvious steam breakthrough is observed in the three production wells, an HTG solution slug (0.1 PV) is injected. Then, this model is kept under the reservoir temperature condition for about 24 h to simulate gelling. After that, steam flooding is reactivated. To maximize EOR, three profile control processes by HTG injection are performed. The results are shown in Table 8.11. The incremental cumulative oil is 294.4 mL and the current oil recovery factor reaches 67.2%. Compared with hybrid steam–surfactant co-injection, the EOR performance of the HTG system is more significant. Because of the high plugging strength of this HTG system, it can provide a good direction for the EOR of the heavy oil reservoir after steam flooding, especially for a steamed heavy oil reservoir with serious steam breakthrough.

TABLE 8.11 Cumulative oil in each profile control process.

	Oil production for each well (mL)			
Stages	**1**	**2**	**4**	**Cumulative oil (mL)**
First HTG injection operation	48.2	39	92.2	179.4
Second HTG injection operation	27.5	25	41	93.5
Third HTG injection operation	5	6.5	10	21.5
Sum	80.7	70.5	143.2	294.4

HTG, high-temperature gel.

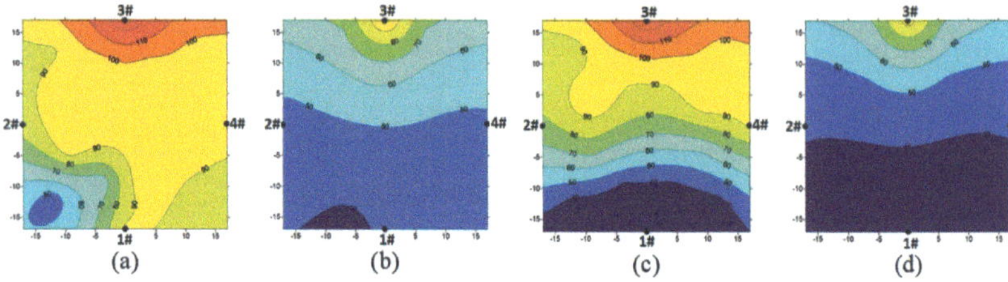

FIGURE 8.16 Temperature distributions after third high-temperature gel injection operation: (A) temperature distribution at the top of reservoir after steam breakthrough is observed in well 1; (B) temperature distribution at the bottom of reservoir after steam breakthrough is observed in well 1; (C) temperature distribution at the top of reservoir after steam breakthrough is observed in well 4; (D) temperature distribution at the bottom of reservoir after steam breakthrough is observed in well 4.

Fig. 8.16 shows the reservoir temperature distribution after the third HTG injection operation. During the third HTG injection, a steam breakthrough path between 3 and 4 is observed, and then the wells of 2 and 1. Compared with the previous steam flooding processes, the sequence of interwell steam breakthrough has changed. Fig. 8.16 shows that after the injection of the HTG system, the previous steam breakthrough path is plugged and the advancement of the steam front in the reservoir is more stable. The steam injection profile control is obvious. Also, during the entire profile control process, the advancement rate of the bottom reservoir temperature is smoother than before the profile control.

On the other hand, a hybrid thermal-chemical process can be applied in the post-SAGD heavy oil reservoir to enhance heavy oil recovery. Typical processes include N_2-foam—enhanced SAGD and surfactant-enhanced SAGD. The former is applied to address nonuniform steam conformance along a horizontal wellbore. Especially for heavy oil reservoirs with serious permeability heterogeneity, N_2-foam—enhanced SAGD will be an important follow-up method at a later stage of SAGD. It can effectively improve a steam injection profile along a horizontal wellbore. On the other hand, because of the high flow resistance of N_2 foam, the vertical expansion rate of a steam chamber is reduced and the lateral expansion rate is increased. In comparison, the latter can be applied to increase the oil production rate by emulsification and wettability alteration. However, considering the stimulation behavior of high-temperature steam, improvements caused by a surfactant are limited.

8.5 Macroscale enhanced oil recovery mechanisms

From a macroscale point of view, oil recovery is a product of sweep efficiency and displacement efficiency. Therefore, to enhance heavy oil at the middle or later stage of steam injection, increasing sweep efficiency or displacement efficiency, or both, by a suitable method is the right direction. For hybrid EOR, because of the difference between injected additives, their focus is also different.

First, for a hybrid thermal–NCG process, based on the 3D experimental results in this chapter, it has the heating effect of steam and also of chemical and gas operations. NCG fractions aim to improve the mobility of heavy oil and recover reservoir energy. For a hybrid process with NCG mixtures, in addition to conventional thermal recovery mechanisms of steam injection (reducing oil viscosity, heat expansion, changing relative permeability, and steam distillation), there are other EOR mechanisms:

(1) CO_2 has high solubility in heavy crude oil. The dissolution of CO_2 can increase oil volume and enhance reservoir expansibility and oil mobility.
(2) The dissolution of CO_2 can further reduce oil viscosity.
(3) With the dissolution of CO_2 in heavy crude oil, foamy oil can be formed in a reservoir, which helps EOR.
(4) The dissolution of CO_2 in formation water can form carbonated water, which can react with carbonate minerals in a reservoir to improve reservoir permeability.
(5) N_2 has low solubility in heavy crude oil and is useful for recovering reservoir pressure, which gives the reservoir high expansibility.

On the other hand, the heat carriers discussed in this book (multithermal fluids) are produced from a combustion reaction in a generator. They are a gas mixture of N_2, CO_2, CH_4, and CO. They are different from the conventional so-called gas mixture of steam and NCGs. Based on the 3D experiments of MFAGD in Section 8.4.1, for EOR mechanisms, because of a high injection rate of NCG, in addition to the heat insulation mechanism, it has the mechanisms of energy recovery, gas dissolution, foamy oil, and auxiliary cleanup.

Second, a hybrid thermal-solvent process possesses the mechanisms of the dissolution and diffusion of a solvent in heavy crude oil. As discussed in Chapter 1, commonly used solvents include CO_2, light hydrocarbons, toluene, and naphtha. Typical processes include expanding solvent SAGD, steam-alternating solvent, and solvent-enhanced steam flooding. Dominant mechanisms include viscosity reduction by dissolution, solvent extraction, and diffusion. For this process, the problem of asphaltene deposition is usually a top concern, especially for a hybrid thermal–CO_2 process, as discussed in Chapter 2. In this chapter, from 3D experimental results of MFAGD, because of the extraction behavior of CO_2, a phenomenon of asphaltene deposition is observed. Compared with the unconsolidated view of original oil sands, after the effect of CO_2 extraction, the oil sands show a solid-like shape under the performance of deposited asphaltene.

Third, a hybrid thermal-chemical process is applied to increase macroscale sweep efficiency and microscale displacement efficiency. From the 3D experimental results in this chapter, for the chemical agents with high plugging strength (e.g., NCG foam, gel, and solid particles), their application can significantly increase macroscale sweep efficiency by plugging steam breakthrough paths. For other chemical agents, including polymer, alkali, and surfactant, their EOR mechanisms are different. The addition of a polymer can increase microscale displacement efficiency by improving the mobility ratio. Alkali can improve the displacement efficiency by changing the wettability of a rock surface. For a surfactant, however, it can enhance emulsification behavior and reduce interface tension between oil and water under formation conditions to improve microscale displacement efficiency. In some cases, application of a surfactant also has the mechanism of flow path blockage through emulsion droplets.

In this chapter, a set of macroscale experiments are performed to illustrate the recovery of hybrid EOR processes in heavy oil reservoirs. Combining this chapter with Chapter 7, hybrid EOR can improve the recovery of steam-based processes under both microscale and macroscale conditions. However, once they are applied at the field scale, some problems may be encountered to adapt their applicability. In the next chapter, challenges and problems during applications of different hybrid EOR processes will be discussed.

References

[1] Dong X, Liu H, Hou J, Zhang Z, Chen Z. Multi-thermal fluid assisted gravity drainage process: a new improved-oil-recovery technique for thick heavy oil reservoir. J Petrol Sci Eng 2015;133:1–11.

[2] Dong X. The development mechanism and method screening for offshore heavy oil reservoirs with multi-thermal fluid. PhD dissertation. Beijing): China University of Petroleum; 2014.

[3] Lu C. The improvement effects of gas and chemical agent during steam injection process for heavy oil reservoirs. PhD dissertation. Beijing: China University of Petroleum; 2015.

[4] Lu C, Zhao W, Liu Y, Dong X. Pore-scale transport mechanisms and macroscopic displacement effects of in-situ oil-in-water emulsions in porous media. ASME J Energy Res Technol 2018;140(10):102904.

[5] Pujol L, Boberg TC. Scaling accuracy of laboratory steam flooding models. In: SPE California regional meeting, Bakersfield, California; 1972.

[6] Kimber KD, Farouq Ali SM, Puttagunta VR. New scaling criteria and their relative merits for steam recovery experiments. J Can Petrol Technol 1988;27(4):86–94.

[7] Bagci S, Aybak T, Shamsul A. A comparative laboratory analysis of steamflooding with horizontal wells in heavy oil reservoirs with bottom water zone. J Can Petrol Technol 1999;38(13):1–8.

[8] Ni J, Zhou X, Yuan Q, Lu X, Zeng F, Wu K. Numerical simulation study on steam-assisted gravity drainage performance in a heavy oil reservoir with a bottom water zone. Energies 2017;10(12):1999.

[9] Tian J, Liu H, Pang Z. A study of scaling 3D experiment and analysis on feasibility of SAGD process in high pressure environment. J Petrol Sci Eng 2017;150:238–49.

[10] Butler RM. Thermal recovery of oil and bitumen. Englewood Cliffs, New Jersey: Prentice Hall; 1991.

[11] Butler RM. Steam-assisted gravity drainage, concept, development, performance and future. J Can Petrol Technol 1994;33(2):44–50.

[12] Dong X, Liu H, Chen Z, et al. Enhanced oil recovery techniques for heavy oil and oilsands reservoirs after steam injection. Appl Energy 2019;239:1190–211.

[13] Hascakir B. How to select the right solvent for solvent-aided steam injection processes. J Petrol Sci Eng 2016;146:746–51.

[14] Yu T. Experimental study on air assisted CSS for medium deep ultra heavy oil reservoir. Special Oil Gas Reservoirs 2013;20(5):77–80.

[15] Mohebati MH. Experimental and numerical investigation of hydrocarbon Co-injection with steam in the steam assisted gravity drainage process for in-situ heavy oil and bitumen recovery. PhD dissertation. University of Calgary; 2014.

[16] Nesse T. Experimental comparison of hot water propane injection to steam propane injection for recovery of heavy oil. Master thesis. Texas A&M University; 2004.

[17] Zheng Q. Adaptability of EOR techniques during steam flooding in heavy oil reservoirs. PhD dissertation. Beijing): China University of Petroleum; 2013.

[18] Liu H, Zheng Q. Experimental and numerical studies on the steam flooding process in heavy oil reservoirs. Technical report. Beijing): China University of Petroleum; 2012.

[19] Liu H, Zhao Y, Dong X. Study on the principle and optimization of nitrogen-assisted production increase in steam huff and puff wells. Technical report. Beijing): China University of Petroleum; 2019.

[20] Dong X, Liu H, Zhang Z, Wang L, Chen Z. Performance of multiple thermal fluids assisted gravity drainage process in post SAGD reservoirs. J Petrol Sci Eng 2017;154:528–36.

Chapter 9

Challenges in application of hybrid enhanced oil recovery processes

9.1 Reservoir adaptability

For the choice of an enhanced oil recovery (EOR) process for heavy oil reservoirs, it is an urgent and difficult job to screen the optimal one. For the EOR processes proposed in this book, each is suitable only for a certain type of reservoir. For screening, the specific reservoir properties and previous recovery history of a reservoir are important, but its economic factor also needs to be considered [1–4]. Moreover, once an optimal process is screened, the next important issue is its reservoir adaptability. This includes an optimal range of reservoir parameters (e.g., reservoir thickness, reservoir permeability, oil saturation, reservoir pressure, and oil viscosity).

9.1.1 Process screening

Considering the difference in their operations, the three hybrid EOR processes can be classified into three modes: cyclic operation, continuous (or plugging) operation, and gravity drainage operation, as shown in Table 9.1 [5]. A cyclic operation can be used as a follow-up technique for a cyclic steam stimulation (CSS) process. A continuous injection (or plugging) operation can be applied as a follow-up technique for steam flooding. A gravity drainage operation can be employed to improve the recovery of steam-assisted gravity drainage (SAGD). The classification in Table 9.1 shows that hybrid EOR processes can be screened for potential heavy oil reservoirs.

During process screening, from the current status of a typical heavy oil reservoir, primary screening can be performed from the classification in Table 9.1, from which a potential process list can be obtained. After that, a series of in-door experiments can be conducted to evaluate the performance of these potential hybrid EOR processes, from which this potential list can be further reduced. Then, advanced reservoir simulators (e.g., Eclipse-E500 or

Hybrid Enhanced Oil Recovery Processes for Heavy Oil Reservoirs.
https://doi.org/10.1016/B978-0-12-823954-4.00008-4

TABLE 9.1 Hybrid enhanced oil recovery processes in heavy oil reservoirs.

Process		Cyclic injection stimulation (CSS)	Continuous injection (plugging) operation (SF)	Steam-assisted gravity drainage (SAGD)
Hybrid thermal- solvent process	Additive	C_xH_y (light hydrocarbons), toluene, naphtha, diesel		
	Technique	(1) Solvent aided CSS (2) LASER (3) SAP (4) CSI	(1) Solvent-steam flooding (2) SESF	(1) SAS (2) ES-SAGD (3) SA-SAGD (4) SC-SAGD (5) VAPEX
Hybrid thermal- noncondensable gas (NCG) process	Additive	N_2, air, CO_2, flue gas, methane		
	Technique	(1) N_2-CSS (2) CO_2-CSS (3) CH_4-CSS (4) Flue gas-CSS (5) Air-CSS (6) HDNS, HDCS, HDAS	(1) Steam-NCG flooding (2) Multithermal fluid flooding	(1) NCG-SAGD (2) SAGP (3) GAGD (4) Multithermal fluid-assisted gravity drainage (5) SAGDOX
Hybrid thermal- chemical process	Additive	Alkali, polymer, surfactant, foam, gel		
	Technique	(1) HSSP (2) NCG-foam (3) High-temperature gel (HTG)	(1) Alkaline-steam flooding (2) SSP (3) Polymer (4) NCG-foam (5) HTG	(1) Surfactant-aided SAGD (2) FA-SAGD (3) CAFA-SAGD

CAFA-SAGD, Chemical additives and foam assisted SAGD; *CSI*, Cyclic solvent injection; *ES-SAGD*, Expanding solvent-SAGD; *FA-SAGD*, Foam-assisted SAGD; *GAGD*, Gas-assisted-gravity-drainage; *HDNS*, Horizontal well, Dissolver, Nitrogen, and Steam; *HDCS*, Horizontal well, Dissolver, CO_2, and Steam; *HDAS*, Horizontal well, Dissolver, Air, and Steam; *HSSP*, Hybrid steam-surfactant process; *LASER*, Liquid addition to steam for enhancing recovery; *SF*, Steam flooding; *SESF*, Solvent enhanced steam flooding; *SAP*, Solvent-aided process; *SAGDOX*, Oxygen-SAGD process; *SSP*, Steam-surfactant process; *SA-SAGD*, Surfactant assisted-SAGD; *SC-SAGD*, Solvent cyclic SAGD; *VAPEX*, Vapor extraction.

CMG-STARS) can be applied to provide a solid foundation for the recommendation. From the experimental and numerical simulation results, an economic evaluation will be the most important step for the final recommendation. This evaluation needs to consider comprehensively the effects of different factors (e.g., an oil price, costs of surface facilities and operation equipment, the source of an additive, and even the costs of water treatment) [5]. Finally, once an optimal process is determined, an operating procedure combing the results of numerical simulation and field tests should be carefully designed to take advantage of every additive for maximum benefit. For a typical hybrid thermal EOR process, an injection sequence of different additives (e.g., steam, noncondensable gas [NCG], or chemical agents) should first be optimized. Steam is used to reduce oil viscosity by heat transfer; NCG is used to improve sweep efficiency or reduce oil viscosity by gas dissolution (CO_2); chemical agents can be used to reduce fluid viscosity (a viscosity reducer [VR]), control steam injection conformance, or plug a steam channeling path (foam, gel). For example, for a post-CSS heavy oil reservoir, the main issue is usually the expansion of a heated area, and a hybrid thermal-NCG process should be the most effective one for follow-up [5–7].

In a field application, screening an optimal EOR process is complicated. Many tests and assessments need to be performed before its application in the field. Only after the in-door experiments are successful and an economic evaluation is profitable does a process have the potential to be applied. On the other hand, although an increase in oil is observed in the laboratory, the performance of a field-scale test can still be poor. This is caused by a great difference between laboratory test conditions and actual field conditions. The recovery of most hybrid EOR processes in field-scale tests is significantly different from that in the laboratory. Consistence between laboratory and field operations is always the highest concern for the selection of EOR processes.

9.1.2 Reservoir adaptability of enhanced oil recovery processes

For a given heavy oil reservoir, once an optimal recovery process is determined, the next important step is to analyze reservoir adaptability. The purpose of this step is to determine a range of reservoir physical parameters accurately. Numerical reservoir simulation is a commonly used method for addressing reservoir adaptability, and the increase in oil recovery and cumulative oil–steam ratio are the most effective assessment indicators in this step. For hybrid EOR, because of the advantages of their recovery mechanisms, they usually have a wider technology application range than conventional pure steam injection. The reservoir adaptabilities of pure steam injection and hybrid steam-NCG injection in different heavy oil reservoirs in Bohai Bay, China National Offshore Oil Corporation were discussed using the method of numerical simulation [8–10] (Table 9.2). As shown, the scope of application of

TABLE 9.2 Adaptability of different recovery processes in different heavy oil reservoirs in Bohai offshore oil field.

Heavy oil reservoir type	Recovery method	Adaptability					
		Formation depth (m)	Reservoir thickness (m)	Permeability (10^{-3} μm^2)	Oil viscosity (cp@50°C)	Water/oil net ratio	Net to gross
Parameter range		500–2000	5–30(boundary) 20–60(thick)	1000–6000	5000–100,000	1–20	0.4–1.0
Heavy oil reservoir with boundary water	Steam injection	<1400	>10	Suited	<30,000	<5	>0.60
	Hybrid steam–noncondensable gas injection	Suited	>10	Suited	Suited	Suited	Suited
Thick heavy oil reservoir	Steam injection	<1400	>30	Suited	<30,000	–	>0.55
	Hybrid steam–noncondensable gas injection	Suited	Suited	Suited	Suited	–	Suited

hybrid steam-NCG injection is wider than that of pure steam injection for both of the heavy oil reservoirs. This is caused by the extra EOR mechanisms of NCGs. Because of the application of hybrid EOR, previously unprofitable heavy oil reservoirs by pure steam injection may now be reconsidered. Hybrid EOR can significantly increase recoverable heavy oil reserves.

9.2 Reservoir lithology

Reservoir lithology, including sandstones and carbonates, is another important factor for screening a suitable EOR process for heavy oil reservoirs. Most enhanced thermal recovery processes have been successfully applied in sandstone heavy oil reservoirs. Fractured carbonate heavy oil reservoirs are characterized by high oil viscosity, low matrix permeability, and unfavorable wettability of the matrix [11]. Therefore, considering the unique reservoir characteristics of fractured carbonate reservoirs, the EOR mechanisms of thermal recovery processes may be different from those in sandstone reservoirs. For steam injection in carbonate heavy oil reservoirs, recovery mechanisms include the conventional steam injection mechanisms (e.g., oil viscosity reduction and thermal expansion) as well as capillary imbibition and gravity drainage [12,13]. In fractured carbonate heavy oil reservoirs, during steam injection, the injected steam first enters a fracture system and displaces the heavy oil in this system. Then, under the effect of a capillary force, the heavy oil accumulated in a matrix system is replaced. A matrix system is the main oil accumulation system in a fractured heavy oil reservoir. Thus, the effect of capillary imbibition is the key issue of the successful development of fractured carbonate heavy oil reservoirs.

Different from capillary imbibition in light oil reservoirs, temperature will be the key factor for the performance of capillary imbibition in fractured heavy oil reservoirs. As the temperature increases, the oil viscosity decreases and oil recovery from an imbibition effect increases. A series of static imbibition tests were performed to study the effect of temperature on capillary imbibition in fractured carbonate heavy oil reservoirs [14]. Table 9.3 shows the experimental

TABLE 9.3 Imbibition recovery at different temperatures.

Heavy oil	Oil viscosity (cp @ 50°C)	Recovery factor @ 21°C (%)	Stage recovery factor @ 60°C (%)	Stage recovery factor @ 85°C (%)	Total imbibition recovery factor (%)
1	66.6	2.12	28.22	15.98	46.32
2	198.8	1.40	19.97	12.61	33.98
3	878.1	0.28	12.91	14.08	27.27

results. As the temperature increased, more oil was recovered under the effect of a capillary force. On the other hand, oil viscosity is another important factor affecting the performance of imbibition. As the oil viscosity increases, each stage recovery factor caused by imbibition behavior is gradually reduced.

For the field performance of steam-based processes in fractured carbonate heavy oil reservoirs, the Lacq Superieur oil field in France was first performed with a steam drive pilot in Oct. 1977 [15]. At the end of Jun. 1980, an increase in oil production was obtained (176 $Mbbl^3$). In China, the CaoGu-1 reservoir in Shengli oil field, the ninth heavy oil block in Karamay oil field, and Zao35 in the Dagang oil field are other successful applications of steam injection in fractured carbonate heavy oil reservoirs [3,16]. In north Alberta, Canada, development of the Grosmont Formation has been challenging. In Grosmont, the processes of CSS, steam flooding, SAGD, in situ combustion (ISC), cyclic solvent-SAGD, and cold solvent injection were all tested [17,18]. Because of severe permeability heterogeneity, CSS showed more promising applications compared with steam flooding and ISC. In addition, during operations, compared with the use of steam, the operation cost without involving steam was lower. Especially for cold solvent injection, the cost was about half that of steam injection [18].

Therefore, based on this discussion, a hybrid thermal-solvent process is a promising technique for fractured carbonate heavy oil reservoirs. A steam-over-solvent injection process has been proposed for EOR for fractured carbonate heavy oil reservoirs [19,20]. This process has three phases: initial steam injection to heat a reservoir (Phase 1), solvent injection to recover matrix oil by a diffusion-imbibition-drainage technique (Phase 2), and steam injection again to recover more oil and retrieve the solvent (Phase 3) [11,21]. The numerical simulation results show that a diffusion coefficient was one of the most important factors affecting the recovery performance of this process. Also, the effect of gravity drainage is nonnegligible [22].

9.3 Offshore versus onshore heavy oil fields

The discovered heavy oil resources in the world occupy a large portion. How to recover offshore heavy oil resources effectively is a major concern for many oil companies, especially when oil production from onshore heavy oil fields is decreasing. Most discovered offshore heavy oil fields in the world are located in the North Sea (United Kingdom and Norway), Mexico Bay (Mexico), Bohai Bay (China), the Middle East (Saudi Arabia), and Latin America (Brazil), as shown in Fig. 9.1 [5,10,23]. There are various reservoir types and oil viscosities of the discovered offshore heavy oil resources. Considering the space limitation of an offshore platform, water and chemical flooding with a simple surface facility are the most commonly used recovery processes. However, for offshore heavy oil reservoirs whose oil viscosity is high, a conventional cold oil production method is no longer effective, and thermal recovery may

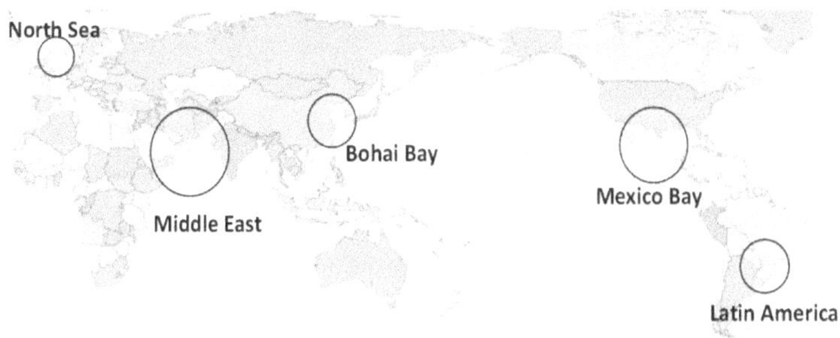

FIGURE 9.1 Distribution of offshore heavy oil resources in the world.

become the only option to unlock highly viscous offshore heavy oil reservoirs. Compared with the operation in an onshore environment, the requirement for offshore sites is stricter. This is because of great differences in production environments and operation conditions between onshore and offshore oil fields. In particular, because of the limited service life of an offshore platform, a higher recovery rate and factor are usually primary requirements for developing offshore heavy oil reservoirs.

Table 9.4 indicates the status of some offshore heavy oil reservoirs in the world. The method of cold oil production (e.g., water and polymer flooding) is the first option for most offshore heavy oil reservoirs. The method of polymer flooding has been successfully applied to recover a heavy oil reservoir, Canada, whose oil viscosity is over 6000 cp [24,25]. On the other hand, the application of thermal oil recovery processes has been evaluated in Mariner and Bressay offshore oil fields [26–28]. Table 9.4 shows that a thermal recovery process was carried out in the Emeraude offshore oil field in Congo (steam flooding), Lake Maracaibo in Venezuela (CSS), and Bohai offshore oil field in China (cyclic multithermal fluid [MTF] injection) [29,30]. Especially for heavy oil reservoirs in the Bohai offshore oil field, oil viscosity reached 10,000 cp. In addition, considering the diversity of heavy oil reservoir types in Bohai Bay, many thermal recovery processes have been experimentally and numerically evaluated, including ISC, steam flooding, and SAGD. In 2010, a cyclic MTFs injection process was performed in NB35-2 heavy oil reservoirs in Bohai Bay, as discussed in Section 1.8.2 in Chapter 1. Considering the successful operation in Emeraude, Lake Maracaibo and Bohai Bay oil fields, thermal recovery will be an important technique for the future development of offshore heavy crude oil.

9.4 Conversion time and operation time

Hybrid EOR are the best potential follow-up recovery processes for heavy oil reservoirs. Based on the discussions in the previous chapters and this one, an

TABLE 9.4 Status of some offshore heavy oil reservoirs in the world.

No.	Offshore oil field	Location	Operation company	Water depth (m)	Oil viscosity (cp @ RC)	Recovery process
1	Captain field	North Sea, UK	ChevronTexaco	104	88	Water flooding; polymer flooding
2	Bently field	North Sea, UK	Statoil; Shell	110	1500	Water flooding; polymer flooding
3	Mariner field	North Sea, UK	Statoil		508 67	Water flooding; hot water flooding, steam flooding, steam-assisted gravity drainage (evaluated)
4	Bressay field	North Sea, UK			1000	Water flooding; polymer flooding; in situ combustion, steam flooding, hot water (evaluated)
5	Grane field	North Sea, Norway	Statoil	123		Water flooding
6	Jubarte field	Brazil	Petrobras	1066	3000 @ 20°C	Water flooding
7	Snorre	Norway	Saga petroleum			Water flooding; noncondensable gas–foam; WAG flooding
8	Tambaredjo field	Suriname	Shell; Esso; Elf		600	Water flooding; cyclic steam stimulation test; polymer flooding; in situ combustion (pilot)
9	Emeraude field	Congo	Elf Acquitaine		100	Steam flooding
10	Lake maracaibo	Venezuela	PDVSA	50	635	Cyclic steam stimulation
11	Bohai oil field	Bohai bay, China	China National Offshore Oil Corporation	15	4580; 10,000	Water flooding; polymer flooding; cyclic multithermal fluids

RC, Reservoir condition; *WAG*, Water alternating gas.

application of hybrid EOR can effectively enhance heavy oil recovery after steam injection. Therefore, an optimal time to convert from pure steam injection to hybrid EOR will be the most important parameter to determine the success of the operation. If conversion is too early, the advantages of hybrid EOR cannot be fully used. If conversion is too late, operation costs will increase. On the other hand, after successful conversion from pure steam injection to hybrid EOR, through a lengthy operation, the next important issue considered is the follow-up methods after hybrid EOR. That is, one needs to consider the operation time of a hybrid EOR process. In this section, these two important issues will be discussed.

9.4.1 Conversion time for hybrid enhanced oil recovery processes

The conversion time depends on the recovery performance of pure steam injection. Once the recovery of pure steam injection is no longer effective, follow-up recovery is required. For CSS recovery, after multiple CSS cycles, a series of serious problems will occur (e.g., steam fingering, channeling, and steam overlap, as discussed in Chapter 2). Fig. 9.2 shows the cyclic oil−steam ratio (OSR) and cyclic cumulative oil production of three typical CSS wells. As the CSS cycle increases, the recovery performance becomes worse and both the cyclic OSR and cumulative oil decrease. As the CSS cycle reaches above 6, changes in the cyclic OSR and cumulative oil production become relatively stable. On the other hand, as the oil viscosity increases, both decrease. For an extra heavy oil reservoir (50,000 cp < μ_o), the cumulative oil production in the first CSS cycle is only about 20% compared with that in a normal heavy oil reservoir (10,000 cp >μ_o). Also, as the CSS cycle increases, under the thermal effect of steam, the recovery performance of a CSS well in an extraheavy oil reservoir can be slightly improved. Fig. 9.2 indicates that oil viscosity and the CSS cycle are two important parameters that affect the

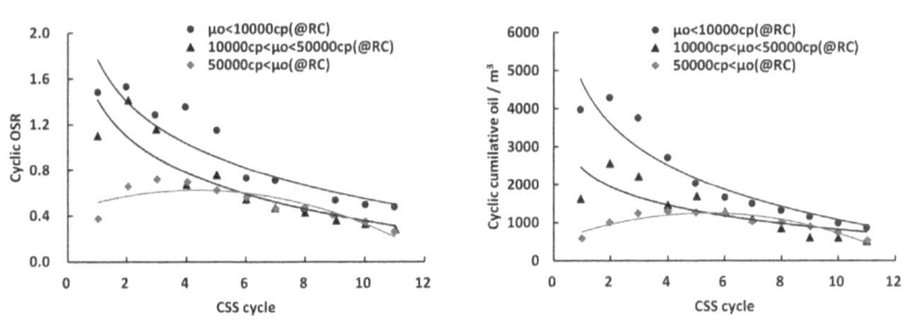

FIGURE 9.2 Recovery performance of CSS wells in heavy oil reservoirs with different oil viscosities. (A) Cyclic OSR and (B) cyclic cumulative oil production. RC in legend refers to reservoir conditions.

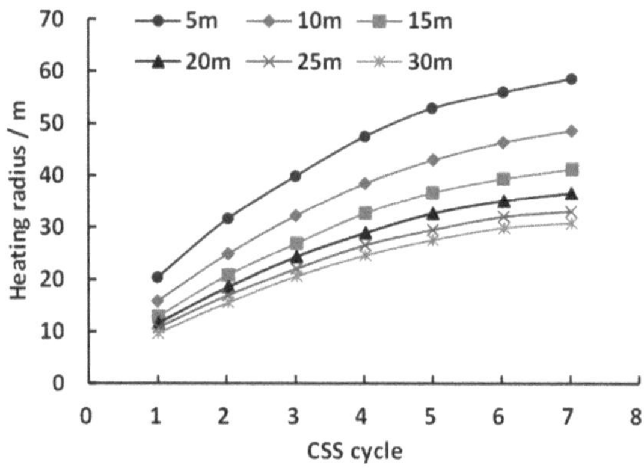

FIGURE 9.3 Heating radius of cyclic steam stimulation (CSS) well in heavy oil reservoir with different reservoir thicknesses. *OSR*, oil–steam ratio.

performance of a CSS well in heavy oil reservoirs. Once the CSS cycle has increased to above six, a follow-up EOR can be considered. However, to convert from a pure steam injection process to a follow-up EOR process, these two quantities should be comprehensively considered.

Besides cyclic OSR and cumulative oil production, a heating radius is another indicator that can be used to evaluate the performance of a CSS recovery process. Fig. 9.3 shows the changes in the heating radius of a CSS well in a typical heavy oil reservoir ($\mu_o < 10{,}000$ cp). As shown, as the CSS cycle increases, the heating radius is increasing. But, once the CSS cycle reaches above five or six, the increasing tendency is relatively stable. It indicates that at this time, as the CSS recovery process continues, the heating radius is not significantly increased. The changing tendencies are not largely different among the results of different reservoir thicknesses. But, under the same steam injection volume and CSS cycle, compared with the heating radius of a thin heavy oil reservoir, the heating radius of a thick heavy oil reservoir is lower. Therefore, from the results of a heating radius, the conversion time is the same as that according to the cyclic OSR and cumulative oil production.

9.4.2 Operation time for hybrid enhanced oil recovery processes

Similar to the discussion in Section 9.4.1, the operation time of a hybrid EOR process refers when follow-up recovery is required it. To evaluate the operation time accurately, the three indicators of cyclic OSR, cyclic cumulative oil production, and the heating radius can be applied. Among them, the OSR is the most commonly used indicator. In this section, a method involving a cyclic OSR and oil recovery under different conditions is proposed to estimate the operation time.

Specifically, N_2-CSS is a typical hybrid thermal-NCG process after pure steam-based CSS recovery. Using the method of numerical simulation, the

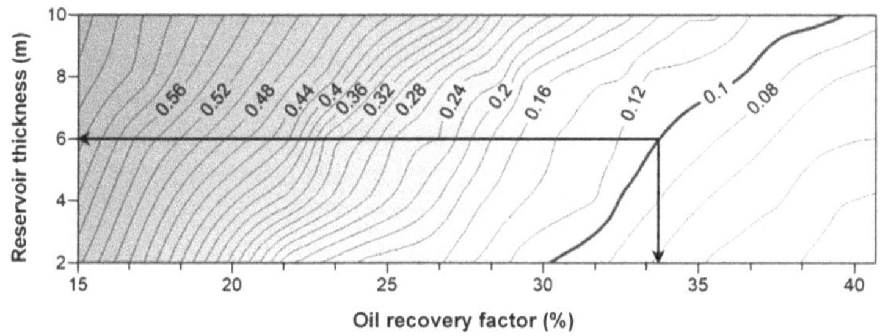

FIGURE 9.4 Plot of operation time of N_2—cyclic steam stimulation (CSS) process.

cyclic OSRs of an N_2-CSS process under different conditions in a typical heavy oil reservoir can be calculated. Then, according to the reservoir thickness and oil recovery, a plot for the operation time of N_2-CSS in heavy oil reservoirs can be obtained, as shown in Fig. 9.4. For a typical heavy oil reservoir developed using N_2-CSS, if the reservoir thickness is given, the operation time can be derived from this plot. For example, for a heavy oil reservoir with a thickness of 6 m, from this plot, the critical oil recovery factor is about 34%. This indicates that if current oil recovery is below 34%, the operation time will not be achieved. However, once oil recovery is above 34%, follow-up recovery after N_2-CSS will be required.

9.5 Formation damage

For heavy oil reservoirs, steam injection is the first operated recovery technique. Based on the discussion in Chapter 2, a series of problems will be encountered after a steam injection process (e.g., fine migration, mineral dissolution, mineral transformation, and clay swelling). These complicated physical and chemical reactions indicate that the reservoir rock has been significantly damaged after a lengthy steam injection operation. However, for hybrid EOR processes, in addition to steam, other additives (NCG, solvent, or chemical agents) are co-injected during recovery. They further enhance reactions between fluids and reservoir rocks. In this section, formation damage caused by hybrid EOR processes will be discussed. Specifically, the discussion includes the adsorption or retention of chemical additives and interactions among rock, brine, and CO_2.

9.5.1 Adsorption and retention of chemical additives in reservoirs

A hybrid thermal-chemical process is a typical hybrid EOR process in heavy oil reservoirs. As shown in Table 9.1, different hybrid thermal-chemical processes have been widely applied to improve the recovery of poststeam injection heavy oil reservoirs. However, during hybrid thermal-chemical processes, because of the interaction between chemical agents and rocks, adsorption loss of chemical additives can be observed. It can have an important influence on the successful operation of a hybrid thermal-chemical process. Mechanisms of adsorption loss of chemical additives in porous media include ion pairing, dispersion, ion exchanging, an electrostatic force, and hydrophobic bonding [31]. During a hybrid thermal-chemical process, the adsorption behavior of chemical additives can highly depend on the type of chemical agent and rock characterization [32,33]. Moreover, changes in the formation water salinity and the rock surface can tremendously affect adsorption behavior between a surfactant and rock. As a typical surfactant, a VR is one of the most commonly used chemical agents in a hybrid thermal-chemical process. From the chemical structure, surfactants are classified as anionic, nonionic, cationic, or zwitterionic [34]. Accurately evaluating the adsorption rate of surfactant has an important role in the successful operation of a hybrid thermal-chemical process. Table 9.5 shows commonly used adsorption models for the adsorption of surfactants in porous media. They can be used to predict the adsorption loss of surfactants conveniently under different conditions.

On the other hand, the adsorption behavior of a surfactant is also a time-dependent process. Therefore, a model is required for the adsorption kinetics of a surfactant in a hybrid thermal-chemical process. Four adsorption kinetic models have been proposed, including a pseudofirst-order rate equation, a pseudosecond-order kinetic model, an intraparticle diffusion model, and Elovich's model [31]. The first model determines the kinetic process of a liquid–solid phase system by a first-order rate equation [39]. Comparatively, the second model is a second-order equation, and it has been widely applied to evaluate the adsorption of herbicides, dyes, oils, and organic materials from aqueous solutions [39,40]. The third model assesses the diffusion mechanism

TABLE 9.5 Commonly used adsorption models for adsorption behavior of surfactants in porous media.

No.	Adsorption model	Formula	Reference
1	Linear isotherm	$q_e = K_{Li}C_e$	[35]
2	Langmuir isotherm	$q_e = \frac{q_o K_{La} C_e}{1+K_{la} C_e}$	[36]
3	Freundlich isotherm	$q_e = K_{Fr} C_e^{1/n}$	[37]
4	Temkin isotherm	$q_e = BlnK_{Te} + BlnC_e$	[38]

q_e refers to equilibrium adsorption density, mg/g; K_{Li}, K_{La}, K_{Fr}, and K_{Te} are constants in the models of linear isotherm, Langmuir isotherm, Freundlich isotherm, and Temkin isotherm, respectively; C_e represents the equilibrium surfactant concentration; q_o is the maximum adsorption density in the Langmuir isotherm model; n is a constant in the Freundlich isotherm model; and B is the slope of a straight line plot in the Temkin isotherm model.

during adsorption [41]. The fourth model was proposed by Zeldowitsch (1934); it can be applied to describe the adsorption of chemisorption kinetics of gas in a solid heterogeneous system [42].

Based on this discussion, adsorption loss can directly affect the performance of a hybrid thermal-chemical process. Therefore, how to reduce the adsorption loss of chemical agents in a reservoir is critical for oil companies. Ionic liquids, nanoparticles and cosurfactants are three typical methods. First, ionic liquids can reduce the adsorption of a surfactant onto formation rocks by a sacrificing mechanism. Then, the main mechanism of an adsorption reduction by nanoparticles is by bonding to surfactant monomers or micelles. Also, sacrificing can be observed under a low concentration nanoparticle condition. Another important method is to employ other additives. Alkali is one of the most commonly used cosurfactants. When alkali is applied, the pH value of a surfactant solution can be increased to reduce the positive charges of rock surfaces, especially carbonate, dolomite, and kaolinite rocks [31].

9.5.2 Corrosion reactions in a rock–brine–CO_2 system

Based on the discussion in previous chapters, CO_2 is a potential additive applied in hybrid EOR processes in heavy oil reservoirs. Compared with other NCGs (e.g., N_2, air, and flue gas), CO_2 can take full advantage of dissolution, extraction, and diffusion. On the other hand, CO_2 can dissolve into a water phase to form an acidic environment under HCO_3^-. Thus, a corrosion reaction in a rock–brine–CO_2 system can have an important role in recovering hybrid EOR, especially for carbonates and limestones. In this section, an experimental method is proposed to explore the corrosion reaction in a rock–brine–CO_2 system under different conditions.

9.5.2.1 Experimental method

First, a high-temperature, high-pressure (HTHP) reactor is applied to test corrosion between a water-saturated rock core and CO_2 under different conditions. Then, using an electron microscope and core slices, the effect of CO_2 corrosion on the pore structure of the rock core can be determined.

The detailed experimental procedures are:

- ➢ A crushed rock core is first saturated with formation water and weighed;
- ➢ The rock core is placed into the HTHP reactor;
- ➢ CO_2 is injected into reactor until the designated pressure is achieved;
- ➢ The reactor temperature is set and maintained at the designated temperature condition for 3 days;
- ➢ The pressure changes are recorded during the whole process until the pressure no longer changes;
- ➢ The experiment is terminated and the rock core is weighed again;
- ➢ A core slice is taken to test changes in the rock pore structure.

With changes in the rock pore structure, the permeability of the rock core also changes. Therefore, another experiment in this section is performed to test changes in rock permeability under the effect of a corrosion reaction in the rock–brine–CO_2 system. In this experiment, a high-temperature core holder is applied to provide a high-temperature environment. The detailed experimental procedures are:

- ≻ A standard rock core is first weighed and saturated with formation water;
- ≻ The core permeability is tested using N_2;
- ≻ CO_2 is injected until the desired pressure condition is achieved;
- ≻ The temperature of the core holder is set and maintained for 3 days;
- ≻ The rock core is taken out and dried in an oven;
- ≻ The core mass and permeability are tested again;
- ≻ Differences in the core mass before and after the experiment are compared.

9.5.2.2 Experimental results

Table 9.6 shows the mass changes of rock after the experiments. After the test in the HTHP reactor, a slight reduction in rock mass can be observed. With an increase in temperature, the corrosion interaction is enhanced, and thus the degree of rock mass reduction is more significant, which confirms the enhancement of a corrosion reaction in the rock–brine–CO_2 system.

Then, bases on these experiments, a series of rock slices are prepared. In addition, an electron microscope is applied to testing the effect of a corrosion reaction on the pore structure of rock cores. The results are shown in Fig. 9.5. The test results show that the rock is a calcareous fine feldspathic quartz sandstone. Fig. 9.5A and B (an original rock core) shows a small amount of intergrain dissolution pores. As shown, once the corrosion rection is induced, the number of intergrain dissolution pores is increased, and some particle dissolution pores can be also observed (Fig. 9.5C and D). Then, with an increase in the test temperature and pressure, the corrosion reaction is enhanced and some microscopic and small fractures can be observed (Fig. 9.5F–H). Under a high-pressure condition, CO_2 can preferentially enter a fracture

TABLE 9.6 Crushed rock mass before and after CO_2 corrosion experiments.

No.	Temperature (°C)	Pressure (MPa)	Mass before experiment (g)	Mass after experiment (g)	Mass reduction ratio (%)
1	50	0.6	4.375	4.372	0.069
2	100	2.0	4.832	4.811	0.435
3	120	8.5	4.571	4.526	0.984

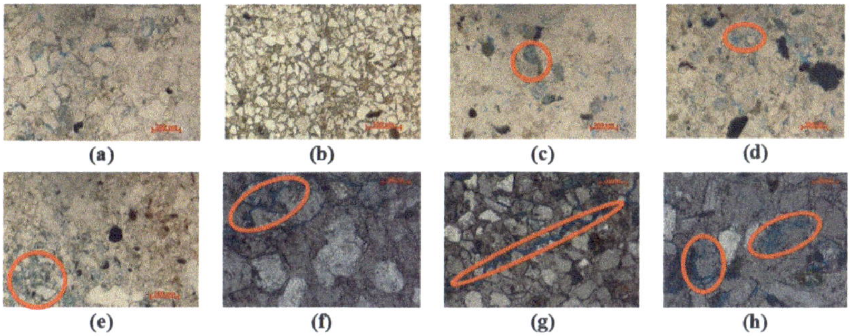

FIGURE 9.5 Changes in rock pore structure with the effect of corrosion reaction of CO_2. (A, B) Test results of original rock core; (C) test results of experiment 1; (E, F) test results of experiment 2; (G, H) test results of experiment 3.

system and diffuse along the fracture surfaces. Thus, a corrosion reaction can be induced between the soluble mineral (calcite and feldspar) and CO_2, and more dissolution pores can be observed in the formation rocks.

Table 9.7 shows changes in core permeability. Like the results in Table 9.6, the rock−brine−CO_2 interaction can have an important effect on rock properties for rock mass as well as permeability. As for CO_2 corrosion, rock permeability is tremendously improved, which further indicates that the injection of CO_2 in hybrid EOR can take advantage of EOR mechanisms and improve fluid flow by increasing rock permeability. Especially for poststeam injection heavy oil reservoirs, the reservoir temperature is usually higher than the original formation temperature. The results in Table 9.7 show that a high-temperature condition can help a CO_2 corrosion reaction and significantly increase rock permeability.

9.6 Methods after hybrid enhanced oil recovery processes

Based on the discussion in Section 9.4.2, once a hybrid EOR process is no longer effective, a follow-up EOR method is required. Table 9.1 shows that all three different hybrid EOR processes (thermal-NCG, thermal-solvent, and thermal-chemical) can be applied to improve the recovery of heavy oil reservoirs after the operation of three different steam injection modes. This indicates that the selection of a suitable hybrid EOR process is not sensitive to the steam injection mode; it highly depends on the reservoir properties and previous recovery operation process. Similarly, the selection of an EOR process after a hybrid EOR process depends on its previous operation. Moreover, once a previous hybrid EOR process is terminated for some special reason (technical or nontechnical), the reservoir properties and conditions should be reevaluated. It is helpful for the follow-up EOR process for heavy oil reservoirs.

TABLE 9.7 Rock core mass and permeability before and after CO_2 corrosion experiments.

No.	Temperature (°C)	Pressure (MPa)	Mass (g)		Permeability (10^{-3} μm^2)		Permeability increase ratio (%)
			Before test	After test	Before test	After test	
1	25	2	61.255	61.230	0.8424	0.9562	13.51↑
2	120	8	59.628	58.130	0.6389	1.217	76.41↑

Specifically, a hybrid steam-N_2 process is the most typical hybrid thermal–NCG process. Table 9.1 shows that it can be applied to improve a post-CSS heavy oil reservoir through cyclic steam-N_2 injection. In addition, it can be applied to improve the recovery of poststeam flooding and post-SAGD heavy oil reservoirs by continuous/slug steam-N_2 injection and hybrid N_2-SAGD (i.e., SAGP). During the field operation of a hybrid steam-N_2 process, after this process is performed for a long time, a large amount of N_2 can accumulate in the top zone of a heavy oil reservoir. Because of its low heat conductivity, this artificial N_2 layer can be considered a cotton comforter to reduce heat loss between a reservoir and its overburden. Therefore, once another EOR process is selected to replace the previous hybrid steam-N_2 process, this N_2 layer will be the top issue to be considered. In the best case, the newly selected EOR process should take advantage of residual heat as well as the current reservoir properties (e.g., current pressure, a fluid distribution, and oil saturation) to help operate this process. For example, a hybrid steam-N_2-VR process has been applied as a follow-up process of a hybrid steam-N_2 process [5]. In addition, the strategy of a variable gas–steam ratio is commonly used to improve the recovery of a hybrid steam-N_2 process. This operation strategy can take advantage of additives.

For a hybrid thermal-solvent process, based on the discussion in Sections 2.8.1 in Chapter 2 and 5.4.2 in Chapter 5, asphaltene deposition should be the priority in a follow-up EOR process after this hybrid EOR process. Because of asphaltene deposition, a fluid flow path is narrow or even blocked. Also, a large amount of asphaltenes accumulate in a formation. Furthermore, because of the high operation cost, the solvent recovery rate is an important issue in this process. It indicates reducing operation costs and addressing the problem of asphaltene deposition are the most important issues for the new recovery process. Similar to the strategy of a variable gas–steam ratio in a hybrid thermal-NCG process, the variable solvent-steam ratio is a potential technology. Because of the reduction in a solvent injection volume, the amount of precipitated asphaltenes can be reduced. In addition, a hybrid thermal-solvent-NCG process is a good choice. In this process, the mechanisms of steam, solvent, and NCG will be fully combined. Moreover, an amount of NCG will be injected to replace the solvent, and thus the operation cost can be reduced.

For a hybrid thermal-chemical process, as discussed in Chapter 1, this process is applied to plug a steam breakthrough path and improve a steam injection profile. During this process, considering its operation cost, a chemical additive is injected by a small slug and then subsequent steam is injected. After a long time, a large amount of chemicals can be found in a formation. For some heavy oil reservoirs, more than one hybrid thermal-chemical process is conducted; therefore, more chemicals can be observed. Among so many chemical additives, a foaming agent and a surfactant are commonly used, and a gel system has minimum application because of its high temperature requirement. For hybrid thermal-chemical processes, once a valid period of chemical additives is achieved (a pressure drop is reduced, or an oil production rate is reduced), it will be terminated and another recovery process will be required. For the follow-up recovery process, a hybrid thermal-chemical process with a higher plugging strength can be a potential process. An NCG foaming system will be a follow-up process for a steam-surfactant co-injection process, and a high-temperature gel system will be a good choice after an NCG foaming system. In addition, in some heavy oil reservoirs with multiple steam breakthrough paths, multiple chemical additive slugs are applied to improve steam injection profiles.

Based on the discussion in Chapters 1–8, it can be concluded that hybrid EOR processes can be used to improve the recovery performance of heavy oil reservoirs effectively. However, once they are applied in the actual field, some challenges may hinder their application and extension. In this chapter, a series of challenges during the application of a hybrid EOR process are discussed. The discussion in this chapter can help their future application in an actual heavy oil field. In the next chapter, some other EOR processes and the future trends of heavy oil EOR processes will be addressed.

References

[1] Butler RM. Thermal recovery of oil and Bitumen. GravDrain's Blackbook; 1997.

[2] Speight James G. Enhanced recovery methods for heavy oil and tar sands. Gulf Publishing Company; 2009.

[3] Liu HQ. Principle and design of thermal oil recovery processes. Petroleum Industry Press; 2013 (In Chinese).

[4] Kalateh R, Ogg L, Charkazova M, Gerogiorgis I. AES-2016-A database and workflow integration methodology for rapid evaluation and selection of Improved Oil Recovery (IOR) technologies for heavy oil fields. Adv Eng Software 2016;100:176–97.

[5] Dong X, Liu H, Chen Z, et al. Enhanced oil recovery techniques for heavy oil and oilsands reservoirs after steam injection. Appl Energy 2019;239:1190–211.

[6] Jha RK, Kumar M, Benson I, Hanzlik E. New insights into steam/solvent-Coinjection process mechanism. SPE J 2013;18(5):867–77.

[7] Farouq Ali SM. Practical heavy oil recovery. Lecture Notes. University of Calgary; 2007.

[8] Xu W, Zhang F, Wang D, Wu T, Dong X, Chen Z, Xu J. CNOOC studies steam recovery in offshore Bohai heavy oil field. Oil Gas J 2018;116(4):46–51.

[9] Dong X. The development mechanism and method screening for offshore heavy oil reservoirs with multi-thermal fluid. PhD Dissertation. Beijing: China University of Petroleum; 2014.

[10] Dong X, Liu H, Hou J, Zhang T, Zhan J, Chen Z, Hong C. The thermal recovery methods and technical limits of Bohai offshore heavy oil reservoirs: a case study. In: OTC 26080 presented at the offshore technology conference Brazil, Rio de Janeiro, Brazil, 27–29 October, 2015; 2015.

[11] Babadagli T. Evaluation of EOR methods for heavy-oil recovery in naturally fractured reservoirs. J Petrol Sci Eng 2003;37:25–37.

[12] Telmadarreie A, Trivedi JJ. New insight on carbonate-heavy-oil recovery: pore-scale mechanisms of post-solvent carbon dioxide foam/polymer-enhanced-foam flooding. SPE J 2016;21(5):1655–68.

[13] Alvarez J.M., Sawatzky R.P., Forster L.M., Coates R.M., Alberta's Bitumen carbonate reservoirs – moving forward with advanced R&D. Paper presented at the second world heavy oil congress, Edmonton, AB, 10–12 March 2008.

[14] Wang Q. The flow mechanism and development strategy for naturally fractured heavy oil reservoirs. PhD Dissertation. Beijing: China University of Petroleum; 2012.

[15] Sahuquet BC, Ferrier JJ. Steam-drive pilot in a fractured carbonated reservoir: Lacq Superieur field. J Petrol Technol 1982;34(4):873–80.

[16] Huo GR, Li XM, Zhang GQ. Thermal oil recovery technologies of heavy oil reservoirs in Shengli oilfield. Petroleum Industry Press; 1999 [In Chinese].

[17] Novak J., Edmunds N., Cimolai M. A history match of CSS recovery in the Grosmont. Paper presented at the Canadian international petroleum conference, Calgary, Alberta, June 12–14, 2007.

[18] Edmunds N, Barrett K, Solanki S, Cimolai M. Prospects for commercial Bitumen recovery from the Grosmont carbonate, Alberta. J Can Petrol Technol 2009;48(9):26–32.

[19] Al Bahlani A.M., Babadagli T. Steam-over-solvent injection in fractured reservoirs (SOS-FR) for heavy-oil recovery experimental analysis of the mechanism. SPE 123568 presented at the SPE Asia Pacific oil and gas conference and exhibition, Jakarta, Indonesia, August 4–6, 2009.

[20] Al Bahlani AM, Babadagli T. Laboratory scale experimental analysis of Steam-Over-Solvent injection in Fractured Reservoirs (SOS-FR) for heavy-oil recovery. J Petrol Sci Eng 2012;84–85:42–56.

[21] Al Bahlani A.M., Babadagli T. Heavy-oil recovery in naturally fractured reservoirs with varying wettability by steam solvent Co-injection. SPE 117626 presented at the international thermal operations and heavy oil symposium, Calgary, Alberta, Canada, 20–23 October, 2008.

[22] Singh R, Babadagli T. Mechanics and upscaling of heavy oil Bitumen recovery by steam-over-solvent injection in fractured reservoirs method. J Can Petrol Technol 2011;50(1):33–42.

[23] Wehunt C.D., Burke N.E., Noonan S.G. and Bard T.R. Technical challenges for offshore heavy oil field developments. Paper OTC 15281 presented at the offshore technology conference, Houston, Texas, USA, 5–8 May 2003.

[24] Bai J., Wassmuth F.R., Jost R.W., Zhao L. Hydrophobically-Modified cellulosic polymer for heavy oil displacement in Saline conditions, SPE 157917 presented at the SPE heavy oil conference Canada, Calgary, Alberta, Canada, 12–14 June, 2012.

[25] Mukherjee S., Süss M.P. Performance evaluation and design aspects of polymer and hot water Co-injection in an offshore heavy oil field under challenging conditions, SPE 169711 presented at the SPE EOR conference at oil and gas West Asia, Muscat, Oman, 31 March-2 April, 2014.

[26] Jayasekera A.J., Goodyear S.G. The development of heavy oil fields in the U.K. Continental shelf: past, present and future, SPE 54623 presented at the SPE Western regional Meeting, May 26–27, 1999, Anchorage, Alaska, USA.

[27] Berg E.A., Resksten K.,Scott A.S., Ibatullin T., Mollerstad H., Aasum Y., Julseth L. Heavy oil offshore UK: recommended mariner reservoir development strategy, SPE 145618 presented at the SPE offshore europe oil and gas conference and exhibition September 6–8, 2011, Aberdeen, UK.

[28] Alvarado V, Manrique EJ. Engineering design challenges and opportunities beyond waterflooding in offshore reservoirs, SPE 37634 presented at the offshore technology conference May 6–9, 2013, Houston, Texas, USA.

[29] Couderc BM, Verpeaux JF, Monfrln D, Quettler LH. Emeraude vapeur: a steam pilot in an offshore environment. SPE Reserv Eng 1990;5(4):508–16.

[30] Tang XX, Ma Y, Sun YT. Research and field test of complex thermal fluid huff and puff technology for offshore viscous oil recovery. China Offshore Oil Gas 2011;23:185–8 [in Chinese)].

[31] Ahmadi M, Chen Z. Challenges and future of chemical assisted heavy oil recovery processes. Adv Colloid Interface Sci 2020;275:102081.

[32] Daoshan L, Shouliang L, Yi L, Demin W. The effect of biosurfactant on the interfacial tension and adsorption loss of surfactant in ASP flooding. Colloids Surf A Physicochem Eng Asp 2004;244:53–60.

[33] Jian G, Puerto MC, Wehowsky A, Dong P, Johnston KP, Hirasaki GJ. Static adsorption of an ethoxylated nonionic surfactant on carbonate minerals. Langmuir 2016;32:10244–52.

[34] Myers D. Surfactant science and technology. John Wiley & Sons; 2005.

[35] Langmuir I. The constitution and fundamental properties of solids and liquids. Part I. solids. J Am Chem Soc 1916;38:2221–95.

[36] Freundlich H. Uber die adsorption in lumngen. Z Phys Chem 1906;57:385–470.

[37] Ruthven DM. Principles of adsorption and adsorption processes. John Wiley & Sons; 1984.

[38] Ahmadi MA, Shadizadeh SR. Experimental investigation of a natural surfactant adsorption on shale-sandstone reservoir rocks: static and dynamic conditions. Fuel 2015;159:15–26.

[39] Ahmadi MA, Shadizadeh SR. Spotlight on the new natural surfactant flooding in carbonate rock samples in low salinity condition. Sci Rep 2018;8:10985.

[40] Ahmadi MA, Shadizadeh SR. Adsorption of novel nonionic surfactant and particles mixture in carbonates: enhanced oil recovery implication. Energy Fuel 2012;26:4655–63.

[41] Low M. Kinetics of chemisorption of gases on solids. Chem Rev 1960;60:267–312.

[42] Yuh-Shan H. Citation review of Lagergren kinetic rate equation on adsorption reactions. Scientometrics 2004;59:171–7.

Chapter 10

Other enhanced oil recovery processes and future trends

10.1 Introduction

As an important petroleum resource, heavy oil has an important role in the world energy supply. The world's proven heavy oil reserves are about 9911.8×10^8 t and the world's heavy oil production is about 5000×10^4 t [1]. When oil prices are low, to enhance heavy oil recovery effectively with a low operation cost is the first option for all oil companies. Through lengthy steam injection, both heavy oil and formation properties have significantly changed compared with their original properties. Thus, the previously applied steam injection processes (cyclic steam stimulation [CSS], steam flooding, or steam-assisted gravity drainage [SAGD]) are no longer the optimal recovery process. In this book, the most commonly used follow-up processes, hybrid EOR methods, are systemically introduced from laboratory-scale experiments, mechanism simulations, and field-scale applications. In the future, hybrid EOR processes will be the most effective recovery processes for steamed heavy oil reservoirs as well as marginal heavy oil reservoirs (including low reservoir thickness, high formation sensitivity, and large active aquifers).

By the application of hybrid EOR processes, heavy oil reserves that cannot be economically recovered have become available. As discussed in Chapter 1, some hybrid EOR processes have been applied to enhance heavy oil recovery. The performance of these hybrid EOR processes can be described through three aspects [2–4]:

(1) Increasing the flowability of heavy crude oil. For heavy oil reservoirs with high oil viscosity, the presence of different agents (a solvent, noncondensable gas (NCG), and a surfactant) can change the fluid state of a fluid in porous media to increase its flowability.

(2) Increasing the flowing distance of heavy crude oil. For heavy oil reservoirs with the recovery process of steam huff-n-puff, the heating radius of a single well is usually limited (Fig. 9.2). Especially for a production well

Hybrid Enhanced Oil Recovery Processes for Heavy Oil Reservoirs.
https://doi.org/10.1016/B978-0-12-823954-4.00009-6

with multiple CSS cycles, the application of a hybrid enhanced oil recovery (EOR) process can increase the heating radius based on steam injection. Thus, the flowing distance of heavy crude oil is increased.

(3) Increasing the flow rate of heavy crude oil. In the middle or later stage of steam injection, because of a reduction in the relative permeability of the oil phase, the effective flow rate of heavy crude oil is low. However, once a hybrid EOR process is considered, under the nonisothermal flow condition, multiphase fluid flow behavior in porous media can effectively reduce the relative permeability of the water phase and increase that of the oil phase. Thus, the flow rate of heavy crude oil is increased.

Although the advantages of hybrid EOR processes are prominent, in some field cases their recovery performance may not be economical. It is caused by geological or reservoir conditions, but it also depends on the operation by petroleum engineers. In addition, economics are important. Besides hybrid EOR processes, some recovery techniques are proposed for heavy oil recovery. Different from conventional EOR processes, most of these techniques have the characteristics of a multidiscipline background. In this chapter, they will be systematically introduced.

10.2 Electrical heating

For conventional steam-based heavy oil recovery processes, the cost of steam generation accounts for most of the operation cost. On the other hand, considering the requirement of a large amount of steam in heavy oil recovery processes, surface oil—water separation treatment is another important issue. If the costs of steam generation and production liquid treatment can be reduced, economic benefits can be extremely impressive. With such a background, the method of electrical heating is proposed. Especially for heavy oil reservoirs whose reservoir temperature is not too low and oil viscosity is not too high (e.g., the Orinoco oil belt in Venezuela), an electrical heating method will be a potential EOR technique. Moreover, for heavy oil reservoirs with low injectivity, a great formation depth, and a high degree of heterogeneity, the electrical heating method is attractive. In particular, because of the effect of wellbore heat loss, a conventional steam injection process cannot be applied in a deep heavy oil reservoir. For electrical heating, however, steam injection is no longer required. This indicates that the issue of heat loss will no longer be a problem. It may be applied to recover a deep reservoir effectively.

For electrical heating, from the frequency of the electrical current used, it can be classified into three categories: low-frequency electric resistive (ohmic) heating (less than 100 Hz), medium-frequency electromagnetic (EM) induction heating (100—300 kHz), and high-frequency (radio-frequency or microwave) EM heating (10—100 MHz (radio-frequency, 100 MHz to 100 GHz, microwave) [5], as shown in Table 10.1. The heating methods of the three

TABLE 10.1 The difference in three electrical heating methods.

Method	Main components	Heating process
Low-frequency electric resistive (ohmic) heating	Electrode assembly, power conditioning unit, power delivery system, grounding system, recording/monitoring system.	This method is dominant because of heating ions via energy transfer from a heater, with more mobile electrons that carry the bulk of the current [2].
Medium-frequency electromagnetic induction heating	Electromagnetic antenna or induction coil	In this method, a conductor is used to induce a magnetic field in a formation. Thus, the magnetic field is changed, which induces a secondary current and circulates in a medium to generate heat [6].
High-frequency (radio-frequency or microwave) electromagnetic heating	Radio-frequency antenna or induction coil	Dipoles are formed by molecules and tend to align them with an electric field. The molecular movement produces heat in a reservoir [7].

electrical heating processes in reservoirs are different. heating has been applied to assist steam injection in SAGD. By combining inductive heating and steam injection, the efficiency of a preheating process can be significantly improved [8].

For a field application, a low-frequency electrical heating field trial in the Rio Panan field, Brazil was performed in Dec. 1987 [9]. After an electrical heating system was turned on, a sudden increase in the oil production rate was observed. After 40 days of heating, the oil production rate increased from 0.2 to 1.0 m^3/day at 20 kW. An electrothermal dynamic stripping process (ETDSP) is another typical electrical resistive heating method. This process can be applied to heavy oil reservoirs whose depth is too shallow for steam injection and too deep for surface mining. In 2007, a field pilot test for the ETDSP process was performed in the McMurray formation by E-T Energy Ltd [10]. An obvious increase in oil was observed.

Comparatively, for a high-frequency electrical heating process, EM heating, the appropriate EM source frequency for heating heavy oil or oil sands reservoirs usually falls in the radio-frequency (RF) band [5]. Since the 1970s, the method of EM heating-based thermal recovery has been attempted. Because of the limitations in an understanding of the complex physics of EM heating, there is scarce information on its successful field applications. In

2012, a pilot-scale implementation of RF heating, known as effective solvent extraction incorporating electromagnetic heating (ESEIEH), successfully demonstrated the application potential of an RF source in a field [7]. For an oil field with a production rate of 10,000 bbl/day, the cost of ESEIEH was estimated to be \$50/bbl, whereas it was \$70/bbl for SAGD [11,12].

10.3 Nanotechnology

As a cutting-edge technology, nanotechnology has been applied in many fields, such as chemistry, biomedicine, physics, space, and engineering. For petroleum engineering, nanomaterials has important application potential to improve exploitation and development processes for petroleum reservoirs. Specifically, through the application of nanomaterials, oil mobility during the life of a petroleum reservoir can be significantly improved. Because of the small size of nanoparticles (NPs) (1−100 nm), they can easily be transported with low retention in porous media. The risk of blocking a pore space is no longer a problem. Importantly, compared with other additives, nanomaterials or NPs have the unique characteristics of a high surface-to-volume ratio, wettability alternation, interfacial tension (IFT) reduction, and viscosity reduction [13,14]. During application, NPs are added to a liquid phase (water); then, prepared nanofluids or smart fluids are injected or co-injected with other fluids to improve oil recovery.

Fig. 10.1 shows relative permeability curves with and without NP conditions. With the mechanism of wettability alteration of NPs, both a crossover point and end points of the relative permeability curves are shifted rightward [15], which improves an oil recovery process.

Second, for an IFT reduction, although it is the main mechanism of surfactant injection, an additional reduction in IFT is observed at the presence of NPs. Many experimental studies have found that adding NPs can reduce the

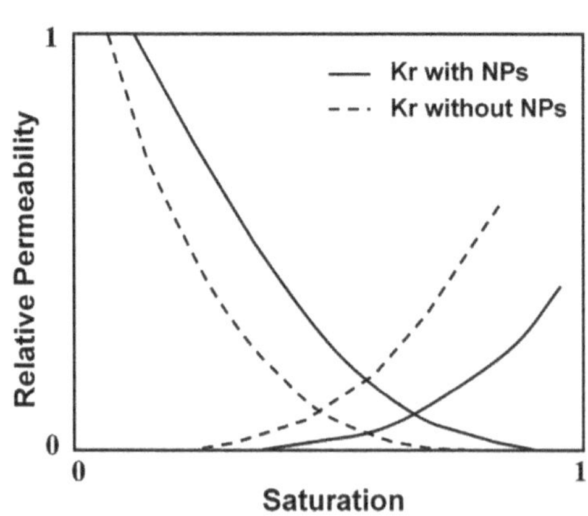

FIGURE 10.1 Relative permeability curves with and without nanoparticles (NPs).

IFT and contact angle between oil and nanofluid–water [14,16]. As shown in Fig. 10.2, the presence of NPs in a three-phase contact region can increase the tendency to form a wedge structure, and thus the wettability of a pore surface shifts toward more water wet. By analyzing internal driving forces, a structural disjoining pressure has an important role in the spreading behavior of NPs, including the van der Waals, electrostatic, and solvation forces on a pore wall surface [16]. A series of tests was performed to discuss the effect of sodium dodecyl sulfate (SDS) surfactant/silica NPs on IFT, a contact angle, and oil recovery. As the NPs concentration increased, both the IFT and the contact angle were reduced, and oil recovery increased [17].

Third, viscosity reduction is caused by the adsorption of heavy components (asphaltene and resin) on the surfaces of NPs. Heavy crude oil has a high asphaltene content and complex rheologic behavior in porous media. Once NPs are injected, the interaction between NPs and asphaltenes can modify the rheologic properties of heavy oil and reduce the size of asphaltene aggregates at the microscale level [18]. Detailed mechanisms include van der Waals, H-bonding, π-π stacking, and acid-based interactions [18]. On the other hand, the catalytic properties of NPs can contribute to catalytic decomposition reactions of heavy crude oil [19]. Therefore, some lighter components are created, and oil viscosity is reduced. Furthermore, compared with other chemical agents for oil viscosity reduction (e.g., surfactant), the perdurability of a viscosity reduction of heavy oil-nanofluids is significantly enhanced [17]. For a heavy oil field operation in Colombia, heavy oil viscosity was reduced by more than 60% during continuous nanofluid injection; the oil production rate increased by about 10% [19].

Types of NPs include metal/metal oxide, silica, and organic. NP-based EOR processes can act as an adsorbent and catalyst to help upgrade heavy oil and enhance recovery. Specifically, for different thermal recovery processes in heavy oil reservoirs, including steam injection and in situ combustion, an application of nanotechnology can decompose asphaltenes to improve heavy oil quality [18]. Even in an electromagnetic heating process, the presence of magnetic NPs can increase recovery through perturbation at a water–oil interface [20].

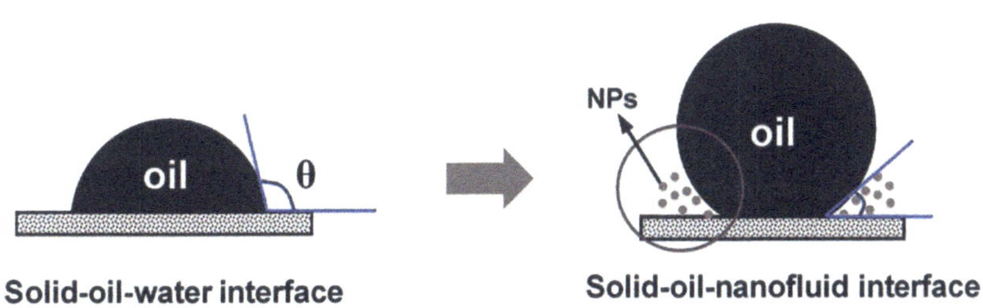

FIGURE 10.2 Effect of nanoparticles (NPs) on wettability of pore surface.

10.4 Ionic liquids

Ionic liquids (ILs) are organic salts composed of organic cations and organic or inorganic anions [21]. They are proposed as a potential alternative to surfactants because of their surface activity. Their properties are dominated by a number of cations and anions, which made ILs a good candidate for many petrochemical industries. Different from the application of ILs in the petrochemical industry (e.g., upgrading [extra] heavy oil and refining), they can also be applied in heavy oil EOR processes. After ILs are injected into a formation, the polar components of heavy oil (asphaltene and resins) can diffuse into ILs. Thus, the viscosity of crude oil is reduced and a reduction in polar components is observed [22,23]. Therefore, during hybrid thermal-solvent processes, an application of ILs can inhibit the occurrence of asphaltene precipitation. A simulation study observed that only a small amount of asphaltene can deposit in a porous medium and the presence of ILs can prevent a pore plugging problem caused by asphaltene deposition [23]. On the other hand, because of electrostatic interactions between sand grains and ILs in porous media, an IL medium can reduce the energy of adhesion more significantly than an aqueous solution [24]. The other mechanisms include emulsification, a reduction in IFT, catalysis, hydrocracking, and hydrogenation [25]. Thus, to some extent, the EOR mechanisms of ILs are similar to those of nanofluids. a type of ILs, eutectic-based ILs or deep eutectic solvents (DES), has attracted wide interest as novel solvents [26]. Mohsenzadeh et al. studied the performance of DES in steam flooding for heavy oil reservoirs [25,27]. In situ upgrading was observed, such as an increase in American Petroleum Institute (API) gravity, a reduction in sulfur (16%) and an increase in saturate hydrocarbons in products.

Although the advantages of ILs in heavy oil EOR processes are significant, a field-scale application is still challenging, mainly because of complicated formation conditions (e.g., high reservoir temperature, high salinity, and serious reservoir heterogeneity) and high operation costs.

10.5 Solar and nuclear energy

For thermal recovery processes for heavy oil reservoirs, the cost of steam generation takes up a high proportion of the entire operation cost for many oil companies. Therefore, how to reduce the cost of steam generation is a top concern. Based on this, solar or nuclear energy-based steam generation technology has gained much attention. It can significantly reduce the amount of natural gas consumption for steam generation in thermal recovery processes. Compared with traditional steam generation facilities, solar and nuclear energy-based methods can make EOR much environmentally friendlier, cleaner, and more effective. The application of solar and nuclear energy can help carbon neutrality. In this section, both solar and nuclear energy steam generation methods are mentioned.

10.5.1 Solar energy

First, for solar energy-based steam generation, a solar collector is applied to heat water and generate steam. It can increase net profits tremendously for heavy oil production [28–30]. Because of the effect of day–night variation on steam output, a conventional fuel-fired steam generator has been used to balance variations in steam output by a solar steam generator. Using this mode, fuel consumption for a steam generator in Kuwait can be reduced by about 75% [31]. Although the application of a solar steam generator is attractive, its application is dominated by many factors, including the duration of sunshine, climactic conditions, and policies regarding encouragement. The effects of climactic conditions include dust, major sandstorm events, wind, erosion, and drifting sand [32].

For application, some solar facilities are under way or planned in field operations in San Joaquin Valley, Oman, and Kuwait [33]. For the Bright Source project in San Joaquin Valley, it is adjacent to Chevron's Coalinga field. From this project, 60% quality steam is generated at 500°F and 700 psi [34]. In Oman, because of its low latitude, large seasonal variations cannot significantly affect solar irradiation; most locations can receive high direct normal irradiance (DNI) (>2000 kW h/a). The Amal field (DNI: 2057 kW h/a) in southern Oman was finally selected as the solar-steam generation pilot site using enclosed-trough technology [35]. Furthermore, Kuwait announced a future national oil production target, with a program of heavy oil development planned to reach 270,000 b/p by 2030 [36]. Solar energy–based steam generation technology is a potential method for thermal recovery for heavy oil reservoirs. For heavy oil fields with a high DNI, the application of solar energy can tremendously reduce operation costs and some carbon emissions to help carbon neutrality.

10.5.2 Nuclear energy

Similar to solar energy, except to generate electricity, nuclear energy can be applied to replace traditional natural gas–fired methods for steam generation. As an important form of clean energy, its application significantly reduces greenhouse gas emissions and reduces operational costs, especially regarding carbon neutrality [37]. In 1977, the application of nuclear energy for heavy oil and bitumen recovery in Alberta was proposed [38]. By using a organic-cooled CANDU reactor(CANDU-OCR) nuclear reactor from Atomic Energy of Canada Limited (AECL), the specifications of a steam generator for in situ application can be met. In 2003, based on the CANDU reactor concept, AECL designed a new reactor, the ACR-700 Advanced CANDU Reactor. For ACR-700, most thermal energy produced by the nuclear reactor could be used for steam production rather than electricity generation. Therefore, higher steam quality was possible for a nuclear configuration because steam was produced by a heat exchanger instead of gas firing. Moreover, regarding the economical profits of an ACR-700 reactor and a natural gas-fired facility to supply steam for SAGD, steam from a nuclear facility was more economical [39].

For a field application, an advanced nuclear power plant was placed for heavy oil recovery in the Orinoco Oil Belt, Venezuela. In this nuclear power plant, three 1200-MW high-temperature gas-cooled reactors were built successively. They can supply sufficient heat energy and steam for oil production and even electricity [40]. To inject steam, the pressure of produced steam can reach 12−17 MPa at saturation temperatures. Furthermore, in China and some other countries, small nuclear power plants have been planned to assist steam generation in thermal EOR. Using clean energy and reducing carbon emissions have become top priorities across the world. However, because of environmental concerns regarding nuclear waste, the application of nuclear energy for EOR is considered as a technology backup.

10.6 Wellbore configurations

Another method to improve thermal recovery performance is to change the conventional steam injection wellbore configuration. For horizontal wells, because of their long horizontal wellbores, the effect of reservoir heterogeneity on steam conformance is significant. However, the presence of a drop in wellbore pressure can also have an important role in steam conformance along a wellbore. To improve the distribution of a steam injection profile and increase the degree of reservoir unlocking along a wellbore, two subsequent wellbore configurations are applied.

10.6.1 Flow control devices

A flow control device (FCD) is a wellbore throttling device that controls inflow (or outflow) rate distribution along a horizontal wellbore. According to the fluid flow direction, it can be classified into an inflow control device (ICD) or an outflow control device (OCD). An ICD can be applied in a steam injection well, and an OCD can be applied in a production well. For heavy oil reservoirs with serious permeability heterogeneity or some adverse geologic factors (e.g., an aquifer, gas cap, and shale barrier), nonuniform distributions of a steam injection profile in an injector and a fluid production profile in a producer can be observed. This indicates that rapid steam breakthrough or water coning can be induced. In such a condition, an FCD can provide an important solution to address injection or production problems. By applying FCDs, an extra pressure drop is attached on a wellbore to limit the fluid flow rate of a high fluid flow interval. Moreover, for a horizontal interval with a low fluid flow rate (caused by reservoir heterogeneity or variable mass fluid flow), an FCD can reduce a drop in pressure and increase the fluid flow rate. Thus, the distribution of a fluid flow rate along an entire horizontal wellbore can be more uniform and flowing uniformity across the wellbore can be improved. For a horizontal well in a heavy oil reservoir, the steam breakthrough time and water invasion time can be delayed, and the thermal recovery performance can be improved.

For a field operation, distributed temperature sensing temperature data can be applied to observe the expansion behavior of a steam chamber along a horizontal wellbore. Once the nonuniform distribution of a steam injection profile along the horizontal wellbore is observed, an FCD is recommended to restrict the fluid flow of a high-permeability interval and improve steam conformance, as shown in Fig. 10.3. On the other hand, in SAGD, some observing wells are placed between adjacent well pairs to monitor temperature distribution. Therefore, from the temperature results of observing wells, steam conformance along a horizontal wellbore can be obtained. They are used to evaluate whether to deploy FCDs. Furthermore, for heavy oil reservoirs with an aquifer, a lean zone, or a gas cap, an application of FCDs can prevent the early breakthrough of an aquifer and gas. FCDs have been widely applied to improve the performance of Horizontal well (HW)-CSS and SAGD [41,42]. The application of FCDs can significantly improve bitumen recovery and decrease Steam-oil ratio (SOR) by improving the conformance of steam injection profiles along horizontal wells.

10.6.2 Dual-pipe well configurations

As in FCDs, a dual-pipe well configuration can be applied to improve the distribution of a steam injection or fluid production profile. For a dual-pipe configuration, there are two pipes in a wellbore, and their relationship can be concentric or parallel, as shown in Fig. 10.4. During field operations, both

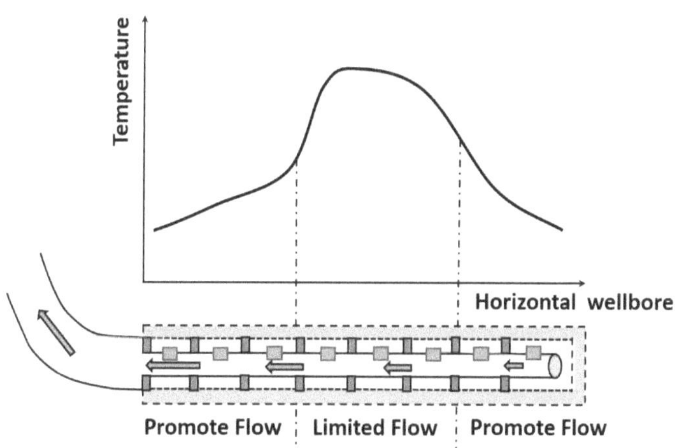

FIGURE 10.3 Schematic for temperature distribution along horizontal wellbore in a heterogeneous heavy oil reservoir.

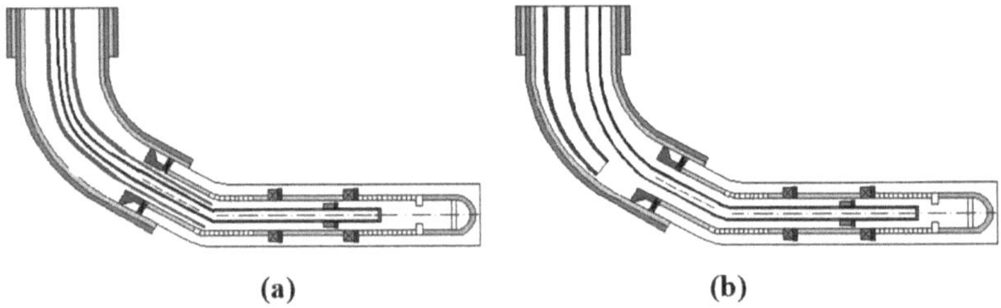

FIGURE 10.4 Concentric and parallel dual-pipe well configurations: (A) concentric dual pipes; (B) parallel dual pipes.

can be used for steam injection or oil production. In SAGD, a dual-pipe well configuration is applied in both a steam injection and production well. By using a dual-pipe well configuration, a steam circulation method can be easily applied to preheat an interwell section. During field operations, a long pipe is applied to inject steam and a short pipe is used for production. Also, in some thermal recovery operations, one pipe can be used for steam injection and the other for oil production.

For a dual-pipe well configuration, one pipe can be used to heat a formation area around a heel-end section, and the other one can be used to heat a formation area around a toe-end section. Moreover, for a specific recovery process, based on actual production characteristics, the locations of the two pipes can be changed. In some cases, one can even be placed in the middle of a wellbore. Therefore, by changing the specific locations of the dual pipes, an entire horizontal wellbore can have a uniform steam injection profile distribution and the effect of permeability heterogeneity can be significantly reduced. However, considering the requirement of a wellbore diameter, a parallel dual-pipe well configuration is usually employed for a shallow reservoir, and a concentric dual-pipe is employed for a deep well. The two types of dual-pipe well configurations have been widely applied for heavy oil recovery processes in heavy oil fields [43–45]. Furthermore, in some shallow wells, a multiple-pipe well configuration has been operated to improve the recovery of heavy oil reservoirs.

In this section, a method of an advanced wellbore configuration is introduced to improve thermal recovery. Therefore, in field operations, to enhance the heavy oil recovery effectively, we can combine the advantages of hybrid EOR processes and advanced wellbore configurations. The combined method will be a potential process for heavy oil reservoirs.

10.7 Future trends

As discussed in Chapter 1, heavy oil reserves are abundant. Global proven heavy oil resources are about 9911.8×10^8 t, and recoverable reserves of heavy oil are estimated at 1267.4×10^8 t [1]. They are mainly distributed in America and the Middle East. Annual heavy oil production is about 5000×10^4 t. It has an important role in balancing the world energy supply and demand. Among so many exploitation and development methods of heavy oil reservoirs, a hybrid EOR method has attracted much attention and has been applied across the world. Although some other processes have also been applied, some are still at the laboratory scale and some have higher operation costs. Hybrid EOR methods are still the most effective and profitable processes. In this section, the future trends of heavy oil recovery are discussed. Three aspects will be discussed in this section.

10.7.1 Innovations in effective and low-cost additives

Based on the discussion in this book, hybrid EOR processes will be an attractive and effective method to enhance heavy oil recovery. For poststeamed heavy oil reservoirs, the application of hybrid EOR processes can take full advantage of reservoir residual heat and minimize the impact of steam breakthrough. However, the most important step for a hybrid EOR process is to screen a suitable additive, including the NCG, solvent, or chemical agent. When oil prices are low, an effective and low-cost additive will be the top concern for oil companies. Additives include commonly used commercial ones (e.g., N_2, CO_2, Air, C_xH_y, hydrolyzed polyacrylamide, and SDS) as well as some proposed and invented ones. Many thermally stable and cost-effective chemical additives have been proposed and tested in thermal recovery for heavy oil reservoirs (e.g., switchable-hydrophilicity tertiary amines, thin film spreading agents, alkali-co-solvent-polymer, and hydroxypropyl methylcellulose) [46–49]. Most are just at the laboratory scale. Their costs and compatibility are two important issues for field operations. For a field application, once an applied target heavy oil reservoir is changed, a previous process may not have a good response. This indicates that before a field application, a systematic assessment of the additive feasibility should be performed. It includes not only laboratory-scale experimental studies but also field-scale simulation studies. An effective and low-cost additive is an important prerequisite for a successful field operation.

10.7.2 Accurate characterization of reservoir and fluid properties

To take advantage of hybrid EOR processes in heavy oil reservoirs, characterizing reservoir and fluid properties accurately is the most important issue. It can help the assessment process of an additive in a preidentified heavy oil reservoir. A hybrid EOR process is applied in a marginal heavy oil reservoir or a poststeamed heavy oil reservoir.

Marginal heavy oil reservoirs (e.g., great reservoir depth, high oil viscosity, thin formation thickness, and strong aquifer energy) are out of the criteria for steam injection. In comparison, a hybrid EOR process can make a previously unprofitable heavy oil reservoir available for further development. To develop marginal heavy oil reservoirs effectively, screening a suitable hybrid EOR process will be a challenging. Based on the discussion in Chapters 1–9, different hybrid EOR processes have different recovery characteristics and purposes (increasing displace efficiency or increasing sweep efficiency). Moreover, accurate characterization of reservoir and fluid properties will be an important step for screening.

Then, for poststeamed heavy oil reservoirs, from the discussion in Chapter 2, a lengthy steam–rock interaction can significantly change the reservoir and fluid properties. Generally, steam is alkaline (pH > 7.0), and once the temperature of steam is above 200°C, the properties of rock (including the pore structure and permeability) will be damaged compared with its original properties. On the other hand, under the effect of steam distillation (usually occurring in steam flooding), the component composition of heavy oil also changes. The heavy components (resin and asphaltene) will deposit in porous media. Thus, the deposited asphaltene will reduce the pore radius and increase flow resistance. Furthermore, for a typical hybrid EOR process, the co-/injection of additives (e.g., CO_2, air, and chemical agents) with steam can further damage the reservoir properties, including corrosion, oxidation, and adsorption. This indicates that the parameters used to evaluate feasibility for a preidentified hybrid EOR process have changed. Therefore, through a series of laboratory experimental tests or well logging tests, evaluating the current reservoir and fluid properties after steam injection operation will be crucial.

10.7.3 Optimization of operation modes of a hybrid enhanced oil recovery process

For hybrid EOR processes, steam and additives (i.e., NCG, solvent, and chemical agents) are simultaneously or sequentially injected into a heavy oil reservoir. An injection mode of different fluids can have an important effect on the recovery of a hybrid EOR process. In general, considering the mechanism differences of injected fluids in heavy oil reservoirs, a sequential injection operation is the most commonly used method.

For a newly found marginal heavy oil reservoir, steam is first injected to reduce oil viscosity and fluid flow resistance around wells. To some extent, the first steam injection operation can increase fluid injectivity. After steam is injected, additives (NCG or a solvent) will be injected. However, for poststeamed heavy oil reservoirs, to take advantage of residual thermal energy after steam injection, additives are directly injected. A chemical agent aims to plug steam breakthrough paths. Therefore, if serious steam breakthrough is observed, chemical agents (e.g., a foam system or high-temperature gel) are injected and then steam is injected to increase the heat sweep efficiency. Moreover, because of the effect of thermal stability, chemical agents and steam are rarely co-injected. In some cases, however, steam and additives (mainly NCG or a solvent) are also co-injected (e.g., SAGP [steam and gas pushing], expanding solvent-SAGD), and solvent enhanced steam flooding.

In addition, for some complicated hybrid EOR processes (e.g., Horizontal well, Dissolver, CO_2, and Steam (HDCS)/ Horizontal well, Dissolver, Nitrogen, and Steam (HDNS) and a multiple hybrid slug process), many different additives or multiple slugs will be injected into a reservoir because of the poor performance of a single additive. This complicated operation mode is commonly used for heavy oil reservoirs with serious steam breakthrough, and single slug cannot effectively improve recovery. Typical multiple slugs used in hybrid EOR include high-/low-temperature foam slugs, high-/low-concentration foam slugs, and NCG/foam slugs.

This chapter provides a discussion on some other EOR processes and the future trends of heavy oil EOR processes. Although some may be only at the laboratory scale because of high operation costs or unstable performance, with advances in science and technology, these processes can provide reliable technical support for continuous EOR for heavy oil reservoirs.

Most heavy oil fields have entered later stages of steam injection. Continuing steam injection is no longer economical and effective for their development. Furthermore, the problems of low sweep efficiency, low thermal use, and serious steam fingering prevent the continuous operation of steam injection. Based on these facts, this book systematically discusses hybrid EOR processes, including their experimental tests, theoretical investigations, simulation studies, and field trials. In addition, challenges during the application of these hybrid processes are addressed. Based on the discussion in Chapters 1−9, they are reliable and effective processes for heavy oil reservoirs. They can take full advantage of steam and additives to enhance heavy oil recovery.

References

[1] Liu Z, Wang H, Blackbourn G, et al. Heavy oils and oil sands: global distribution and resource assessment. Acta Geol Sin 2019;93(1):199−212.

[2] Dong X, Liu H, Chen Z, et al. Enhanced oil recovery techniques for heavy oil and oilsands reservoirs after steam injection. Appl Energy 2019;239:1190−211.

[3] Alvarez JM, Han S. Current overview of cyclic steam injection process. J Pet Sci Res 2013;2(3):116−27.

[4] Guo K, Li H, Yu Z. In-situ heavy and extra-heavy oil recovery: a review. Fuel 2016;185:886−902.

[5] Ji D, Harding T, Chen Z, Dong M, Liu H. Modelling of electromagnetic heating process and its applications in oil sands reservoirs. SPE 193905 presented at the SPE reservoir simulation conference, Galveston, Texas, USA, April 10−11, 2019.

[6] Hassanzadeh H, Harding T. Analysis of conductive heat transfer during in-situ electrical heating of oil sands. Fuel 2016;178:290−9.

[7] Saeedfar A, Lawton D, Osadetz K. Directional RF heating for heavy oil recovery using antenna array beamforming. SPE 180695 presented at SPE Canada heavy oil technical conference, Calgary, Alberta, Canada, June 7–9, 2006.
[8] Ghannadi S, Irani M, Chalaturnyk R. Overview of performance and analytical modeling techniques for electromagnetic heating and applications to steam-assisted-gravity-drainage process startup. SPE J 2016;21(2):311–33.
[9] Plzarro JOS, Trevlsan OV. Electrical heating of oil reservoirs: numerical simulation and field test results. J Petrol Technol 1990;42(10):1320–6.
[10] McGee BCW. Electro-thermal pilot in the athabasca oil sands: theory versus performance. Paper presented at the Canadian international petroleum conference, Calgary, Alberta, June 17–19, 2008.
[11] Wise S, Patterson C. Reducing supply cost with Eseieh™ pronounced easy. In: SPE 180729 presented at SPE Canada heavy oil technical conference, Calgary, Alberta, Canada, June 7–9, 2016; 2016.
[12] Sadeghi A, Hassanzadeh H, Harding T. Modeling of desiccated zone development during electromagnetic heating of oil sands. J Petrol Sci Eng 2017;154:163–71.
[13] Rezk MY, Allam NK. Impact of nanotechnology on enhanced oil recovery: a mini-review. Ind Eng Chem Res 2019;58:16287–95.
[14] Medina OE, Olmos C, Lopera SH, Cortes FB, Franco CA. Nanotechnology applied to thermal enhanced oil recovery processes: a review. Energies 2019;12:4671.
[15] Alomair O, Alajmi A. Experimental study for enhancing heavy oil recovery by nanofluid followed by steam flooding NFSF. SPE 184117 presented at the SPE heavy oil conference and exhibition, Kuwait City, Kuwait, December 6–8, 2016.
[16] Chengara A, Nikolov AD, Wasan DT, Trokhymchuk A, Henderson D. Spreading of nanofluids driven by the structural disjoining pressure gradient. J Colloid Interface Sci 2004;280:192–201.
[17] Montes D. Upscaling process for nanofluids usage for heavy oil mobility enhancement: experimental design and field trial application. SPE 204273 presented at the SPE annual technical conference and exhibition, virtual, October 26–29, 2020.
[18] Montes D, Orozco W, Taborda EA, et al. Development of nanofluids for perdurability in viscosity reduction of extra-heavy oils. Energies 2019;12(6):1068.
[19] Franco CA, Zabala R, Cortés FB. Nanotechnology applied to the enhancement of oil and gas productivity and recovery of Colombian fields. J Petrol Sci Eng 2017;157:39–55.
[20] Al-Shehri A, Ellis E, Servin JMF, Kosynkin DV, Kanj MY, Schmidt HK. Illuminating the reservoir: magnetic nanomappers. In Proceedings of the SPE Middle East oil and gas show and conference, Manama, Bahrain, March 10–13, 2013.
[21] Hanamertani AS, Pilus RM, Manan NA, Ahmed S. Surface and interfacial tension behavior in the use of ionic liquids as additives for surfactant-based enhanced oil recovery. SPE 192278 presented at the SPE kingdom of Saudi arabia annual technical symposium and exhibition, Dammam, Saudi Arabia, April 23–26, 2018.
[22] Fan ZX, Wang TF, He YH. Upgrading and viscosity reducing of heavy oils by [BMIM][$AlCl_4$] ionic liquid. J Fuel Chem Technol 2009;37(6):690–3.
[23] Bin Dahbag MS, Hossain ME. Simulation of ionic liquid flooding for chemical enhance oil recovery using CMG STARS software. SPE 182836 presented at the SPE kingdom of Saudi arabia annual technical symposium and exhibition, Dammam, Saudi Arabia, April 25–28, 2009.
[24] Hogshead CG, Manias E, Williams P, et al. Studies of bitumen-silica and oil-silica interactions in ionic liquids. Energy Fuel 2011;25(1):293–9.

[25] Mohsenzadeh A, Al-Wahaibi Y, Jibril A, Al-Hajri R, Shuwa S. The novel use of deep eutectic solvents for enhancing heavy oil recovery. J Petrol Sci Eng 2015;130:6–15.

[26] Abbott AP, Barron JC, Ryder KS, Wilson D. Eutectic-based ionic liquids with metal-containing anions and cations. Chem Eur J 2007;13(22):6495–501.

[27] Mohsenzadeh A, Al-Wahaibi Y, Al-Hajri R, Jibril A, Mosavat N. Sequential deep eutectic solvent and steam injection for enhanced heavy oil recovery and in-situ upgrading. Fuel 2017;187:417–28.

[28] Sandler J, Fowler G, Cheng K, Kovscek AR. Solar-generated steam for oil recovery: reservoir simulation, economic analysis, and life cycle assessment. SPE 153806 presented at the SPE western regional meeting, Bakersfield, California, USA, March 21–23, 2012.

[29] Anderson T. Economic analysis of solar-based thermal enhanced oil recovery. SPE 173466 presented at the SPE annual technical conference and exhibition, Amsterdam, The Netherlands, October 27–29, 2014.

[30] Akhmedzhanov TK, Nuranbayeva BM, Gussenov IS, Ismagilova LT. Enhanced oil recovery and natural bitumen production through the use of sinusoidal wells and solar thermal method. J Petrol Sci Eng 2017;159:506–12.

[31] Avaritsiotis J. Solar energy enhanced oil recovery: a case study for Kuwait. SPE 193801 presented at the SPE international heavy oil conference and exhibition, Kuwait City, Kuwait, December 10–12, 2018.

[32] Palmer D, O'Donnell J, Walter B. Solar enhanced oil recovery application to Kuwait's heavy oil fields. SPE 175290 presented at the SPE Kuwait oil and gas show and conference, Mishref, Kuwait, October 11–14, 2015.

[33] Kovscek AR. Emerging challenges and potential futures for thermally enhanced oil recovery. J Petrol Sci Eng 2012;98–99:130–43.

[34] Moritis G. Chevron starts California demo of solar-to-steam enhanced recovery. Oil Gas J 2011;109:86–9.

[35] Palmer D, O'Donnell J. Pilot steam generator uses solar energy successfully for EOR operations in Oman, SPE 169745 presented at the SPE EOR conference at OGWA, Muscat, Oman, 31 March-2 April, 2014.

[36] Wang J, Brandt AR, O'Donnell J. Potential for solar energy use in the global petroleum sector. SPE 187578 presented at the SPE Kuwait oil and gas show and conference, Kuwait city, Kuwait, 15–18 October, 2017.

[37] Carvajal-Osorio H. Nuclear power in heavy oil extraction and upgrading. IAEA Bull 1989;3:50–5.

[38] Puitagunta VR, Sochaski RO, Robertson RFS. A role for nuclear energy in the recovery of oil from the tar sands of Alberta. J Can Petrol Technol 1977;16(3):28–49.

[39] Dunbar RB, Sloan TW. Does nuclear energy have a role in the development of Canada's oil sands? J Can Petrol Technol 2004;43(9):19–22.

[40] Carvajal-Osorio H. An advanced nuclear power plant for heavy oil exploitation in the Venezuelan Orinoco oil belt. Nucl Eng Des 1992;136(1–2):219–27.

[41] Buell RS, Gurton R, Sims J, Wells M, Adnyana GP, Shirdel M, Muharam C, Gorham T, Riege E, Dulac GB. Design and operational experience with horizontal steam injectors in kern river field, California, USA. SPE 187258 presented at the SPE annual technical conference and exhibition, San Antonio, Texas, USA, October 9–11, 2017.

[42] Banerjee S, Hascakir B. Flow control devices in SAGD completion design enhanced heavy oil bitumen recovery through improved thermal efficiencies. SPE 185703 presented at the SPE western regional meeting, Bakersfield, California, USA, April 23, 2017.

[43] Zhang FL, Zhao HY. Steam based recovery technologies of heavy oil reservoirs in Liaohe oilfield. Petroleum Industry Press; 2007 [in Chinese].
[44] Liu HQ. Principle and design of thermal oil recovery processes. Petroleum Industry Press; 2013 [In Chinese].
[45] Dong X, Liu H, Zhang Z, Wang C. The flow and heat transfer characteristics of multi-thermal fluid in horizontal wellbore coupled with flow in heavy oil reservoirs. J Petrol Sci Eng 2014;122:56–68.
[46] Pratama RA, Babadagli T. New formulation of tertiary amines for thermally stable and cost-effective chemical additive: synthesis procedure and displacement tests for high-temperature tertiary recovery in steam Applications. SPE 201769 presented at the SPE annual technical conference and exhibition, virtual, October 26–29, 2020.
[47] Dang G, Nghiem L, Nguyen N, Yang C, Shrivastava V, Mirzabozorg A, Li H, Chen Z. Modeling of CoSolvent assisted chemical flooding for enhanced oil recovery in heavy oil reservoirs. Paper SPE-190196-MS, SPE improved oil recovery conference, Tulsa, Oklahoma, USA, April 14–16, 2018.
[48] He H, Wang Y, Zhao M, Cheng L, Liu P. Laboratory evaluation of thermoreversible gel for in-depth conformance control in steam-stimulated wells. SPE 157871 presented at the SPE heavy oil conference Canada, Calgary, Alberta, Canada, June 12–14, 2012.
[49] Srivastava P, Castro L. Successful field application of surfactant additives to enhance thermal recovery of heavy oil. SPE 140180 presented at the SPE Middle East oil and gas show and conference, Manama, Bahrain, September 25–28, 2011.

Index

'*Note*: Page numbers followed by "t" indicate tables.'

S

T

FSC
www.fsc.org
MIX
Papier | Fördert gute Waldnutzung
FSC® C083411

Kolibri 360 GmbH
Ferdinand-Jühlke-Straße 7
99095 Erfurt, Deutschland
produktsicherheit@kolibri360.de